**THE NORTHEAST
AEGEAN ISLANDS**
Pages 124–157

THE DODECANESE

D1330174

THE
NORTHEAST
AEGEAN
ISLANDS

THE
SPORADES AND
EVVOIA

CRETE
Pages 244–281

THE CYCLADES

THE
DODECANESE

CRETE

EYEWITNESS TRAVEL

THE GREEK ISLANDS

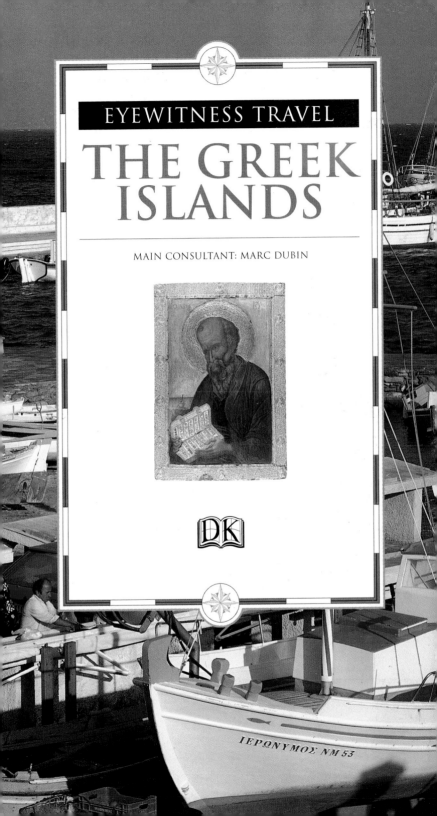

EYEWITNESS TRAVEL

THE GREEK ISLANDS

MAIN CONSULTANT: MARC DUBIN

DK

LONDON, NEW YORK,
MELBOURNE, MUNICH AND DELHI
www.dk.com

PROJECT EDITOR Jane Simmonds
ART EDITOR Stephen Bere
EDITORS Isabel Carlisle, Michael Ellis, Simon Farbrother,
Claire Folkard, Marianne Petrou, Andrew Szudek
DESIGNERS Jo Doran, Paul Jackson, Elly King, Marisa Renzullo
MAP CO-ORDINATORS Emily Green, David Pugh
VISUALIZER Joy Fitzsimmons
LANGUAGE CONSULTANT Georgia Gotsi

CONTRIBUTORS AND CONSULTANTS
Rosemary Barron, Marc Dubin, Stephanie Ferguson, Mike Gerrard, Andy
Harris, Lynette Mitchell, Colin Nicholson, Robin Osborne, Barnaby
Rogerson, Paul Sterry, Tanya Tsikas

MAPS
Gary Bowes, Fiona Casey, Christine Purcell (ERA-Maptec Ltd)

PHOTOGRAPHERS
Max Alexander, Joe Cornish, Paul Harris, Rupert Horrox,
Rob Reichenfeld, Linda Whitwam, Francesca Yorke

ILLUSTRATORS
Stephen Conlin, Steve Gyapay, Maltings Partnership, Chris Orr &
Associates, Mel Pickering, Paul Weston, John Woodcock

Reproduced by Colourscan (Singapore)
Printed and bound in China by L. Rex Printing Co. Ltd

First published in Great Britain in 1997
by Dorling Kindersley Limited
80 Strand, London WC2R 0RL

**Reprinted with revisions 1998, 1999, 2000,
2001, 2002, 2003, 2004, 2006, 2007, 2009**

Copyright 1997, 2009 © Dorling Kindersley Limited, London
A Penguin Company

ISBN: 978 1 40531 187 8

Front cover main image: blue domed church near Thíra, Santoríni

**The information in this
DK Eyewitness Guide is checked regularly.**
Every effort has been made to ensure that this book is as up-to-
date as possible at the time of going to press. Some details,
however, such as telephone numbers, opening hours, prices,
gallery hanging arrangements and travel information are liable to
change. The publishers cannot accept responsibility for any
consequences arising from the use of this book, nor for any
material on third party websites, and cannot guarantee that any
website address in this book will be a suitable source of travel
information. We value the views and suggestions of our readers
very highly. Please write to: Publisher, DK Eyewitness Travel
Guides, Dorling Kindersley, 80 Strand, London WC2R 0RL.

The harbour at Réthymno, Crete

CONTENTS

The Turkish Prince Cem arriving in
Rhodes (15th century)

INTRODUCING THE
GREEK ISLANDS

◁ Fishermen unloading their catch at Mýkonos harbour in the Cyclades

Family on a
scooter

Gorgon's head from Evvoia

Garides saganáki, prawns with
feta in a tomato sauce

Kámpos beach on Ikaría in the
Northeast Aegean Islands

Néa Moní on
Chíos, Northeast
Aegean Islands

HOW TO USE THIS GUIDE

This guide helps you to get the most from your visit to the Greek Islands. *Introducing the Greek Islands* maps the country in its historical and cultural context, including a quick comparison chart with *Choosing Your Island*. *Ancient Greece* gives a background to the many remains and artifacts to be seen. The seven regional chapters, plus *A Short Stay in Athens*, describe important sights, with maps and illustrations. Restaurant and hotel recommendations can be found in *Travellers' Needs*. The *Survival Guide* has tips on everything from the Greek telephone system to transport networks.

THE GREEK ISLANDS AREA BY AREA

The islands have been divided into six groups, each of which has a separate chapter. Crete has a chapter on its own. A map of these groups can be found inside the front cover of the book. Each island group is colour coded for easy reference.

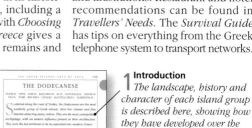

1 Introduction
The landscape, history and character of each island group is described here, showing how they have developed over the centuries and what they offer to the visitor today.

Each island group can be quickly identified by its colour coding.

A locator map shows you where you are in relation to other island groups.

2 Regional Map
This shows all the islands covered in the chapter. Main ferry routes are marked and there are useful tips on getting around the islands.

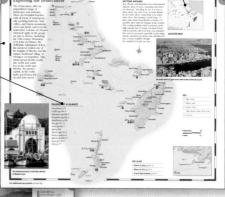

Islands at a Glance lists the islands alphabetically. Each island has a cross reference to its entry.

The main ferry routes, roads and transport points are marked on each map.

A locator map shows you where you are in relation to other islands in the group.

3 Detailed information
Most of the islands are described individually. Within each island entry there is detailed information on all the sights. Major islands have an island map showing all the main towns, villages, sights and beaches.

Story boxes highlight special or unique aspects of a particular sight.

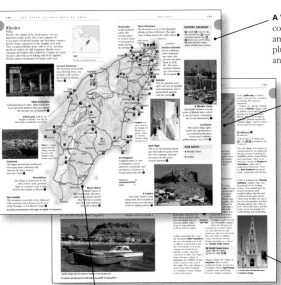

A Visitors' Checklist gives contact points for tourists and transport information, plus details of market days and local festival dates.

4 Greece's top islands
An introduction covers the history, character and geography of the island. The main sights are numbered and plotted on the map. They are described in more detail on the following pages.

Following pages describe the islands in more detail.

The main ferry routes, roads, transport points and recommended beaches are marked on the map.

5 Street-by-Street Map
Towns, or districts, of special interest to visitors are shown in detailed 3D, giving a bird's-eye view.

Stars indicate the sights that no visitor should miss.

A Visitors' Checklist provides the practical information you will need to plan your visit.

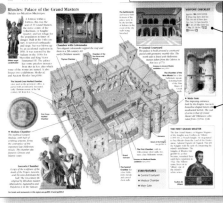

6 Greece's top sights
These are given one or more full pages. Historic buildings are dissected to reveal their interiors. Plans and reconstructions of ancient sites are provided.

INTRODUCING THE GREEK ISLANDS

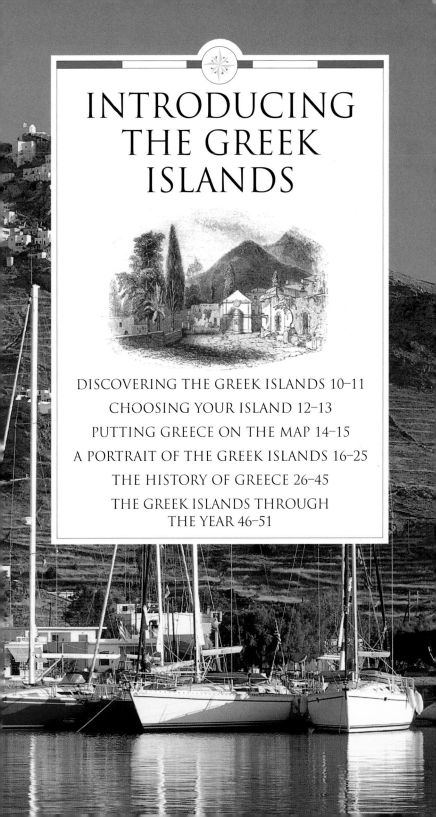

DISCOVERING THE GREEK ISLANDS

At first glance, the hundreds of islands scattered in the seas around Greece may seem similar, with landscapes of terraced fields, olive groves, vineyards and barren goat pastures, but each is distinct in character even from its nearest neighbour. From party resorts and sybaritic hotels to artists'

Monkey Orchid, Crete

retreats, temples and hillsides ablaze with wild flowers, these glorious, gold-fringed islands offer holidays in all guises.

Divided into six groups, plus Crete, the largest of all, this section introduces the main groups, highlighting their top attractions. See also *Choosing Your Island*, pages 12–13.

Agios Nikítas beach and harbour on Lefkáda, Ionian Islands

THE IONIAN ISLANDS

- **Majestic island scenery**
- **Buzzing resorts and nightlife**
- **Corfu's Venetian architecture**

With golden sands, green-shadowed mountainsides and bright blue water, it is no wonder these film-set islands attract attention.

Relive the myth hiking Odysseus' rugged island, **Ithaca** *(see pp86–7)*, take a boat ride around **Zákynthos** *(see pp90–1)* striking Blue Grotto or get underground in the subterrannean caves of mountainous **Kefalloniá** *(see pp88–9)*, the island famous for its star-turn in *Captain Corelli's Mandolin*.

Corfu *(see pp68–79)*, the largest and perhaps most scenic of the Ionians, throngs in peak season but has wide beaches, lively nightlife and elegant cafés amid Venetian architecture in Corfu Town.

Close by, tiny **Paxós** *(see p84)* is a haven of peace and

pretty villages, while **Lefkáda** *(see p85)* offers sheltered bays busy with windsurfers.

THE ARGO-SARONIC ISLANDS

- **Chic harbour towns**
- **Kýthira's deserted beauty**
- **Daytrips to Athens**

The rocky Argo-Saronics once contained some of the wealthiest seaports in Greek waters and the harbours of **Aígina** *(see pp92–5)* , **Póros** *(see p100)* and car-free **Ydra** *(see pp100–1)*, are still lined with dignified Neo-Classical mansions – picturesque reminders of that era.

Kýthira *(see pp102–3)*, lies far to the south, just off the tip of the Peloponnese. Its deserted beaches, rugged coastline and sleepy hilltop villages offer welcome respite from the summer crowds of its noisier siblings.

Just a short hop from the mainland, frequent ferry

routes mean the Acropolis and other fabled sights of Athens are only hours away.

THE SPORADES AND EVVOIA

- **First-class beachlife**
- **Island tradition on Skýros**
- **Unspoilt hinterland of Evvoia**

The popular islands of **Skiáthos** *(see pp108–9)* and **Skópelos** *(see pp112–13)* attract many visitors with their endless sparkling bays, crystal water and colourful harbours filled with glossy yachts. Watersports, boat hire, plus a vast choice of tavernas and bars do a roaring trade here.

Lonely **Skýros** *(see pp116–17)* is an artists' retreat offering traditional island culture in costumes, old-fashioned villages and herds of wild ponies, while **Evvoia** *(see pp118–19)* is an unsung hero of deserted coastline and wild, mountainous space.

Ydra harbour, surrounded by 18th-century ship-owners' mansions

The fortified towers of the Palace of the Master, on Rhodes

THE NORTHEAST AEGEAN ISLANDS

- Uncrowded beaches
- Ancient and medieval sites
- Eastern flavour of Lésvos

Surprisingly untouched by tourism, this dispersed cluster of islands is rich with ancient sites, natural charms and some superb beaches for the crowd-weary.

Take your pick from mastic villages and the Byzantine monastery, Néa Moní, on **Chíos** *(see pp146–53)*, ancient temples, woods and waterfalls on **Samothráki** *(see pp132–3)*, **Ikaría**'s *(see p153)* rocky coastline and lush valleys, or pretty villages on **Thásos** *(see pp128–31)* and wine-tasting on **Sámos** *(see pp154–7)*.

Lésvos *(see pp136–45)* with its Ottoman domes and lively bazaar has an eastern feel. Be sure to take a plunge in a natural thermal spa.

THE DODECANESE

- Ideal for island-hopping
- Monastery of St-John
- Rhodes' sun, sand and sights

The sizzling Dodecanese, the hottest of all the islands, are perfect, in their variety and mutual proximity, for a combination holiday. Hop by ferry or hydrofoil from the large island of **Kos** *(see*

pp170–73) to **Pátmos** *(see pp164–5)*, the "Jeruselem of the Aegean", and the 11th-century Monastery of St John. Then on to the utter tranquillity of dots on the map like **Lipsí** *(see p166)*, **Symi** *(see pp178–9)* and **Tílos** *(see p177)*. There is even a bubbling semi-active volcanic crater on **Nísyros** *(see p176)*.

Rhodes *(see pp180–95)* rewards a longer visit for its world-class sights including the hilltop acropolis at Líndos and the fortified Palace of the Masters in Rhodes Old Town. Happily for some, history comes combined with fabulous beaches, raucous nightlife and 300 sunny days a year.

THE CYCLADES

- Pretty hilltop villages
- Chic bars and nightclubs
- Ancient temples on Delos

The Cyclades, a volcanic archipelago of wide horizons and whitewashed villages, pretty with windmills and blue-domed churches, typify the Greek Islands ideal – and their variety. Sophisticated hedonists flock to the stylish hotels and cosmopolitan nightclubs of **Mýkonos** *(see pp214–15)*, **Amorgós** *(see pp233)* and **Santoríni** *(see pp238–41)*, with its sea-filled volcanic caldera, while nature-lovers will find excellent hiking on **Naxos** *(see pp230–1)* and snorkling on **Páros** *(see pp226–9)*. Central to them all, minute

Delos *(see pp218–19)* is one of the most important archaeological sites in Greece and one huge outdoor museum.

Minoan Palace of Knosós, Crete, built around 1700 BC

CRETE

- Spring flowers and wildlife
- Minoan palaces
- Hiking the Samariá Gorge

Sprawling Crete, Greece's largest island, attracts return visitors for its wealth of beaches, natural beauty – in spring, the hillsides burst with wild flowers and birdsong – and excellent facilities. Relics such as the Minoan palaces of **Knosós** *(see pp272–5)* and **Phaestos** *(see pp266–7)* wait to be explored as do busy port towns and museums.

The **Samariá Gorge** *(see pp250–1)* is one of Crete's top sights. Hikers will relish the tortuous 18-km (11-mile) route and the reward of glorious mountain scenery.

The crystal waters of Síkinos, typical of the Cyclades

Choosing Your Island

One great appeal of the Greek Islands is the sheer variety of attractions and activities on offer. Choosing the right island for the type of holiday you want – whether it be action-packed, historical or lazy (or a combination) – can be a bewildering decision, however. This chart gives a quick reference point to the strengths, charms and facilities of each island covered in this guide.

KEY

- ☐ The Ionian Islands
- ☐ The Argo-Saronic Islands
- ☐ The Sporades and Evvoia
- ☐ The Northeast Aegean Islands
- ☐ The Dodecanese
- ☐ The Cyclades
- ☐ Crete

Kayaks for hire in Skiáthos, The Sporades and Evvoia

KEY

★ Excellent

• Available

	DIVING	SNORKELLING	DAY TRIPS	WATERSPORTS
THE IONIAN ISLANDS				
CORFU (see pp72–83)	•	•	★	★
PAXOS (see p84)				
LEFKADA (see p85)	•	•	★	★
ITHACA (see pp86–7)				
KEFALLONIA (see pp88–9)	•	•	•	•
ZAKYNTHOS (see pp90–1)	•	•		•
THE ARGO-SARONIC ISLANDS				
SALAMINA (see p96)			•	
AIGINA (see pp96–9)	•			
POROS (see p100)	•			
YDRA (see pp100–1)	•			
SPETSES (see p101)	•			•
KYTHIRA (see pp102–3)	•			
THE SPORADES AND EVVOIA				
SKIATHOS (see pp108–9)	•	•		•
SKOPELOS (see pp112–13)	•	•		•
ALONNISOS (see p114)				
SKYROS (see pp116–17)				
EVVOIA (see pp118–23)	•	•	•	•
THE NORTHEAST AEGEAN ISLANDS				
THASOS (see pp128–31)		•		
SAMOTHRAKI (see pp132–3)				
LIMNOS (see pp134–5)				
LESVOS (see pp136–45)				•
CHIOS (see pp146–53)				•
IKARIA (see p153)				
SAMOS (see pp154–7)			•	•
THE DODECANESE				
PATMOS (see pp162–5)		•	•	•
LIPSI (see p166)				
LEROS (see pp166–7)				
KALYMNOS (see pp168–9)		•	•	•
KOS (see pp170–3)			•	•
ASTYPALAIA (see p174)				
NISRYOS (see pp174–6)				
TILOS (see p177)				
SYMI (see pp178–9)		•	•	
RHODES (see pp180–97)	•	•	★	•
CHALKI (see pp198–9)				
KASTELLORIZO (see p199)		★		
KARPATHOS (see pp202–3)				•
THE CYCLADES				
ANDROS (see pp208–11)				
TINOS (see pp212–13)				
MYKONOS (see pp214–15)	•	•		•
DELOS (see pp218–19)				
SYROS (see pp220–23)				
KEA (see p223)				
KYTHNOS (see p224)				
SERIFOS (see pp224–5)				•
SIFNOS (see p225)		•		
PAROS (see pp226–9)	★	★	•	•
NAXOS (see pp230–33)				
AMORGOS (see p233)	•	•		
IOS (see p234)				
SIKINOS (see pp234–5)				
FOLEGANDROS (see p235)				
MILOS (see pp236–3)				
SANTORINI (see pp238–41)	•	•	•	•
CRETE (see pp244–81)	★	★	★	•

On the map: BULGARIA, FYR OF MACEDONIA, ALBANIA, MAINLAND GREECE, AEGEAN SEA, TURKEY

CAMPING	HORSERIDING	BEACH SPORTS	SWIMMING	HIKING	FISHING	BIRDWATCHING	SAILING/BOAT HIRE	MARINE LIFE	CYCLING	KAYAKING	GOLF	ARCHAEOLOGY	FAMILY BEACHES	NUDIST BEACHES	NIGHTLIFE	UNSPOILT ISLANDS	GAY SCENE	RESTAURANTS	WINE	SHOPPING	SCENIC	SPA/LUXURY HOTEL	ART SCENE	MUSEUMS	MAJOR AIRPORT
★	•	•	•	★	•	•	•		•	•	★		★	•	★			★	•	•	•	•	•	•	•
				•					•			•	•												
•		•	•	•	★		•		•	•			•	•											
	•		•	•						•		•													
•	•	•	•	•			•		•	•			•	•	•				•		•				
•		•	•				•	•					•	•	•										
			•																						
•												•						•							
									•				•												
														•		•	•		•				★		
			•										•								•				
		•	•	•			•		•				•						•						
•	•	•	•	•			•		•				•	•	•		•	•		•					
		•	•	•			•		•				•	•											
		•	•	•		•		•	•				•								•		•		
													•								•				
•		•	•	•	•	•	•		•				•	•				•			•				
•	•	•	•	•			•		•				•	•	•				•		•				
			•	•			•		•			•			•				•		•				
		•	•	•			•		•	•			•	•		★				•			•	•	•
•		•	•	•		★	•		•	•			•	•					•		•			★	•
			•						•				•	•		•									
		•	•	•			•		•				•	•				•		•				•	•
•		•	•	•	•		•		•				•	•	•						•				
			•	•			•		•					•							•				
•			•				•							•											
•			•						•					•											
		•	•	•			•						•	•	•										•
•			•											•			•								
			•						•					•			•								
			•	•					•			•	•			•			•						
•	•	•	•	•			•		•	•	•	•	•	•	★		•	•	•	★		•		•	•
			•										•			•									
			•													•				•					
		•	•	•	•		•	•	•	•			•			•					★				
•			•	★					★	•			•	★											
•			•	•										•				•						•	
•		•	•				•					•	★	★		★	•		★		★	•		•	
•										★				★											
		•	•				•		•				•		•										•
•			•	•			•			•			•		•										
•			•	•									•		•										
•		•	•	★			•						•		•										
•			•										•			•									
•	•	•	•	•	•		•		•	•		•	•	★		•	•								•
•		•	•	•					•			•	•	•							•				•
•			•	★									•	★							★				
•		•	•				•						•												
•			•												★										
•			•	•																					
•			•						•	★			•		•					★					•
•		•	•				•				★		★			★	•	•	★	★	★	•	★	•
★	★	•	•	★	•	★	•	•	★	•	•	★	★	★	★			•	★	★	•	★	★	★	•

Putting Greece on the Map

Occupying the southernmost tip of the Balkan peninsula, Greece divides into over 2,000 islands stretching from the Ionian Sea in the west to the Aegean Sea in the east. The mainland has borders with Albania, Bulgaria, Turkey and the FYR of Macedonia. Of Greece's 10.9 million people, over 10 per cent live on the islands, while a third live in Athens.

KEY

- Main international ferry service
- International airport
- Motorway, dual-carriageway
- Major road
- Railway line
- National boundary

A PORTRAIT OF THE GREEK ISLANDS

Greece is one of the most visited European countries, but also one of the least known. At a geographical crossroads, the modern Greek state dates only from 1830, and combines elements of the Balkans, Middle East and Mediterranean.

Of the thousands of Greek islands, large and small, only about a hundred are today permanently inhabited. Barely 10 per cent of the country's population of just under eleven million lives on the islands, and for centuries a large number of Greek islanders have lived abroad: currently there are over half as many Greeks outside the country as in. The proportion of their income sent back to relatives significantly bolsters island economies. Recently there has been a trend for reverse immigration, with expatriate Greeks returning home to influence the architecture and cuisine on many islands.

Islands lying within sight of each other can have vastly different

Greek priest

histories. Most of the archipelagos along sea lanes to the Levant played a crucial role between the decline of Byzantium and the rise of modern Greece. Crete, the Ionian group and the Cyclades were occupied by the Venetians and exposed to the influence of Italian culture.

The Northeast Aegean and Dodecanese islands were ruled by Genoese and Crusader overlords in medieval times, while the Argo-Saronic isles were completely resettled by Albanian Christians.

Island and urban life in contemporary Greece were transformed in the 20th century despite years of occupation and war, including a civil war, which only ended after

Fishermen mending their nets on Páros in the Cyclades

◁ **A backstreet in Anógeia on Crete**

A village café on Crete's Lasíthi Plateau

the 1967–74 colonels' Junta. Recently, based on the revenues from tourism and the EU, there has been a rapid transformation of many of the islands from backwater status to prosperity. Until the 1960s most of the Aegean Islands lacked paved roads and basic utilities. Even larger islands boasted just a single bus and only a few taxis as transport and emigration, either to Athens or overseas, increased.

Frescoed saint from monastery of St John, Pátmos

RELIGION, LANGUAGE AND CULTURE
During the centuries of domination by Venetians and Ottomans *(see pp40–41)* the Greek Orthodox church preserved the Greek language, and

with it Greek identity, through its liturgy and schools. The query *Eísai Orthódoxos* (Are you Orthodox?) is virtually synonymous with *Ellinas eísai* (Are you Greek?). Today, the Orthodox Church is still a powerful force, despite the secularizing reforms of the first democratically elected PASOK government of 1981–5. While no self-respecting couple would dispense with church baptisms for their children, civil marriages are now as valid in law as the religious service. Sunday Mass is popular, particularly with women, who often socialize there as men do at *kafeneía* (cafés).

Many parish priests, recognizable by their tall stovepipe hats and long beards, marry and have a second trade (a custom that helps keep up the numbers of entrants to the church). However, there has also been a recent renaissance in celibate monastic life, perhaps as a reaction to postwar materialism.

The beautiful and subtle Greek language, that other hallmark of national identity, was for a long time

Traditional houses by the sea on Kefalloniá, the Ionian Islands

Stepped streets and pastel colours at Oía on Santorini in the Cyclades

DEVELOPMENT AND DIPLOMACY

While compared to most of its Balkan neighbours Greece is a wealthy and stable country, by Western economic indicators Greece languishes at the bottom of the EU league table and will be a net EU beneficiary for several years to come. The country's persistent negative trade deficit is aggravated by the large number of luxury goods imported on the basis of *xenomanía* – the belief that goods from abroad are of a superior quality to those made at home. Cars are the most conspicuous of these imports, since Greece is one of the very few European countries not to manufacture any of its own.

A family in Kos on their scooter

a field of conflict between the written *katharévousa*, an artificial form hastily devised around the time of Independence, and the slowly evolved everyday speech, or *dimotikí* (demotic Greek).

Today's prevalence of the more supple *dimotikí* was perhaps a foregone conclusion in an oral culture. Storytelling is still as prized in Greece as in Homer's time, with conversation pursued for its own sake in *kafeneía*. The bardic tradition is alive with poet-lyricists such as Mános Eleftheríou, Níkos Gátsos and Apóstolos Kaldáras. Collaborations such as theirs have produced accessible works which have played an important role keeping *dimotikí* alive from the 19th century until today.

During recent times of censorship under dictatorship or foreign rule, writers and singers have been a vital source of news and information.

Greece still bears the hallmarks of a developing economy, with profits from the service sector and agriculture accounting for two-thirds of its GNP. With EU membership since 1981, and an economy that is more capitalist than not, Greece has lost its economic similarity to Eastern Europe before the fall of the Iron Curtain. Recent years have seen many improvements: loss-making enterprises

A beach at Plakiás on Crete

Windmills at Olympos on the island of Kárpathos, in the Dodecanese

have been sold off by the state, inflation has dipped to single figures for the first time since 1973, interest rates are falling and Greece was accepted as a member of the EU monetary union. The euro has been its sole currency since March 2002.

Tourism ranks as the largest hard currency earner, compensating for the depression in world shipping and the fact that Mediterranean agricultural products are duplicated within the EU. Now the lifeblood of many islands,

Children dressed for a festival in Koskinoú village, Rhodes

tourism has only been crucial since the late 1960s. The unprepossessing appearance of many island tourist facilities owes much to a megadevelopment ethos and permit-granting policy formulated under the Junta. More recent developments have an appearance that is more in harmony with their natural surroundings. Planners hope that traditional high-volume and low-spending package tourism will defer to the new rich of central Europe, pan-Orthodox pilgrimages and special-interest tourism. To attract higher spenders the infra-structure of the islands is being upgraded, with plans for spas, yacht marinas, new airports and tele-communication links.

The fact that the Greek state is less than 200 years old and in the years since 1922 has been politically unstable means that Greeks have very little faith in government insti-tutions. Everyday life operates on networks

Festival bread from Chaniá's covered market on Crete

Threshing with donkeys in the Cyclades

of personal friendships and official contacts. The classic political designations of Right and Left have only acquired their conventional meanings in Greece since the 1930s. Among politicians, the dominant figure of the early 20th century was the anti-royalist Liberal Elefthérios Venizélos, who came from Crete. The years since World War II have been over-shadowed by two politicians: the late Andréas Papandréou, three times premier as head of the Panhellenic Socialist Movement (PASOK), and the late conservative premier Konstantínos Karamanlís, who died in 1998.

Thriving Pythagóreio harbour on the island of Sámos

With the Cold War over, Greece looks more than likely to assert its underlying Balkan identity. Relations with its nearest neighbours, and particularly with Albania, have improved considerably since the fall of the Communist regime there in 1990. Greece is already the number-one investor in neighbouring Bulgaria, and after a recent rapprochement with Skopje, (formerly Yugoslavian Macedonia) it seems as if Greece is now poised to become a significant regional power.

HOME LIFE

The family is still the basic Greek social unit. Under traditional island land distribution and agricultural practices, one family could sow, plough and reap its own fields, without the help of cooperative work parties. Today's family-run businesses are still the norm in the many port towns. Arranged marriages and granting of dowries, though not very common, persist; most single young people live with their parents or another relative until marriage; and outside the largest university towns, such as Rhodes town, Irákleio or Mytilíni, few couples dare to cohabit "in sin". Children from the smaller islets board with a relative while attending secondary school on the larger islands. Despite the renowned Greek love of children, Greece has a very low birth rate – in Europe, only Italy's is lower. Currently, the Greek birth rate is less than half of pre-World War II levels.

Fish at Crete's Réthymno market

Macho attitudes persist on the islands and women often forgo any hope of a career in order to look after the house and children. Urban Greek women are seeing a rise in status as new imported attitudes have started to creep in. However, no amount of outside influence is likely to jeopardize the essentially Greek way of life, which remains vehemently traditional.

A man with his donkey in Mýkonos town in the Cyclades

Vernacular Architecture on the Greek Islands

Greek island architecture varies greatly, even between neighbouring islands. Yet despite the fact that the generic island house does not exist, there are shared characteristics within and between island groups. The Venetians in Crete, the Cyclades, Ionian Islands and Dodecanese, and the Ottomans in the Northeast Aegean strongly influenced the indigenous building styles developed by vernacular builders.

View of the town of Chóra on Astypálaia in the Dodecanese, with the kástro above

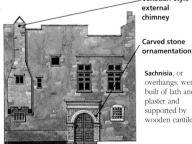

Venetian-style external chimney

Carved stone ornamentation

Sachnísia, or overhangs, were built of lath and plaster and supported by wooden cantilevers.

Venetian-style town houses *on Crete date from Venice's 15th- to 17th-century occupation. Often built around a courtyard, the ground floor was used for storage.*

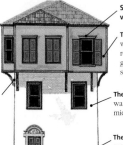

Sash windows with shutters

The top floor was for receiving guests and sleeping.

The kitchen was on the middle storey.

The stone ground floor housed animals and tools.

Arcade on ground floor supporting veranda

Lesvian pýrgoi *are fortified tower-dwellings at the centre of a farming estate. First built in the 18th century, most surviving examples are 19th century and found near Mytilíni town.*

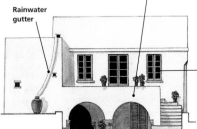

Rainwater gutter

Double "French" windows of the parlour

Sífnos archontiká *or town houses are found typically in Kástro, Artemónas and Katavatí. They are two storeyed, as opposed to the one-storey rural cottage.*

KASTRO ARCHITECTURE

The kástro or fortress dwelling of Antíparos dates from the 15th century. It is the purest form of a Venetian pirate-safe town plan in the Cyclades.

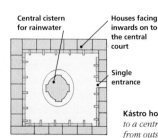

Central cistern for rainwater

Houses facing inwards on to the central court

Single entrance

Plan of a courtyard kástro

Chimneypot from broken urn

Stairway to central court

Plaster and whitewash surface

Kástro housefronts, *with their right-angled staircases, face either on to a central courtyard or a grid of narrow lanes with limited access from outside. The seaward walls have tiny windows. Kástra are found on Síkinos, Kímolos, Sífnos, Antíparos and Folégandros.*

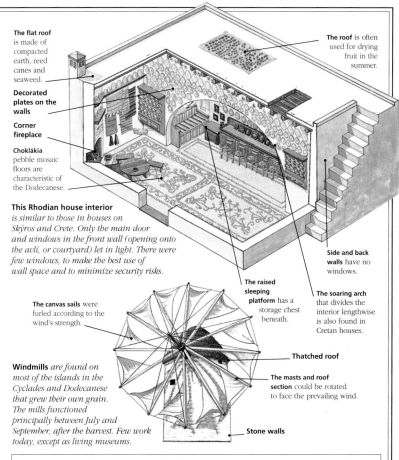

The flat roof is made of compacted earth, reed canes and seaweed.

The roof is often used for drying fruit in the summer.

Decorated plates on the walls

Corner fireplace

Choklákia pebble mosaic floors are characteristic of the Dodecanese.

This Rhodian house interior *is similar to those in houses on Skýros and Crete. Only the main door and windows in the front wall (opening onto the avlí, or courtyard) let in light. There were few windows, to make the best use of wall space and to minimize security risks.*

Side and back walls have no windows.

The raised sleeping platform has a storage chest beneath.

The soaring arch that divides the interior lengthwise is also found in Cretan houses.

The canvas sails were furled according to the wind's strength.

Windmills *are found on most of the islands in the Cyclades and Dodecanese that grew their own grain. The mills functioned principally between July and September, after the harvest. Few work today, except as living museums.*

Thatched roof

The masts and roof section could be rotated to face the prevailing wind.

Stone walls

LOCAL BUILDING METHODS AND MATERIALS

Lava masonry is found on the volcanic islands of Lésvos, Límnos, Nísyros and Mílos. The versatile and easily split schist is used in the Cyclades, while lightweight lath and plaster indicates Ottoman influence and is prevalent on Sámos, Lésvos, the Sporades and other northern islands. Mud-and-rubble construction is common on all the islands for modest dwellings, as is the *dóma* or flat roof of tree trunks supporting packed reed canes overlaid with seaweed and earth.

Unmortared wall of schist slabs

Masoned volcanic boulders

Slate (or "fish-scale") roof

Pantiled roof, found in the Dodecanese

Flat earthen roof or *dóma*

Arched buttresses for earthquake protection

Marine Life

By oceanic standards, the Mediterranean and Aegean are small, virtually landlocked seas with a narrow tidal range. This means that relatively little marine life is exposed at low tide, although coastal plants and shoreline birds are often abundant. However, if you snorkel close to the shore or dive below the surface of the azure coastal waters, a wealth of plant and animal life can be found. The creatures range in size from myriad shoals of tiny fish and dainty sea slugs to giant marine turtles, huge fish and imposing spider crabs.

Triton shell

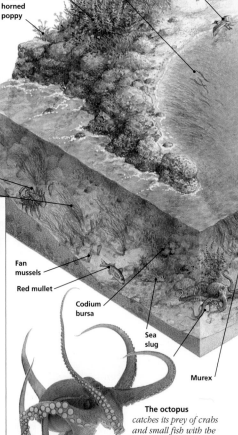

The great pipefish's *elongated body is easily mistaken for a piece of drifting seaweed. It lives among rocks, pebbles and weed, often in rather shallow water, and can be spotted when snorkelling.*

Mediterranean gull

Masked crab

Sea spurge

Tama risk

Yellow-horned poppy

The spiny spider crab *is ungainly when removed from water but agile and surprisingly fast-moving in its element. The long legs allow it to negotiate broken, stony ground easily.*

Neptune grass (*Posidonia*)

TOP SNORKELLING AREAS

Snorkelling can be enjoyed almost anywhere around the Greek coast, although remoter areas are generally more rewarding.

• Kefalloniá and Zákynthos: you may find a rare loggerhead turtle *(see p91)* off the east coast.
• Rhodes: wide variety of fish near Líndos on the sheltered east coast.
• Evvoia: the sheltered waters of the west coast harbour sponges.
• Santoríni: the volcanic rock of the caldera has sharp drop-offs to explore.

Fan mussels

Red mullet

Codium bursa

Sea slug

Murex

The octopus *catches its prey of crabs and small fish with the rows of powerful suckers along each of its eight legs. It can also change its colour and squeeze through the tiniest of crevices.*

The sea turtle, *or loggerhead, needs sandy beaches to lay its eggs and has been badly affected by the intrusion of tourists. The few remaining nesting beaches are now given a degree of protection from disturbance.*

This jellyfish, *called a "by-the-wind-sailor", uses a buoyant float to catch the wind and skim across the sea. Storms will often wash them up on to the beach. Swimmers beware: even the detached threadlike tentacles of some species can inflict painful stings.*

Redshank

Sea balls

Sea horses *are surprisingly common in the seas around Greece. They often live among beds of seagrass and curl their tails around the plants to provide a firm anchorage. Unusually for fish, they show parental care, the male having a brood pouch in which he incubates his offspring.*

Pilchard

Bath sponge

Moray eel

Red gurnard

Violet sea snail

Shore crab

A John Dory *is a majestic sight as it patrols among offshore rocks. It has a flattened, oval-shaped body and long rays on its dorsal fin. Where the species is not persecuted or exploited, some individuals can become remarkably confident and even inquisitive.*

The swimming crab *is one of the most aggressive of all crabs and can inflict a painful nip. It can swim using the flattened, paddlelike tips of its back legs.*

SAFETY TIPS FOR SNORKELLING

• Mediterranean storms can arrive out of nowhere so seek local advice about weather and swimming conditions before you go snorkelling.
• Do not go snorkelling if jellyfish are in the area.
• Take your own snorkel and mask with you to ensure you use one that fits properly.
• Never snorkel unaccompanied.
• Wear a T-shirt or wet suit to avoid sunburn.
• Avoid swimming near river mouths and harbours. The waters will be cloudy and there may be risks from boats and pollution.
• Always stick close to the shore and check your position from time to time.

THE HISTORY OF GREECE

The history of Greece is that of a nation, not of a land: the Greek idea of nationality is governed by language, religion, descent and customs, not so much by location. Early Greek history is the story of internal struggles, from the Mycenaean and Minoan cultures of the Bronze Age to the competing city-states that emerged in the 1st millennium BC.

Alexander the Great, by the folk artist Theófilos

After the defeat of the Greek army by Philip II of Macedon at Chaironeia in 338 BC, Greece became absorbed into Alexander the Great's empire. With the defeat of the Macedonians by the Romans in 168 BC, Greece became a province of Rome. As part of the Eastern Empire she was ruled from Constantinople and became a powerful element within the new Byzantine world.

In 1453, when Constantinople fell to the Ottomans, Greece disappeared as a political entity. The Venetian republic quickly established fortresses on the coast and islands in order to compete with the Ottomans for control of the important trade routes in the Ionian and Aegean seas. Eventually the realization that it was the democracy of Classical Athens that had inspired so many revolutions abroad gave the Greeks themselves the courage to rebel and, in 1821, to fight the Greek War of Independence. In 1830 the Great Powers that dominated Europe established a protectorate over Greece, marking the end of Ottoman rule.

After almost a century of border disputes, Turkey defeated Greece in 1922. This was followed by the dictatorship of Metaxás, and then by the war years of 1940–4, during which half a million people were killed. The present boundaries of the Greek state have only existed since 1948, when Italy returned the Dodecanese. Now an established democracy and member of the European Union, Greece's fortunes seem to have come full circle after 2,000 years of foreign rule.

A map of Greece from the 1595 Atlas of Abraham Ortelius called *Theatrum Orbis Terrarum*

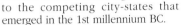
◁ The Knights of the Order of St John from a 15th-century history of the siege of Rhodes

Prehistoric Greece

During the Bronze Age three separate civilizations flourished in Greece: the Cycladic, during the 3rd millennium; the Minoan, based on Crete but with an influence that spread throughout the Aegean Islands; and the Mycenaean, which was based on the mainland but spread to Crete in about 1450 BC when the Minoans went into decline. Both the Minoan and Mycenaean cultures found their peak in the Palace periods of the 2nd millennium when they were dominated by a centralized religion and bureaucracy.

Mycenaean gold brooch

PREHISTORIC GREECE

Areas settled in the Bronze Age

Neolithic Head *(3000 BC)*
This figure was found on Alónnisos in the Sporades. It probably represents a fertility goddess who was worshipped by farmers to ensure a good harvest. These figures ind- icate a certain stability in early communities.

The town is unwalled, showing that inhabitants did not fear attack.

Cycladic Figurine
Marble statues such as this, produced in the Bronze Age from about 2800 to 2300 BC, have been found in a number of tombs in the Cyclades.

Multistorey houses

Minoan Bathtub Sarcophagus
This type of coffin, dating to 1400 BC, is found only in Minoan art. It was probably used for a high-status burial.

TIMELINE

200,000 BC	5000 BC	4000 BC	3000 BC	2000 B
	7000 Neolithic farmers in northern Greece	**3200** Beginnings of Bronze-Age cultures in Cyclades and Crete	**2000** Arrival of first Greek-speakers on mainland Greece	
200,000 Evidence of Palaeolithic civilization in northern Greece and Thessaly		*"Frying Pan" vessel from Sýros (2500– 2000 BC)*	**2800–2300** Kéros-Sýros culture flourishes in Cyclades	
			2000 Building of palaces begins in Crete, initiating First Palace period	

Mycenaean Death Mask

Large amounts of worked gold were discovered in the Peloponnese at Mycenae, the ancient city of Agamemnon. Masks like this were laid over the faces of the dead.

Forested hills

The inhabitants are on friendly terms with the visitors.

WHERE TO SEE PREHISTORIC GREECE

The Museum of Cycladic Art in Athens *(see p291)* has the leading collection of Cycladic figurines in Greece. In the National Archaeological Museum *(p286)* Mycenaean gold and other prehistoric artifacts are on display. Akrotíri *(p241)* on Santoríni in the Cyclades has Minoan buildings surviving up to the third storey. The city of Phylakopi on Mílos *(p237)* also has Mycenaean walls dating to 1500 BC. Crete, the centre of Minoan civilization, has the palaces of Knosós *(pp272–5)*, Phaestos *(pp266–7)* and Agía Triáda *(p263)*.

Cyclopean Walls

Mycenaean citadels, as this one at Tiryns in the Peloponnese, were encircled by walls of stone so large that later civilizations believed they had been built by giants. It is unclear whether the walls were used for defence or just to impress.

Oared sailing ships

MINOAN SEA SCENE

The wall paintings on Santoríni *(see pp238–41)* were preserved by the volcanic eruption at the end of the 16th century BC. This section shows ships departing from a coastal town. In contrast to the warlike Mycenaeans, Minoan art reflects a more stable community which dominated the Aegean through trade, not conquest.

Mycenaean Octopus Jar

This 14th-century BC vase's decoration follows the shape of the pot. Restrained and symmetrical, it contrasts with relaxed Minoan prototypes.

1750–1700 Start of Second Palace period and golden age of Minoan culture in Crete	**1525** Volcanic eruption on Santoríni devastates the region	**1250–1200** Probable destruction of Troy, after abduction of Helen *(see p54)*	*Helen of Troy*
		1450 Mycenaeans take over Knosós; use of Linear B script	
1800 BC	**1600 BC**	**1400 BC**	**1200 BC**
1730 Destruction of Minoan palaces; end of First Palace period		*Minoan figurine of a snake goddess, 1500 BC*	**1200** Collapse of Mycenaean culture
1600 Beginning of high period of Mycenaean prosperity and dominance			**1370–50** Palace of Knosós on Crete destroyed for second time

The Dark Ages and Archaic Period

Silver coin from Athens

In about 1200 BC, Greece entered a period of darkness. There was widespread poverty, the population decreased and many skills were lost. A cultural revival in about 800 BC accompanied the emergence of the city-states across Greece and inspired new styles of warfare, art and politics. Greek colonies were established as far away as the Black Sea, present-day Syria, North Africa and the western Mediterranean. Greece was defined by where Greeks lived.

Kouros *(530 BC)*
Koúroi were early monumental male nude statues. Idealized representations rather than portraits, they were inspired by Egyptian statues, from which they take their frontal, forward-stepping pose.

Bronze breastplate

MEDITERRANEAN AREA, 479 BC

▨ *Areas of Greek influence*

The double flute player kept the men marching in time.

Bronze greaves protected the legs.

Solon *(640–558 BC)*
Solon was appointed to the highest magisterial position in Athens. His legal, economic and political reforms heralded democracy.

HOPLITE WARRIORS

The "Chigi" vase from Corinth, dating to about 750 BC, is one of the earliest clear depictions of the new style of warfare that evolved at that period. This required rigorously trained and heavily armed infantrymen called hoplites to fight in a massed formation or phalanx. The rise of the city-state may be linked to the spirit of equality felt by citizen hoplites fighting for their own community.

TIMELINE

Vase fragment showing bands of distinctive geometric line patterns

900 Appearance of first Geometric pottery

1100 BC	1000 BC	900 BC

1100 Migrations of different peoples throughout the Greek world

1000–850 Formation of the Homeric kingdoms

6th-Century Vase
*This bowl (krater)
for mixing wine
and water at
elegant feasts is an
early example of the
art of vase painting.
It depicts mythological
and heroic scenes.*

Bronze helmets
for protection

Spears were used for
thrusting.

The phalanxes
shoved and
pushed, aiming to
maintain an
unbroken shield
wall, a successful
new technique.

Gorgon's head
decoration

Characteristic
round shields

WHERE TO SEE ARCHAIC GREECE

Examples of *koúroi* can be found in the National Archaeological Museum *(see p286)* and in the Acropolis Museum *(p290)*, both in Athens. The National Archaeological Museum also houses the national collection of Greek Geometric, red-figure and black-figure vases. Old *koúroi* lie in the old marble quarry on Náxos *(pp230–33)*. Sámos boasts the impressive Efpalíneio tunnel *(p155)* and a collection of *koúroi (p154)*. Delos has a terrace of Archaic lions *(pp218–19)* and the Doric temple of Aphaia on Aígina is well preserved *(pp98–9)*. Palaiókastro on Nísyros has huge fortifications *(p175)*.

**Hunter Returning
Home** *(500 BC)
Hunting for hares,
deer, or wild boar was
an aristocratic sport
pursued by Greek nobles
on foot with dogs, as
depicted on this cup.*

Darius I *(ruled 521–486 BC)
This relief from Persepolis shows
the Persian king who tried to
conquer the Greek mainland,
but was defeated at the Battle
of Marathon in 490.*

776 Traditional date for the
first Olympic Games

675 Lykourgos initiates
austere reforms in Sparta

600 First Doric
columns built at
Temple of
Hera,
Olympia

*Doric
capital*

490 Athenians defeat
Persians at Marathon

800 BC   **700 BC**   **600 BC**   **500 BC**

750–700 Homer re-
cords epic tales of the
Iliad and *Odyssey*

770 Greeks start founding
colonies in Italy, Egypt
and elsewhere

*Spartan
votive
figurine*

546 Persians gain control
over Ionian Greeks; Athens
flourishes under the tyrant,
Peisistratos, and his sons

630 Poet Sappho
writing in Lésvos

480 Athens destroyed by
Persians who defeat
Spartans at Thermopylae;
Greek victory at Salamis

479 Persians annihilated at Plataiai
by Athenians, Spartans and allies

Classical Greece

The Classical period has always been considered the high point of Greek civilization. Around 150 years of exceptional creativity in thinking, writing, theatre and the arts produced the great tragedians Aeschylus, Sophocles and Euripides as well as the great philosophical thinkers Socrates, Plato and Aristotle. This was also a time of warfare and bloodshed, however. The Peloponnesian War, which pitted the city-state of Athens and her allies against the city-state of Sparta and her allies, dominated the 5th century BC. In the 4th century Sparta, Athens and Thebes struggled for power only to be ultimately defeated by Philip II of Macedon in 338 BC.

Trading amphora

CLASSICAL GREECE, 440 BC

- Athens and her allies
- Sparta and her allies

Fish Shop
This 4th-century BC Greek painted vase comes from Cefalù in Sicily. Large parts of the island were inhabited by Greeks who were bound by a common culture, religion and language.

Theatre used in Pythian Games

Temple of Apollo

Siphnian Treasury

Perikles
This great democratic leader built up the Greek navy and masterminded the extensive building programme in Athens between the 440s and 420s, including the Acropolis temples.

THE SANCTUARY OF DELPHI

The sanctuary in central Greece, shown in this 1894 reconstruction, reached the peak of its political influence in the 5th and 4th centuries BC. Of central importance was the Oracle of Apollo, whose utterances influenced the decisions of city-states such as Athens and Sparta. Rich gifts dedicated to the god were placed by the states in treasuries that lined the Sacred Way.

TIMELINE

Detail of the Parthenon frieze

462 Ephialtes's reforms pave the way for radical democracy in Athens

431–404 Peloponnesian War, ending with the fall of Athens and start of 33-year period of Spartan dominance

c.424 Death of Herodotus, historian of the Persian Wars

475 BC	450 BC	425 BC

478 With the formation of the Delian League, Athens takes over leadership of Greek cities

451–429 Perikles rises to prominence in Athens and launches a lavish building programme

447 Construction of the Parthenon begins

Bust of Herodotus, probably of Hellenistic origin

Gold Oak Wreath from Vergína
By the mid-4th century BC, Philip II of Macedon dominated the Greek world through diplomacy and warfare. This wreath comes from his tomb.

WHERE TO SEE CLASSICAL GREECE

Athens is dominated by the Acropolis and its religious buildings, including the Parthenon, erected as part of Perikles's mid-5th-century BC building programme *(see pp288–90)*. The island of Delos, the mythological birthplace of Artemis and Apollo, was the centre for the Delian League, the first Athenian naval league. The site contains examples of 5th-century BC sculpture *(pp218–19)*. On Rhodes, the 4th-century Temple of Athena at Líndos *(pp196–7)* is well preserved.

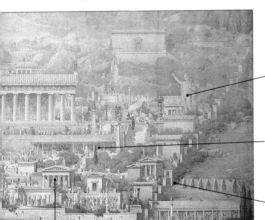

Votive of the Rhodians

Stoa of the Athenians

Sacred Way

Athenian Treasury

Athena Lemnia
This Roman copy of a statue by Pheidias (c.490–c.430 BC), the sculptor-in-charge at the Acropolis, depicts the goddess protector of Athens in an ideal rather than realistic way, typical of the Classical style in art.

Slave Boy *(400 BC)*
Slaves were fundamental to the Greek economy and used for all types of work. Many slaves were foreign; this boot boy came from as far as Africa.

387 Plato founds Academy in Athens

Sculpture of Plato

337 Foundation of the the League of Corinth legitimizes Philip II's control over the Greek city-states

359 Philip II becomes King of Macedon

400 BC

375 BC

350 BC

399 Trial and execution of Socrates

338 Greeks defeated by Philip II of Macedon at Battle of Chaironeia

371 Sparta defeated by Thebes at Battle of Leuktra, heralding a decade of Theban dominance in the area

336 Philip II is assassinated at Aigai and is succeeded by his son, Alexander

Hellenistic Greece

Alexander the Great of Macedon fulfilled his father Philip's plans for the conquest of the Persians. He went on to create a vast empire that extended to India in the east and Egypt in the south. The Hellenistic period was extraordinary for the dispersal of Greek language, religion and culture throughout the territories conquered by Alexander. It lasted from after Alexander's death in 323 BC until the Romans began to dismantle his empire in the mid-2nd century BC. For Greece, Macedonian domination was replaced by that of Rome in AD 168.

Alexander the Great

Relief of Hero-Worship (c.200 BC)
Hero-worship was part of Greek religion. Alexander, however, was worshipped as a god in his lifetime.

Pélla was the birthplace of Alexander and capital of Macedon.

The Mausoleum of Halicarnassus was one of the Seven Wonders of the Ancient World.

Issus, in modern Turkey, was the site of Alexander's victory over the Persian army in 333 BC.

BLACK SEA

Pélla

Athens

Mausoleum of Halicarnassus

Issus

ASIA MINOR

MEDITERRANEAN SEA

Ishtar Gate i Babylon

Ammon

Lighthouse at Alexandria

EGYPT

Alexander died in Babylon in 323 BC.

RED SEA

ARABIA

The Ammon oracle declared Alexander to be divine.

Alexandria, founded by Alexander, replaced Athens as the centre of Greek culture.

Alexander Defeats Darius III
This Pompeiian mosaic shows the Persian leader overwhelmed at Issus in 333 BC. Macedonian troops are shown carrying their highly effective long pikes.

Terracotta Statue
This 2nd-century BC statue of two women gossiping is typical of a Hellenistic interest in private rather than public individuals.

KEY

— — — Alexander's route

☐ Alexander's empire

☐ Dependent regions

TIMELINE

333 Alexander the Great defeats the Persian king, Darius III, and declares himself king of Asia	**301** Battle of Ipsus, between Alexander's rival successors, leads to the break-up of his empire into three kingdoms	**268–261** Chremonidean War, ending with the capitulation of Athens to Macedon
323 Death of Alexander, and of Diogenes		

325 BC	300 BC	275 BC	250 BC

322 Death of Aristotle

287–275 "Pyrrhic victory" of King Pyrros of Epirus who defeated the Romans in Italy but suffered heavy losses

331 Alexander founds Alexandria after conquering Egypt

Diogenes, the Hellenistic philosopher

Fusing Eastern and Western Religion
This plaque from Afghanistan shows the Greek goddess Nike, and the Asian goddess Cybele, in a chariot pulled by lions.

Susa, capital of the Persian Empire, was captured in 331 BC. A masswedding of Alexander's captains to Asian brides was held in 324 BC.

Alexander chose his wife, Roxane, from among Sogdian captives in 327 BC.

WHERE TO SEE HELLENISTIC GREECE

The Aegean was ruled by the Ptolemies in the 3rd and 2nd centuries BC from ancient Thíra on Santoríni, where there are Hellenistic remains: the Sanctuary of Artemídoros of Perge, the Royal Portico, and the Temple of Ptolemy III *(see p240),* houses a collection of Hellenistic sculpture. In Rhodes town, the Hospital of the Knights, now the Archaeological Museum *(p184),* houses a collection of Hellenistic sculpture. The Asklepieion on Kos *(p172)* was the seat of an order of medical priests. The Tower of the Winds *(p287),* in Athens, was built by the Macedonian astronomer Andronikos Kyrrestes.

Roxane
• Alexandropolis

SOGDIANI

• Taxil

Battle elephants were used against the Indian King Poros in 326 BC.

Alexander's army turned back at the River Beas.

BACTRIA

PERSIA

• Susa

War elephant

Sculpture from Persepolis

INDIA

GEDROSIA

CASPIAN SEA

PERSIAN GULF

ARABIAN SEA

Beas

The Persian religious centre of Persepolis, in modern Iran, fell to Alexander in 330 BC.

Alexander's army suffered heavy losses in the Gedrosia desert.

The Death of Archimedes
Archimedes was the leading Hellenistic scientist and mathematician. This mosaic from Renaissance Italy shows his murder in 212 BC by a Roman.

ALEXANDER THE GREAT'S EMPIRE
In forming his empire Alexander covered huge distances. After defeating the Persians in Asia he moved to Egypt, then returned to Asia to pursue Darius, and then his murderers, into Bactria. In 326 his troops revolted in India and refused to go on. Alexander died in 323 in Babylon.

227 Colossus of Rhodes destroyed by earthquake

Colossus of Rhodes

197 Romans defeat Philip V of Macedon and declare Greece liberated

146 Romans sack Corinth and Greece becomes a province of Rome

| 225 BC | 200 BC | 175 BC | 150 BC |

222 Macedon crushes Sparta

217 Peace of Náfpaktos: a call for the Greeks to settle their differences before "the cloud in the west" (Rome) settles over them

168 Macedonians defeated by Romans at Pydna

Roman coin (196 BC) commemorating Roman victory over the Macedonians

Roman Greece

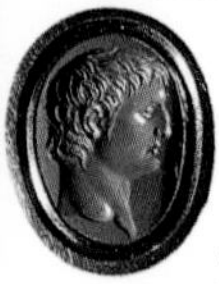

Mark Antony

After the Romans gained control of Greece with the sack of Corinth in 146 BC, Greece became the cultural centre of the Roman Empire. The Roman nobility sent their sons to be educated in the schools of philosophy in Athens. The end of the Roman civil wars between leading Roman statesmen was played out on Greek soil, finishing in the Battle of Actium in Thessaly in 31 BC. In AD 323 the Emperor Constantine founded the new eastern capital of Constantinople; the empire was later divided into the Greek-speaking East and the Latin-speaking West.

THRACIA
MACEDONIA
EPIRUS
ASIA
ACHAIA
CRETE

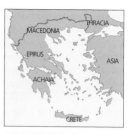

ROMAN PROVINCES, AD 211

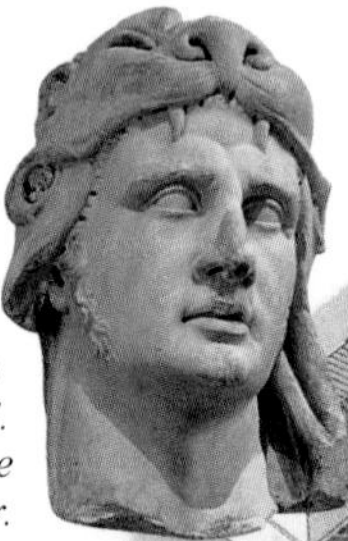

Mithridates
In a bid to extend his territory, this ruler of Pontus, on the Black Sea, led the resistance to Roman rule in 88 BC. He was forced to make peace three years later.

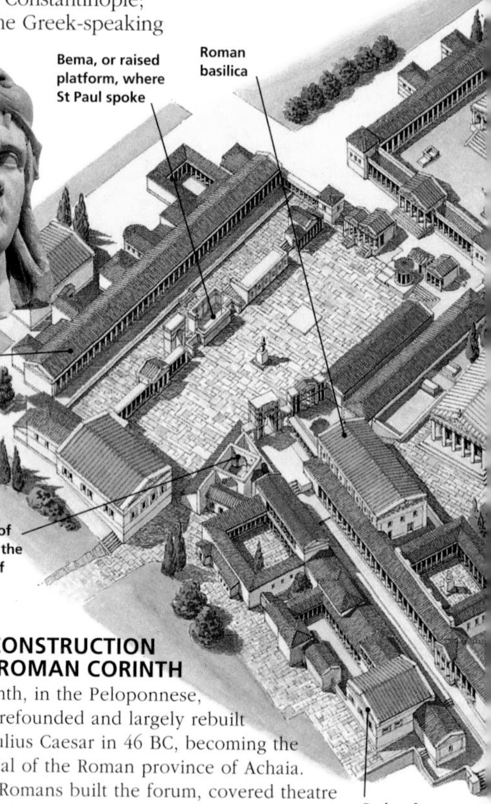

Bema, or raised platform, where St Paul spoke
Roman basilica
Bouleuterion
Springs of Peirene, the source of water
Baths of Eurycles

RECONSTRUCTION OF ROMAN CORINTH
Corinth, in the Peloponnese, was refounded and largely rebuilt by Julius Caesar in 46 BC, becoming the capital of the Roman province of Achaia. The Romans built the forum, covered theatre and basilicas. St Paul visited the city in AD 50–51, working as a tent maker.

ACHAIA

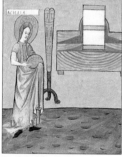

Notitia Dignitatum *(AD 395)*
As part of the Roman Empire, Greece was split into several provinces. The proconsul of the province of Achaia used this insignia.

TIMELINE

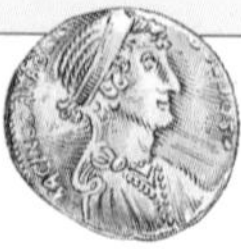

A coin of Cleopatra, Queen of Egypt

49–31 BC Rome's civil wars end with the defeat of Mark Antony and Cleopatra at Actium, in Greece

AD 49–54 St Paul preaches Christianity in Greece

AD 124–131 Emperor Hadrian oversees huge building programme in Athens

100 BC | AD 1 | AD 10

86 BC Roman commander, Sulla, captures Athens

46 BC Corinth refounded as Roman colony

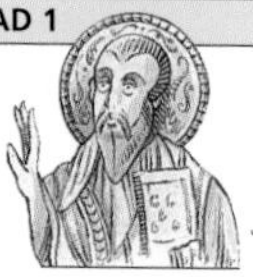

St Paul preaching

AD 66–7 Emperor Nero tours Greece

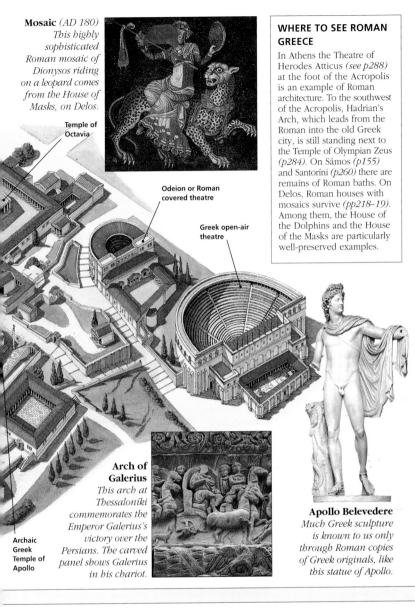

Mosaic *(AD 180) This highly sophisticated Roman mosaic of Dionysos riding on a leopard comes from the House of Masks, on Delos.*

Temple of Octavia

Odeion or Roman covered theatre

Greek open-air theatre

WHERE TO SEE ROMAN GREECE

In Athens the Theatre of Herodes Atticus *(see p288)* at the foot of the Acropolis is an example of Roman architecture. To the southwest of the Acropolis, Hadrian's Arch, which leads from the Roman into the old Greek city, is still standing next to the Temple of Olympian Zeus *(p284)*. On Sámos *(p155)* and Santoríni *(p260)* there are remains of Roman baths. On Delos, Roman houses with mosaics survive *(pp218–19)*. Among them, the House of the Dolphins and the House of the Masks are particularly well-preserved examples.

Archaic Greek Temple of Apollo

Arch of Galerius *This arch at Thessaloníki commemorates the Emperor Galerius's victory over the Persians. The carved panel shows Galerius in his chariot.*

Apollo Belevedere *Much Greek sculpture is known to us only through Roman copies of Greek originals, like this statue of Apollo.*

170 Pausanias completes Guide to Greece for Roman travellers

267 Goths pillage Athens

324 Constantine becomes sole emperor of Roman Empire and establishes his capital in Constantinople

395 Goths devastate Athens and Peloponnese

381 Emperor Theodosius I makes Christianity state religion

AD 200

AD 300

Coin of the Roman Emperor Galerius

293 Under Emperor Galerius, Thessaloníki becomes second city to Constantinople

393 Olympic games banned

395 Death of Theodosius I; formal division of Roman Empire into Latin West and Byzantine East

Byzantine and Crusader Greece

Byzantine court dress arm band

Under the Byzantine Empire, which at the end of the 4th century succeeded the old Eastern Roman Empire, Greece became Orthodox in religion and was split into administrative *themes*. When the capital, Constantinople, fell to the Crusaders in 1204 Greece was again divided, mostly between the Venetians and the Franks. Constantinople and Mystrás were recovered by the Byzantine Greeks in 1261, but the Turks' capture of Constantinople in 1453 marked the final demise of the Byzantine Empire. It left a legacy of hundreds of churches and a wealth of religious art.

BYZANTINE GREECE IN THE 10TH CENTURY

Watch-tower of Tsimiskis

Chapel

Refectory

Two-Headed Eagle
In the Byzantine world, the emperor was also patriarch of the church, a dual role represented in this pendant of a two-headed eagle.

GREAT LAVRA

This monastery is the earliest (AD 963) and largest of the religious complexes on Mount Athos in northern Greece. Many parts have been rebuilt, but its appearance remains essentially Byzantine. The monasteries became important centres of learning and religious art.

Defence of Thessaloníki
The fall of Thessaloníki to the Saracens in AD 904 was a blow to the Byzantine Empire. Many towns in Greece were heavily fortified against attack from this time.

TIMELINE

578–86 Avars and Slavs invade Greece

Gold solidus of the Byzantine Empress Irene, who ruled AD 797–802

400	600	800
529 Aristotle's and Plato's schools of philosophy close as Christian culture supplants Classical thought	**680** Bulgars cross Danube and establish empire in northern Greece	**726** Iconoclasm introduced by Pope Leo III (abandoned in 843) **841** Parthenon becomes a cathedral

Constantine the Great

The first eastern emperor to recognize Christianity, Constantine founded the city of Constantinople in AD 324. Here he is shown with his mother, Helen.

Cypress tree of Agios Athanásios

Christ Pantokrátor

This 14th-century fresco of Christ as ruler of the world is in the Byzantine city and monastic centre of Mystrás.

WHERE TO SEE BYZANTINE AND CRUSADER GREECE

In Athens, the Benáki Museum *(see p291)* contains icons, metalwork, sculpture and textiles. On Pátmos, the treasury of the Monastery of St John, founded in 1088 *(pp164–5)*, is the richest outside Mount Athos. The 11th-century convent of Néa Moní on Chíos *(pp150–51)* has magnificent gold-ground mosaics. The medieval architecture of the Palace of the Grand Masters and the Street of the Knights on Rhodes *(pp186–9)* is particularly fine. Buildings by the Knights on Kos *(pp170–73)* are also worth seeing. The Venetian castle on Páros *(p219)* dates from 1260.

Chapel of Agios Athanásios, founder of Great Lávra

Combined library and treasury

Fortified walls

The katholikón, the main church in Great Lávra, has the most magnificent Byzantine murals on Mount Athos.

Venetian and Ottoman Greece

Venetian lion of St Mark

Following the Ottomans' momentous capture of Constantinople in 1453, and their conquest of almost all the remaining Greek territory by 1460, the Greek state effectively ceased to exist for the next 350 years. Although the city became the capital of the vast Ottoman Empire, it remained the principal centre of Greek population and the focus of Greek dreams of resurgence. The small Greek population of what today is modern Greece languished in an impoverished and underpopulated backwater, but even there rebellious bands of brigands and private militias were formed. The Ionian Islands, Crete and a few coastal enclaves were seized for long periods by the Venetians – an experience more intrusive than the inefficient tolerance of the Ottomans, but one which left a rich cultural and architectural legacy.

GREECE IN 1493

- Areas occupied by Venetians
- Areas occupied by Ottomans

Cretan Painting
This 15th-century icon is typical of the style developed by Greek artists in the School of Crete, active until the Ottomans took Crete in 1669.

Battle of Lepanto *(1571)*
The Christian fleet, under Don John of Austria, decisively defeated the Ottomans off Náfpaktos, halting their advance westwards.

ARRIVAL OF TURKISH PRINCE CEM ON RHODES

Prince Cem, Ottoman rebel and son of Mehmet II, fled to Rhodes in 1481 and was welcomed by the Christian Knights of St John *(see pp188–9)*. In 1522, however, Rhodes fell to the Ottomans after a siege.

TIMELINE

1453 Mehmet II captures Constantinople which is re-named Istanbul and made capital of the Ottoman Empire

1503 Ottoman Turks win control of the Peloponnese apart from Monemvasía

1571 Venetian and Spanish fleet defeats Ottoman Turks at the Battle of Lepanto

1500	1550	1600

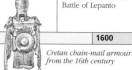

Cretan chain-mail armour from the 16th century

1460 Turks capture Mystrás

1456 Ottoman Turks occupy Athens

1522 The Knights of St John forced to cede Rhodes to the Ottomans

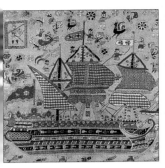

Shipping

Greek merchants traded throughout the Ottoman Empire. By 1800 there were merchant colonies in Constantinople and as far afield as London and Odessa. This 19th-century embroidery shows the Turkish influence on Greek decorative arts.

WHERE TO SEE VENETIAN AND OTTOMAN ARCHITECTURE

The Ionian Islands are particularly rich in buildings dating from the Venetian occupation. The old town of Corfu (*see pp74–7*) is dominated by its two Venetian fortresses. The citadel in Zákynthos (*p90*) is also Venetian. Crete has a number of Venetian buildings: the old port of Irákleio (*pp268–9*) and some of the back streets of Chaniá (*pp252–3*) convey an overwhelming feeling of Venice. Irákleio's fort withstood the Great Siege of 1648–69. Some Ottoman-era houses survive on Thásos (*p131*). Several mosques and other Ottoman buildings, including a library and *hammam* (baths), can be seen in Rhodes old town (*pp182–91*).

The Knights of St John defied the Turks until 1522.

The massive fortifications eventually succumbed to Turkish artillery.

The Knights supported Turkish rebel, Prince Cem.

Dinner at a Greek House in 1801

Nearly four centuries of Ottoman rule profoundly affected Greek culture, ethnic composition and patterns of everyday life. Greek cuisine incorporates Turkish dishes still found throughout the old Ottoman Empire.

1687 Parthenon seriously damaged during Venetian artillery attack on Turkish magazine

1715 Turks reconquer the Peloponnese

Ali Pasha (1741–1822), a governor of the Ottoman Empire

1814 Britain gains possession of Ionian Islands

| 1650 | 1700 | 1750 | 1800 |

Parthenon blown up

1684 Venetians reconquer the Peloponnese

1778 Ali Pasha becomes Vizier of Ioánnina and establishes powerful state in Albania and northern Greece

1801 Frieze on Parthenon removed by Lord Elgin

1814 Foundation of *Filikí Etaireía*, Greek liberation movement

The Making of Modern Greece

Flag with the symbols of the *Filiki Etaireía*

The Greek War of Independence marked the overthrow of the Ottomans and the start of the "Great Idea", an ambitious project to bring all Greek people under one flag *(Enosis)*. The plans for expansion were initially successful, and during the 19th century the Greeks succeeded in doubling their national territory and reasserting Greek sovereignty over many of the islands. However, an attempt to take the city of Constantinople by force after World War I ended in disaster: in 1922 millions of Greeks were expelled from Smyrna in Turkish Anatolia, ending thousands of years of Greek presence in Asia Minor.

THE EMERGING GREEK STATE

  Greece in 1832

  Areas gained 1832–1923

Klephts (mountain brigands) were the basis of the Independence movement.

Massacre at Chíos
This detail of Delacroix's shocking painting Scènes de Massacres de Scio *shows the events of 1822, when Turks took savage revenge for an earlier killing of Muslims.*

Weapons were family heirlooms or donated by philhellenes.

Declaration of the Constitution in Athens
Greece's Neo-Classical parliament building in Athens was the site of the Declaration of the Constitution in 1843. It was built as the Royal Palace for Greece's first monarch, King Otto, in the 1830s.

TIMELINE

1824 The poet Lord Byron dies of a fever at Mesolóngi

1831 President Kapodístrias assassinated

1832 Great Powers establish protectorate over Greece and appoint Otto, Bavarian prince, as king

1834 Athens replaces Náfplio as capital

German archaeologist Heinrich Schliemann

1830	1840	1850	1860	187

1827 Battle of Navaríno

1828 Ioánnis Kapodístrias becomes first President of Greece

King Otto (ruled 1832–62)

1862 Revolution drives King Otto from Greece

1874 Heinrich Schliemann begins excavation of Mycenae

1821 Greek flag of independence raised on 25 March; Greeks massacre Turks at Tripolitsá in Morea

1864 New constitution makes Greece a "crowned democracy"; Greek Orthodoxy made the state religion

Life in Athens
By 1836 urban Greeks still wore a mixture of Greek traditional and Western dress. The Ottoman legacy had not totally disappeared and is visible in the fez worn by men.

WHERE TO SEE 19TH-CENTURY GREECE

In Crete, Moní Arkadíou *(see p260)* is the site of mass suicide by freedom fighters in 1866; the tomb of Venizélos is at Akrotíri *(p251)*. The harbour and surrounding buildings at Sýros *(p220)* are evidence of the importance of Greek seapower in the 19th century.

FLAG RAISING OF 1821 REVOLUTION

In 1821, the Greek secret society *Filikí Etaireía* was behind a revolt by Greek officers which led to anti-Turk uprisings throughout the Peloponnese. Tradition credits Archbishop Germanós of Pátra with raising the rebel flag near Kalávryta in the Peloponnese on 25 March. The struggle for independence had begun.

Corinth Canal
This spectacular link between the Aegean and Ionian seas opened in 1893.

Elefthérios Venizélos
This great Cretan politician and advocate of liberal democracy doubled Greek territory during the Balkan Wars (1912–13) and joined the Allies in World War I.

1880	1890	1900	1910	1920

1893 Opening of Corinth Canal

1896 First Olympics of modern era, held in Athens

1908 Crete united with Greece

1921 Greece launches offensive in Asia Minor

1917 King Constantine is deposed; Greece joins World War I

1922 Turkish burning of Smyrna signals end of the "Great Idea"

Spyrídon Louis, Marathon winner at the first modern Olympics

1899 Arthur Evans begins excavations at Knosós

1912–13 Greece extends its borders during the Balkan Wars

1920 Treaty of Sèvres gives Greece huge gains in territory

1923 Population exchange agreed between Greece and Turkey at Treaty of Lausanne. Greece loses previous gains

Twentieth-Century Greece

The years after the 1922 defeat by Turkey were terrible ones for Greek people. The influx of refugees contributed to the political instability of the interwar years. The dictatorship of Metaxás was followed by invasion in 1940, then Italian, German and Bulgarian occupation and, finally, the Civil War between 1946 and 1949, with its legacy of division. After experiencing the Cyprus problem of the 1950s and the military dictatorship of 1967 to 1974, Greece is now an established democracy and became a member of the European Economic and Monetary Union in 2000.

1947 Internationally acclaimed Greek artist, Giánnis Tsaroúchis, holds his first exhibition of set designs, in the Romvos Gallery, Athens

1938 Death of sculptor Giannoúlis Chalepás, best known for his *Sleeping Girl* funerary statue

1946 Government institutes "White Terror" against Communists

1945 Níkos Kazantzákis publishes *Zorba the Greek*, later made into a film

1933 Death of Greek poet, Constantine (C P) Cavafy

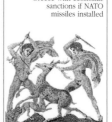

1958 USSR threatens Greece with economic sanctions if NATO missiles installed

1957 Mosaics found by chance at Philip II's 300-BC palace at Pélla

1967 Right-wing colonels form Junta, forcing King Constantine into exile

1925	1935	1945	1955	19

1925	1935	1945	1955	19

1939 Greece declares neutrality at start of World War II

1932 Aristotle Onassis purchases six freight ships, the start of his shipping empire

1951 Greece enters NATO

1948 Dodecanese becomes part of Greece

1955 Greek Cypriots start campaign of violence in Cyprus against British rule

1925 Mános Chatzidákis, who wrote music for the film *Never on Sunday*, is born

1946–9 Civil War between Greek government and the Communists who take to the mountains

1960 Cyprus declared independent

ΟΙ ΗΡΩΙΔΕΣ ΤΟΥ 1940

1940 Italy invades Greece. Greek soldiers defend northern Greece. Greece enters World War II

1944 Churchill visits Athens to show his support for Greek government against Communist Resistance

1963 Geórgios Papandréou's centre-left government voted into power

1973 University students in Athens rebel against dictatorship and are crushed by military forces. Start of decline in power of dictatorship

1988 Eight million visitors to Greece; tourism continues to expand

1993 Andréas Papandréou wins Greek general election for the third time

ΕΛΛΗΝΙΚΗ ΔΗΜΟΚΡΑΤΙΑ 60 HELLAS

1981 Melína Merkoúri appointed Minister of Culture. Start of campaign to restore Elgin Marbles to Greece

2002 Drachma replaced by the Euro at the beginning of March

1994 Because of the choking smog *(néfos)* central Athens introduces traffic restrictions

2004 The Olympic Games take place in Athens

1974 Cyprus is partitioned after Turkish invasion

1975	1985	1995	2005	2010

1975	1985	1995	2005	2010

1975 Death of Aristotle Onassis

1990 New Democracy voted into power; Konstantínos Karamanlís becomes President

2004 Greece win Euro 2004 Football Championship

1974 Fall of Junta; Konstantínos Karamanlís elected Prime Minister

1981 Andréas Papandréou's left-wing PASOK party forms first Greek Socialist government

1998 Karamanlís dies. Kostis Stefanopoulos succeeds him

1973 Greek bishops give their blessing to the short-lived presidency of Colonel Papadópoulos

1997 Athens is awarded the 2004 Olympics

1994 European leaders meet in Corfu under Greek presidency of the EU

1996 Andréas Papandréou dies; Kóstas Simítis succeeds him

THE GREEK ISLANDS THROUGH THE YEAR

May Day wreath

Greek island life revolves around the seasons, and is punctuated by saints' days and colourful religious festivals, or *panigýria*. Easter is the most important Orthodox festival of the year, but there are lively pre-Lenten carnivals on some islands as well. The Greeks mix piety and pleasure, with a great enthusiasm for their celebrations, from the most important to the smallest village fair. There are also festivals that have ancient roots in pagan revels. Other festivals celebrate harvests of local produce, such as grapes, olives and corn, or re-enact various victories for Greece in its struggle for Independence.

SPRING

The Greek word for spring is *ánoixi* (the opening), and it heralds the beginning of the tourist season on the islands. After wintering in Athens or Rhodes, hoteliers and

Children in national dress, 25 March

shopkeepers head for the smaller islands to open up. The islands in spring are at their most beautiful, carpeted with red poppies, camomile and wild cyclamen. Fruit trees are in blossom, fishing boats and houses are freshly painted and people are at their most welcoming. Orthodox Easter is the main spring event, preceded in late February or March with pre-Lenten carnivals. While northern island groups can be showery, by late April, Crete, the Dodecanese and east Aegean islands are usually warm and sunny.

MARCH

Apókries, or Carnival Sunday *(first Sun before Lent)*. There are carnivals on many islands for three weeks leading up to this date, the culmination of pre-Lenten festivities. Celebrations are exuberant at Agiásos on Lésvos and on Kárpathos, while a goat dance is performed on Skýros.
Kathari Deftéra, or Clean Monday *(seven Sundays before Easter)*. This marks the start of Lent. Houses are spring-cleaned and the unleavened bread *lagána* is baked. Clean Monday is also the day for a huge kite-flying contest that takes place in Chalkída on Evvoia.

CELEBRATING EASTER IN GREECE

Greek Orthodox Easter can fall up to three weeks either side of Western Easter. It is the most important religious festival in Greece, and Holy Week is a time for Greek families to reunite. It is also a good time to visit Greece, to see the processions and church services and to sample the Easter food. The ceremony and symbolism is a direct link with Greece's Byzantine past, as well as with earlier more primitive beliefs. The festivities reach a climax at midnight on Easter Saturday when, as priests intone "Christ is risen", fireworks explode to usher in a Sunday of feasting, music and dancing. The Sunday feasting on roast meat marks the end of the Lenten fast, and a belief in the renewal of life in spring. Particularly worthwhile visiting for the Holy Week processions and the Friday and Saturday night services are Olympos on Kárpathos, Ydra, Pátmos and just about any village on Crete.

Priests in robes at the Easter parade of icons

Christ's bier, *decorated with flowers and containing His effigy, is carried in solemn procession through the streets at dusk on Good Friday.*

Candle lighting *takes place at the end of the Easter Saturday Mass. In pitch darkness, a single flame is used to light the candles held by worshippers.*

A workers' rally in Athens on Labour Day, 1 May

Independence Day and **Evangelismós** *(25 Mar)*. A national holiday, with parades and dances nationwide to celebrate the 1821 revolt against the Ottoman Empire. The religious festival, one of the Orthodox church's most important, marks the Archangel Gabriel's announcement to the Virgin Mary that she was to become the Holy Mother. Name day for Evángelos and Evangelía.

APRIL

Megáli Evdomáda, Holy Week *(Apr or May)*, including *Kyriakí ton Vaïón* (Palm Sunday), *Megáli Pémpti* (Maundy Thursday), *Megáli Paraskeví* (Good Friday),

Megálo Sávvato (Easter Saturday) and the most important date in the Orthodox calendar, *Páscha* (Easter Sunday).

Agios Geórgios, St George's Day *(23 Apr)*. A day for celebrating the patron saint of shepherds. This date traditionally marks the beginning of the grazing season in Greece.

Kite-flying competition in Chalkida, Evvoia

MAY

Protomagiá, May Day or Labour Day *(1 May)*. Traditionally, wreaths made with wild flowers and garlic are hung up to ward off evil. In major towns and cities, the day is marked by workers' demonstrations and rallies.

Agios Konstantínos kai Agía Eléni *(21 May)*. A nationwide celebration for the saint and his mother, the first Orthodox Byzantine rulers.

Análipsi, Ascension *(40 days after Easter, usually in May)*. An important Orthodox feast day, celebrated across the nation.

Easter dancing, *for young and old alike, continues the outdoor festivities after the midday meal on Sunday.*

Easter biscuits *celebrate the end of Lent. Another Easter dish, mayerítsa soup, is made of lamb's innards and is eaten in the early hours of Easter Sunday.*

Egg loaves, *made of sweet plaited dough, contain eggs with shells dyed red to symbolize the blood of Christ. Red eggs are also traditionally given as presents on Easter Sunday.*

Lamb roasting *is traditionally done in the open air on giant spits over charcoal, for lunch on Easter Sunday. The first retsina wine from last year's harvest is opened and for dessert there are sweet cinnamon-flavoured pastries.*

Harvesting barley in July, on the island of Folégandros

SUMMER

With islands parched and sizzling, the tourist season is now in full swing. Villagers with rooms to let meet backpackers from the ferries, and prices go up. The islands are sometimes cooled by the strong, blustery *meltémi*, a northerly wind from the Aegean, which can blow up at any time to disrupt ferry schedules and delight windsurfers.

In June, the corn is harvested and cherries, apricots and peaches are at their best. In July herbs are gathered and dried, and figs begin to ripen. August sees the mass exodus from Athens to the islands, especially for the festival of the Assumption on 15 August. By late summer the first of the grapes have ripened, while temperatures soar.

Consecrated bread for religious festivals

JUNE

Pentikostí, Pentecost, or Whit Sunday *(seven weeks after Orthodox Easter)*. An important Orthodox feast day, celebrated throughout Greece.
Agíou Pnévmatos, Feast of the Holy Spirit, or Whit Monday *(the following day)*. A national holiday.
Athens Festival *(mid-Jun to mid-Sep)*, Athens. A cultural festival with modern and ancient theatre and music.
Klídonas *(24 Jun)* Chaniá, Crete *(see pp244–5)*. A festival celebrating the custom of water-divining for a husband. An amusing song is sung while locals dance.
Agios Ioánnis, St John's Day *(24 Jun)*. On some ' islands bonfires are lit on the evening before. May wreaths are consigned to the flames and youngsters jump over the fires.
Agioi Apóstoloi Pétros kai Pávlos, Apostles Peter and Paul *(29 Jun)*. There are festivals at dedicated churches, such as St Paul's Bay, Líndos, Rhodes *(see p197)*.
Agioi Apóstoloi, Holy Apostles *(30 Jun)*. This time the celebrations are for anyone named after one of the 12 apostles.

JULY

Agios Nikódimos *(14 Jul)*, Náxos town. A small folk festival and procession for the town's patron saint.

Festivities on Tínos for Koímisis tis Theotókou, 15 August

Agía Marína *(17 Jul)*. This day is widely celebrated in rural areas, with feasts to honour this saint. She is revered as an important protector of crops and healer of snakebites. There are festivals throughout Crete and at the town of Agía Marína, Léros.
Profítis Ilías, the Prophet Elijah *(18–20 Jul)*. There are high-altitude celebrations in the Cyclades, Rhodes and on Evvoia at the mountain-top chapels dedicated to him. The chapels were built on former sites of Apollo temples.
Agíou Panteleïmonos Festival *(25–28 Jul)*, Tílos *(see p177)*. Three days of song and dance at Moní Agíou Panteleïmonos, culminating in "Dance of the Koupa", or Cup, at Taxiárchis, Megálo Chorió. There are also celebrations at Moní Panachrántou, Andros *(see p209)*.
Simonídeia Festival *(1–19 Aug)*, Kea. A celebration of the work of the island's famous lyric poet, Simonides (556–468 BC), with drama, exhibitions and dance.
Réthymno Festival *(Jul and Aug)*, Réthymno, Crete. The event includes a wine festival and Renaissance fair.

AUGUST

Ippokráteia, Hippocrates Cultural Festival *(throughout Aug)*, Kos *(see p170)*. Art exhibitions are combined with concerts and films, plus the ceremony of the Hippocratic Oath at the Asklepieíon.

One of the many local church celebrations during summer, Pátmos

Dionysía Festival *(first week of Aug)*, Náxos town. A festival of folk dancing in traditional costume, with free food and plenty of wine. **Metamórfosi**, Transfiguration of Christ *(6 Aug)*. An important day in the Orthodox calendar, celebrated throughout Greece. It is a fun day in the Dodecanese, and particularly on the island of Chálki, where you may get pelted with eggs, flour, yoghurt and squid ink. **Koímisis tis Theotókou**, Assumption of the Virgin Mary *(15 Aug)*. A national holiday, and the most important festival in the Orthodox calendar after Easter, and the name day for Maria, Panayiota (female), Panayiotis (male)

AUTUMN

The wine-making months of September and October are still very warm in the Dodecanese, Crete and the Cyclades, although they can be showery further north, and the sea can be rough. October sees the "little summer of St Dimitrios", a pleasant heatwave when the first wine is ready to drink. The shooting season begins and hunters take to the hills in search of pigeon, partridge and other game. The main fishing season begins, with fish such as bream and red mullet appearing on restaurant

The year's first wine

menus. By the end of October many islanders are heading for Athens, packing the ferries and wishing each other

and Despina. Following the long liturgy on the night of the 14th, the icon of the Madonna is paraded and kissed. Then the celebrations proceed, and continue for days, providing an excellent

Women in ceremonial costume, Kárpathos

Kaló Chimóna (good winter). But traditional island life goes on: olives are harvested and strings of garlic, onions and tomatoes are hung up to dry for the winter; flocks of sheep are brought down from the mountains; and fishing nets are mended.

SEPTEMBER

Génnisis tis Theotókou, birth of the Virgin Mary *(8 Sep)*. An important feast day in the Orthodox church calendar. Also on this day, there is a re-enactment of the Battle of Spétses (1822) in the town's harbour *(see p101)*, followed by a fireworks display and feast. **Ypsosis tou Timíou Stavroú**, Exaltation of the True Cross *(14 Sep)*. Though in autumn, this is regarded as the last of Greece's summer festivals. The festivities are celebrated with fervour on Chálki.

opportunity to experience traditional music and spontaneous dance. There are celebrations at Olympos on Kárpathos *(see p203)* and at Panagía Evangelístria on Tínos *(see pp212–3)*.

OCTOBER

Agios Dimítrios *(26 Oct)*. A popular and widely celebrated name day. It is also traditionally the day when the first wine of the year is ready to drink.
Ochi Day *(28 Oct)*. A national holiday, with patriotic parades in the cities, and plenty of dancing. The day commemorates the famous reply by Greece's prime minister of the time, Metaxás, to Mussolini's 1940 call for Greek surrender: an emphatic no *(óchi)*.

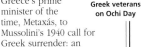
Greek veterans on Ochi Day

NOVEMBER

Ton Taxiarchón Michaíl kai Gavriíl, *(8 Nov)*. Ceremonies at many monasteries named after Archangels Gabriel and Michael, such as at Panormítis, on Sými *(see p179)*. This is an important name day throughout Greece.
Eisódia tis Theotókou, Presentation of the Virgin in the Temple *(21 Nov)*. A religious feast day, and one of the most important for the Orthodox church. Name day for María, Máry.

Strings of tomatoes hanging out to dry in the autumn sunshine

Diving for the cross at Epiphany, 6 Jan

WINTER

Lashed by wild winds and high seas, the islands can be bleak in winter. *Kafeneía* are steamed up and full of men playing cards or backgammon. Women often embroider or crochet, and cook warming stews and soups. Fishermen celebrate Agios Nikólaos, their patron saint, and then preparations get underway for Christmas. The 12-day holiday begins on Christmas Eve, when the wicked goblins, *kallikántzaroi*, are about causing mischief, until the Epiphany in the new year, when they are banished. Pigs are slaughtered for Christmas pork, and cakes representing the swaddling clothes of the infant Christ are made. The Greek Father Christmas comes on New Year's Day and special cakes, called *vasilópita*, are baked with coins inside to bring good luck to the finder.

DECEMBER

Agios Nikólaos *(6 Dec).* This is a celebration for the patron saint of sailors. *Panigýyria* (religious ceremonies)

are held at harbourside churches, and decorated boats and icons are paraded on beaches.

Agios Spyrídon *(12 Dec),* Corfu (*see pp74–9*). A celebration for the patron saint of the island, with a parade of his relics.

Christoúgenna, Christmas *(25 Dec).* A national holiday. Though less significant than Easter in Greece, Christmas is still an important feast day.

Sýnaxis tis Theotókou, meeting of the Virgin's entourage *(26 Dec).* A religious celebration nationwide, and a national holiday. The next day *(27 Dec)* is a popular name day for Stéfanos and Stefanía, commemorating the saint Agios Stéfanos.

JANUARY

Agios Vasíleios, also known as *Protochroniá (1 Jan).* A national holiday to celebrate this saint. The day combines with festivities for the arrival of

Almond biscuits eaten at Christmas and Easter

the new year. Gifts are exchanged and the new year greeting is *Kalí Chroniá.*

Theofánia, or Epiphany *(6 Jan).* A national holiday and an important feast day throughout Greece. There

are special ceremonies to bless the waters at coastal locations throughout many of the islands. A priest at the harbourside throws a crucifix into the water. Young men then dive into the sea for the honour of retrieving the cross.

FEBRUARY

Ypapantí, Candlemas *(2 Feb).* An important Orthodox feast day throughout Greece. This festival celebrates the presentation of the infant Christ at the temple.

Priests in ceremonial robes at Ypapantí, 2 February

MAIN PUBLIC HOLIDAYS

These are the dates when museums and public sites are closed nationwide.

Agios Vasíleios (1 Jan).
Evangelismós (25 Mar).
Protomagiá (1 May).
Megáli Paraskeví (Good Friday).
Páscha (Easter Sunday).
Christoúgenna (25 Dec).
Sýnaxis tis Theotókou (26 Dec).

NAME DAYS

In the past, most Greeks did not celebrate their birthdays past the age of about 12. Instead they celebrated their name days, or *giortí*, the day of the saint after whom they were named at their baptism. Choice of names is very important in Greece, and children are usually named after their grandparents – though in recent years it has become fashionable to give children names from Greece's history and mythology. On St George's day or St Helen's day (21 May) the whole nation seems to celebrate, with visitors dropping in, bearing small gifts, and being given cakes and liqueurs in return. On a friend's name day you may be told *Giortázo símera* (I'm celebrating today) – the traditional reply is *Chrónia pollá* (many years). Today, most people also celebrate their birthdays, regardless of their age.

The Climate of the Greek Islands

Throughout the islands, the tendency is for long, dry summers and mild but rainy winters. The Dodecanese, Cyclades and the Cretan coast are buffeted by a dry north wind called the *meltémi*, which can blow up at any time between June and September, moderating the high temperatures.

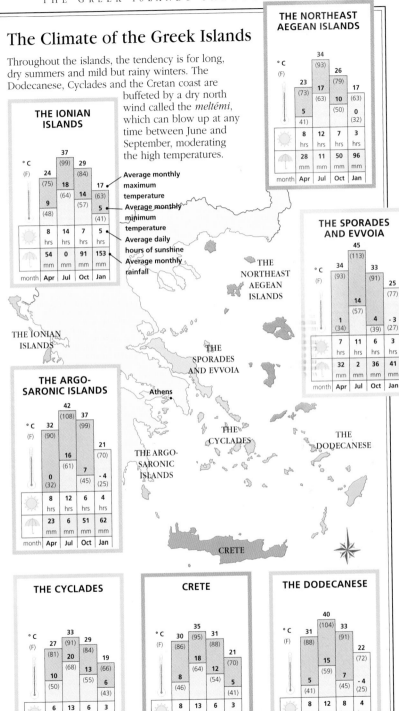

THE NORTHEAST AEGEAN ISLANDS

°C (F)			
23 (73)	34 (93)	26 (79)	17 (63)
5 (41)	17 (63)	10 (50)	0 (32)
8 hrs	12 hrs	7 hrs	3 hrs
28 mm	11 mm	50 mm	96 mm
month Apr	Jul	Oct	Jan

THE IONIAN ISLANDS

°C (F)			
24 (75)	37 (99)	29 (84)	17 (63)
9 (48)	18 (64)	14 (57)	5 (41)
8 hrs	14 hrs	7 hrs	5 hrs
54 mm	0 mm	91 mm	153 mm
month Apr	Jul	Oct	Jan

Average monthly maximum temperature
Average monthly minimum temperature
Average daily hours of sunshine
Average monthly rainfall

THE SPORADES AND EVVOIA

°C (F)			
34 (93)	45 (113)	33 (91)	25 (77)
1 (34)	14 (57)	4 (39)	-3 (27)
7 hrs	11 hrs	6 hrs	3 hrs
32 mm	2 mm	36 mm	41 mm
month Apr	Jul	Oct	Jan

THE ARGO-SARONIC ISLANDS

°C (F)			
32 (90)	42 (108)	37 (99)	21 (70)
0 (32)	16 (61)	7 (45)	-4 (25)
8 hrs	12 hrs	6 hrs	4 hrs
23 mm	6 mm	51 mm	62 mm
month Apr	Jul	Oct	Jan

THE CYCLADES

°C (F)			
27 (81)	33 (91)	29 (84)	19 (66)
10 (50)	20 (68)	13 (55)	6 (43)
6 hrs	13 hrs	6 hrs	3 hrs
19 mm	2 mm	45 mm	91 mm
month Apr	Jul	Oct	Jan

CRETE

°C (F)			
30 (86)	35 (95)	31 (88)	21 (70)
8 (46)	18 (64)	12 (54)	5 (41)
8 hrs	13 hrs	6 hrs	3 hrs
26 mm	1 mm	64 mm	95 mm
month Apr	Jul	Oct	Jan

THE DODECANESE

°C (F)			
31 (88)	40 (104)	33 (91)	22 (72)
5 (41)	15 (59)	7 (45)	-4 (25)
8 hrs	12 hrs	8 hrs	4 hrs
25 mm	3 mm	61 mm	149 mm
month Apr	Jul	Oct	Jan

THE NORTHEAST AEGEAN ISLANDS

THE IONIAN ISLANDS

THE SPORADES AND EVVOIA

Athens

THE CYCLADES

THE DODECANESE

THE ARGO-SARONIC ISLANDS

CRETE

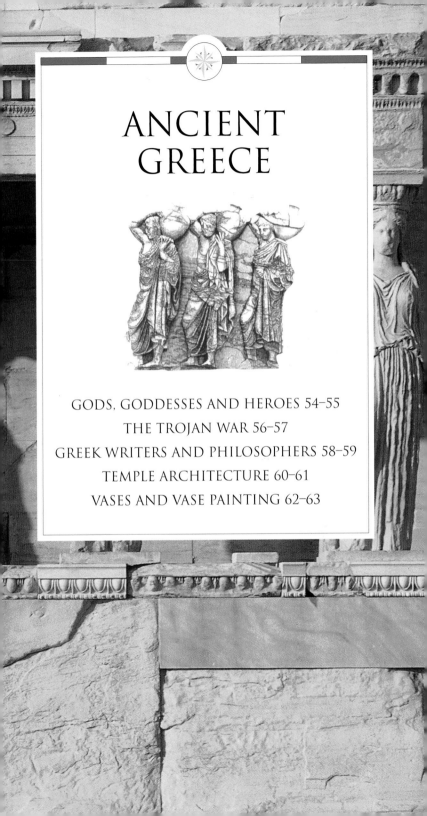

ANCIENT GREECE

Gods, Goddesses and Heroes

The Greek myths that tell the stories of the gods, goddesses and heroes date back to the Bronze Age when they were told aloud by poets. They were first written down in the early 6th century BC and have lived on in Western literature. Myths were closely bound up with Greek religion and gave meaning to the unpredictable workings of the natural world. They tell the story of the creation and the "golden age" of gods and mortals, as well as the age of semimythical heroes, such as Theseus and Herakles, whose exploits were an inspiration to ordinary men. The gods and goddesses were affected by human desires and failings and were part of a divine family presided over by Zeus. He had many offspring, both legitimate and illegitimate, each with a mythical role.

Hades and Persephone *were king and queen of the Underworld (land of the dead). Persephone was abducted from her mother Demeter, goddess of the harvest, by Hades. She was then only permitted to return to her mother for three months each year.*

Zeus was the father of the gods and ruled over them and all mortals from Mount Olympos.

Aris was the goddess of strife.

Clymene, a nymph and daughter of Helios, was mother of Prometheus, creator of mankind.

Poseidon, *one of Zeus's brothers, was given control of the seas. The trident is his symbol of power, and he married the sea-goddess Amphitrite, to whom he was not entirely faithful. This statue is from the National Archaeological Museum in Athens (see p286).*

Hera, sister and wife of Zeus, was famous for her jealousy.

Athena was born from Zeus's head in full armour.

Paris was asked to award the golden apple to the most beautiful goddess.

Paris's dog helped him herd cattle on Mount Ida where the prince grew up.

Dionysos, *god of revelry and wine, was born from Zeus's thigh. In this 6th-century BC cup, painted by Exekias, he reclines in a ship whose mast has become a vine.*

A DIVINE DISPUTE

This vase painting shows the gods on Mount Ida, near Troy. Hera, Athena and Aphrodite, quarrelling over who was the most beautiful, were brought by Hermes to hear the judgment of a young herdsman, the Trojan prince, Paris. In choosing Aphrodite, he was rewarded with the love of Helen, the most beautiful woman in the world. Paris abducted her from her husband Menelaos, King of Sparta, and thus the Trojan War began (*see pp56–7*).

◁ **The statues of the Caryatids on the Erechtheion, Athens**

Artemis, *the chaste goddess of the hunt, was the daughter of Zeus and sister of Apollo. She can be identified by her bow and arrows, hounds and group of nymphs with whom she lived in the forests. Artemis was also the goddess of childbirth.*

Happiness, here personified by two goddesses, waits with gold laurel leaves to garland the winner. Wreaths were the prizes in Greek athletic and musical contests.

Helios, the sun god, drove his four-horse chariot (the sun) daily across the sky.

Hermes was the gods' messenger.

Aphrodite, the goddess of love, was born from the sea. Here she has her son Eros (Cupid) with her.

Apollo, *son of Zeus and brother of Artemis, was god of healing, plague and also music. Here he is depicted holding a lyre. He was also famous for his dazzling beauty.*

THE LABOURS OF HERAKLES

Herakles (Hercules to the Romans) was the greatest of the Greek heroes, and the son of Zeus and Alkmene, a mortal woman. With superhuman strength he achieved success, and immortality, against seemingly impossible odds in the "Twelve Labours" set by Eurystheus, King of Mycenae. For his first task he killed the Nemean lion, and wore its hide ever after.

Killing the Lernaean hydra *was the second labour of Herakles. The many heads of this venomous monster, raised by Hera, grew back as soon as they were chopped off. As in all his tasks, Herakles was helped by Athena.*

The huge boar *that ravaged Mount Erymanthus was captured next. Herakles brought it back alive to King Eurystheus who was so terrified that he hid in a storage jar.*

Destroying the Stymfalian birds *was the sixth labour. Herakles rid Lake Stymfalia of these man-eating birds, which had brass beaks, by stoning them with a sling, having first frightened them off with a pair of bronze castanets.*

The Trojan War

The story of the Trojan War, first narrated in the *Iliad*, Homer's 8th-century BC epic poem, tells how the Greeks sought to avenge the capture of Helen, wife of Menelaos, King of Sparta, by the Trojan prince, Paris. The Roman writer Virgil takes up the story in the *Aeneid*, where he tells of the sack of Troy and the founding of Rome. Recent archaeological evidence of the remains of a city identified with ancient Troy in modern Turkey suggests that the myth may have a basis in fact. Many of the ancient sites in the Peloponnese, such as Mycenae and Pylos, are thought to be the cities of some of the heroes of the Trojan War.

Ajax carrying the body of the dead Achilles

Achilles binding up the battle wounds of his friend Patroklos

GATHERING OF THE HEROES

When Paris *(see p54)* carries Helen back to Troy, her husband King Menelaos summons an army of Greek kings and heroes to avenge this crime. His brother, King Agamemnon of Mycenae, leads the force; its ranks include young Achilles, destined to die at Troy.

At Aulis their departure is delayed by a contrary wind. Only the sacrifice to Artemis of Iphigeneia, the youngest of Agamemnon's daughters, allows the fleet to depart.

FIGHTING AT TROY

The Iliad opens with the Greek army outside Troy, maintaining a siege that has already been in progress for nine years. Tired of fighting, yet still

hoping for a decisive victory, the Greek camp is torn apart by the fury of Achilles over Agamemnon's removal of his slave girl Briseis. The hero takes to his tent and refuses adamantly to fight.

Deprived of their greatest warrior, the Greeks are driven back by the Trojans. In desperation, Patroklos persuades his friend Achilles to let him borrow his armour. Achilles agrees and Patroklos leads the Myrmidons, Achilles' troops, into battle. The tide is turned, but Patroklos is killed in the fighting by Hector, son of King Priam of Troy, who mistakes him for Achilles. Filled with remorse at the news of his friend's death, Achilles returns to battle, finds Hector, and kills him in revenge.

King Priam begging Achilles for the body of his son

PATROKLOS AVENGED

Refusing Hector's dying wish to allow his body to be ransomed, Achilles instead hitches it up to his chariot by the ankles and drags it round the walls of Troy, then takes it back to the Greek camp. In contrast, Patroklos is given the most elaborate funeral possible with a huge pyre, sacrifices of animals and Trojan prisoners and funeral games. Still unsatisfied, for 12 days Achilles drags the corpse of Hector around Patroklos's funeral mound until the gods are forced to intervene over his callous behaviour.

PRIAM VISITS ACHILLES

On the instructions of Zeus, Priam sets off for the Greek camp holding a ransom for the body of his dead son. With the help of the god Hermes he reaches Achilles' tent undetected. Entering, he pleads with Achilles to think of his own father and to show mercy. Achilles relents and allows Hector to be taken back to Troy for a funeral and burial.

Although the Greek heroes were greater than mortals, they were portrayed as fallible beings with human emotions who had to face universal moral dilemmas.

Greeks and Trojans, in bronze armour, locked in combat

ACHILLES KILLS THE AMAZON QUEEN

Penthesileia was the Queen of the Amazons, a tribe of warlike women reputed to cut off their right breasts to make it easier to wield their weapons. They come to the support of the Trojans. In the battle, Achilles finds himself face to face with Penthesileia and deals her a fatal blow. One version of the story has it that as their eyes meet at the moment of her death, they fall in love. The Greek idea of love and death would be explored 2,000 years later by the psychologists Jung and Freud.

Achilles killing the Amazon Queen Penthesileia in battle

THE WOODEN HORSE OF TROY

As was foretold, Achilles *(see p83)* is killed at Troy by an arrow in his heel from Paris's bow. With this weakening of their military strength, the Greeks resort to guile.

Before sailing away they build a great wooden horse, in which they conceal some of their best fighters. The rumour is put out that this is a gift to the goddess Athena and that if the horse enters Troy, the city can never be taken. After some doubts, but swayed by supernatural omens, the Trojans drag the horse inside the walls. That night, the Greeks sail back, the soldiers creep out of the horse and Troy is put to the torch. Priam, with many others, is murdered. Among

An early image of the Horse of Troy, from a 7th-century BC clay vase

the Trojan survivors is Aeneas who escapes to Italy and founds the race of Romans: a second Troy. The next part of the story (the *Odyssey*) tells of the heroes' adventures on their way home *(see p87)*.

DEATH OF AGAMEMNON

Klytemnestra, the wife of Agamemnon, had ruled Mycenae in the ten years that he had been away fighting in Troy. She was accompanied by Aigisthos, her lover. Intent on vengeance for the death of her daughter Iphigeneia, Klytemnestra receives her husband with a triumphal welcome and then brutally murders him, with the help of Aigisthos. Agamemnon's fate was a result of a curse laid on his father, Atreus, which was finally expiated by the murder of both Klytemnestra and Aigisthos by her son Orestes and daughter Elektra. In these myths, the will of the gods both shapes and overrides that of heroes and mortals.

GREEK MYTHS IN WESTERN ART

From the Renaissance onwards, the Greek myths have been a powerful inspiration for artists and sculptors. Kings and queens have had themselves portrayed as gods and goddesses with their symbolic attributes of love or war. Myths have also been an inspiration for artists to paint the nude or Classically draped figure. This was true of the 19th-century artist Lord Leighton, whose depiction of the human body reflects the Classical ideals of beauty. His tragic figure of Elektra is shown here.

Elektra mourning the death of her father Agamemnon at his tomb

Greek Writers and Philosophers

Playwrights Aristophanes and Sophocles

The literature of Greece began with long epic poems, accounts of war and adventure, which established the relationship of the ancient Greeks to their gods. The tragedy and comedy, history and philosophical dialogues of the 5th and 4th centuries BC became the basis of Western literary culture. Much of our knowledge of the Greek world is derived from Greek literature. Pausanias's *Guide to Greece,* written in the Roman period and used by Roman tourists, is a key to the physical remains.

of the life of a very competitive elite. Since *symposia* were an almost exclusively male domain, there is a strong element of misogyny in much of this poetry. In contrast, the fragments of poems discovered by the poet Sappho, who lived on the island of Lésvos, are exceptional for showing a woman competing in a literary area in the male-dominated society of ancient Greece, and for describing with great intensity her passions for other women.

Hesiod with the nine Muses who inspired his poetry

nothing reliable is known. Hesiod, whose most famous poems include the *Theogony,* a history of the gods, and the *Works and Days,* on how to live an honest life, also lived around 700 BC. Unlike Homer, Hesiod is thought to have written down his poems, although there is no firm evidence available to support this theory.

EPIC POETRY

As far back as the 2nd millennium BC, before even the building of the Mycenaean palaces, poets were reciting the stories of the Greek heroes and gods. Passed on from generation to generation, these poems, called *rhapsodes,* were never written down but were changed and embellished by successive poets. The oral tradition culminated in the *Iliad* and *Odyssey* (*see p87*), composed around 700 BC. Both works are traditionally ascribed to the same poet, Homer, of whose life

PASSIONATE POETRY

For private occasions, and particularly to entertain guests at the cultivated drinking parties known as *symposia,* shorter poetic forms were developed. These poems were often full of passion, whether love or hatred, and could be personal or, often, highly political. Much of this poetry, by writers such as Archilochus, Alcaeus, Alcman, Hipponax and Sappho, survives only in quotations by later writers or on scraps of papyrus that have been preserved by chance from private libraries in Hellenistic and Roman Egypt. Through these fragments we can gain glimpses

HISTORY

Until the 5th century BC little Greek literature was composed in prose – even early philosophy was in verse. In the latter part of the 5th century, a new tradition of lengthy prose histories, looking at recent or current events, was established with Herodotus's account of the great war between Greece and Persia (490–479 BC). Herodotus put the clash between Greeks and Persians into a context, and included an ethnographic account of the vast Persian Empire. He attempted to record objectively what people said about the past.

Thucydides took a narrower view in his account of the long years of the Pelopon- nesian war

Herodotus, the historian of the Persian Wars

between Athens and Sparta (431–404 BC). He concentrated on the political history, and his aim was to work out the "truth" that lay behind the events of the war. The methods of Thucydides were adopted by later writers of Greek history, though few could match his acute insight into human nature.

An unusual vase painting of a *symposion* for women only

The orator Demosthenes in a Staffordshire figurine of 1790

ORATORY

Public argument was basic to Greek political life even in the Archaic period. In the later part of the 5th century BC, the techniques of persuasive speech began to be studied in their own right. From that time on some orators began to publish their speeches. In particular, this included those wishing to advertise their skills in composing speeches for the law courts, such as Lysias and Demosthenes. The texts that survive give insights into both Athenian politics and the seamier side of Athenian private life. The verbal attacks on Philip of Macedon by Demosthenes, the 4th-century BC Athenian politician, became models for Roman politicians seeking to defeat their opponents. With the 18th-century European revival of interest in Classical times, Demosthenes again became a political role model.

DRAMA

Almost all the surviving tragedies come from the hands of the three great 5th-century BC Athenians: Aeschylus, Sophocles and Euripides. The latter two playwrights developed an interest in individual psychology (as in Euripides' *Medea*). While 5th-century comedy is full of direct references to contemporary life and dirty jokes, the "new" comedy developed in the 4th century BC is essentially situation comedy employing character types.

Vase painting of two costumed actors from around 370 BC

GREEK PHILOSOPHERS

The Athenian Socrates was recognized in the late 5th century BC as a moral arbiter. He wrote nothing himself but we know of his views through the "Socratic dialogues", written by his pupil, Plato, examining the concepts of justice, virtue and courage. Plato set up his academy in the suburbs of Athens.

His pupil, Aristotle, founded the Lyceum, to teach subjects from biology to ethics, and helped to turn Athens into one of the first university cities. In 1508–11 Raphael painted this vision of Athens in the Vatican.

Aristotle, author of the Ethics, had a genius for scientific observation.

Euclid laid the rules of geometry in around 300 BC.

Plato saw "the seat of ideas" in heaven.

Epicurus advocated the pursuit of pleasure.

Socrates taught by debating his ideas.

Diogenes, the Cynic, lived like a beggar.

Temple Architecture

Temples were the most important public buildings in ancient Greece, largely because religion was a central part of everyday life. Often placed in prominent positions, temples were also statements about political and divine power. The earliest temples, in the 8th century BC, were built of wood and sun-dried bricks. Many of their features were copied in marble buildings from the 6th century BC onwards.

Pheidias, sculptor of the Parthenon, at work

TEMPLE CONSTRUCTION

This drawing is of an idealized Doric temple, showing how it was built and used.

The pediment, triangular in shape, often held sculpture.

The cella, or inner sanctum, housed the cult statue.

The cult statue was of the god or goddess to whom the temple was dedicated.

A ramp led up to the temple entrance.

The stepped platform was built on a stone foundation.

Fluting on the columns was carved *in situ*, guided by that on the top and bottom drums.

The column drums were initially carved with bosses for lifting them into place.

TIMELINE OF TEMPLE CONSTRUCTION

700 BC	600 BC	500 BC	400 BC	300 BC
	522 Temple of Hera, Sámos (Ionic; see p156)	**477–390** Athenian Temple of Apollo, Delos (see pp218–19)	**447–405** Temples of the Acropolis, Athens: Athena Nike (Ionic), Parthenon (Doric), Erechtheion (Ionic) (see pp288–90)	
			Detail of the Parthenon pediment	
	490 Temple of Aphaia, Aígina (Doric; see pp98–9)		**4th century BC** Temple of Lindian Athena, Líndos Acropolis, Rhodes (Doric; see pp196–7)	**Late 4th century BC** Sanctuary of the Great Gods, Samothráki (Doric; see pp132–3)

The gable ends of the roof were surmounted by statues, known as *akroteria*, in this case of a Nike or "Winged Victory". Almost no upper portions of Greek temples survive.

The roof was supported on wooden beams and covered in rows of terracotta tiles, each ending in an upright antefix.

THE DEVELOPMENT OF TEMPLE ARCHITECTURE

Greek temple architecture is divided into three styles, which evolved chronologically, and are most easily distinguished by the column capitals.

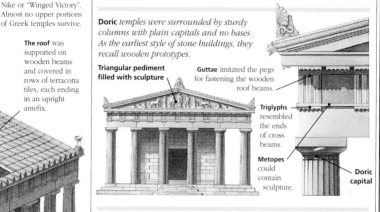

Doric *temples were surrounded by sturdy columns with plain capitals and no bases. As the earliest style of stone buildings, they recall wooden prototypes.*

Triangular pediment filled with sculpture

Guttae imitated the pegs for fastening the wooden roof beams.

Triglyphs resembled the ends of cross beams.

Metopes could contain sculpture.

Doric capital

Stone blocks were smoothly fitted together and held by metal clamps and dowels: no mortar was used in the temple's construction.

The ground plan was derived from the megaron of the Mycenaean house: a rectangular hall with a front porch supported by columns.

Ionic *temples differed from Doric in their tendency to have more columns, of a different form. The capital has a pair of volutes, like ram's horns, front and back.*

Akroteria, at the roof corners, could look Persian in style.

The frieze was a continuous band of decoration.

The Ionic architrave was subdivided into projecting bands.

The Ionic frieze took the place of Doric *triglyphs* and *metopes*.

Ionic capital

Caryatids, *or figures of women, were used instead of columns in the Erechtheion at Athens' Acropolis. In Athens' Agora (see p287), tritons (half-fish, half-human creatures) were used.*

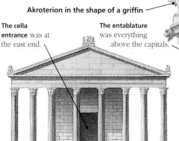

Corinthian *temples in Greece were built under the Romans and only in Athens. They feature columns with slender shafts and elaborate capitals decorated with acanthus leaves.*

The pediment was decorated with a variety of mouldings.

Akroterion in the shape of a griffin

The cella entrance was at the east end.

The entablature was everything above the capitals.

Acanthus leaf capital

Vases and Vase Painting

Donkey cup

The history of Greek vase painting continued without a break from 1000 BC to Hellenistic times. The main centre of production was Athens, which was so successful that by the early 6th century BC it was sending its high-quality black- and red-figure wares to every part of the Greek world. The Athenian potters' quarter of Kerameikós, in the west of the city, can still be visited today. Beautiful works of art in their own right, the painted vases are the closest we can get to the vanished wall paintings with which ancient Greeks decorated their houses. Although vases could break during everyday use (for which they were intended), a huge number still survive intact or in reassembled pieces.

This 6th-century BC black-figure vase *shows pots being used in an everyday situation. The vases depicted are* hydriai. *It was the women's task to fill them with water from springs or public fountains.*

The naked woman holding a *kylix* is probably a flute girl or prostitute.

The white-ground lekythos *was developed in the 5th century BC as an oil flask for grave offerings. They were usually decorated with funeral scenes, and this one, by the Achilles Painter, shows a woman placing flowers at a grave.*

THE SYMPOSION

These episodes of mostly male feasting and drink-ing were also occasions for playing the game of *kottabos*. On the exterior of this 5th-century BC *kylix* are depictions of men holding cups, ready to flick out the dregs at a target.

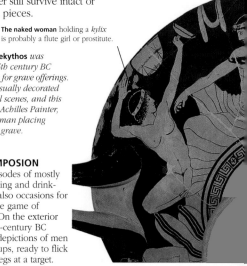

THE DEVELOPMENT OF PAINTING STYLES

Vase painting reached its peak in 6th- and 5th-century BC Athens. In the potter's workshop, a fired vase would be passed to a painter to be decorated. Archaeologists have been able to identify the varying styles of many individual painters of both black-figure and red-figure ware.

The body of the dead man is carried on a bier by mourners.

The geometric design is a proto-type of the later "Greek key" pattern.

Chariots and warriors form the funeral procession.

Geometric style *characterizes the earliest Greek vases, from around 1000 to 700 BC, in which the decoration is in bands of figures and geometric patterns. This 8th-century BC vase, placed on a grave as a marker, is over 1 m (3 ft) high and depicts the bier and funeral rites of a dead man.*

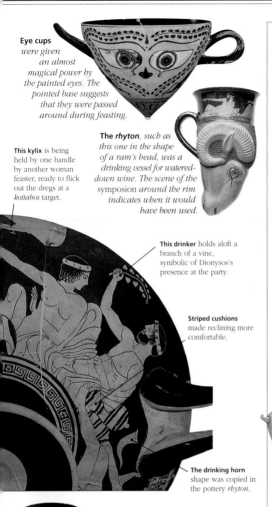

Eye cups *were given an almost magical power by the painted eyes. The pointed base suggests that they were passed around during feasting.*

The rhyton, *such as this one in the shape of a ram's head, was a drinking vessel for watered-down wine. The scene of the symposion around the rim indicates when it would have been used.*

This kylix is being held by one handle by another woman feaster, ready to flick out the dregs at a *kottabos* target.

This drinker holds aloft a branch of a vine, symbolic of Dionysos's presence at the party.

Striped cushions made reclining more comfortable.

The drinking horn shape was copied in the pottery *rhyton*.

Black-figure style *was first used in Athens around 630 BC. The figures were painted in black liquid clay on to the iron-rich clay of the vase which turned orange when fired. This vase is signed by the potter and painter Exekias.*

Red-figure style *was introduced in c.530 BC. The figures were left in the colour of the clay, silhouetted against a black glaze. Here a woman pours from an oinochoe (wine-jug).*

VASE SHAPES

Almost all Greek vases were made to be used; their shapes are closely related to their intended uses. Athenian potters had about 20 different forms to choose from. Below are some of the most commonly made shapes and their uses.

The amphora *was a two-handled vessel used to store wine, olive oil and foods preserved in liquid such as olives. It also held dried foods.*

This krater *with curled handles or "volutes" is a wide-mouthed vase in which the Greeks mixed water with their wine before drinking it.*

The hydria *was used to carry water from the fountain. Of the three handles, one was vertical for holding and pouring, two horizontal for lifting.*

The lekythos *could vary in height from 3 cm (1 in) to nearly 1 m (3 ft). It was used to hold oil both in the home and as a funerary gift to the dead.*

The oinochoe, *the standard wine jug, had a round or trefoil mouth for pouring, and just one handle.*

The kylix, *a two-handled drinking cup, was one shape that could take interior decoration.*

THE GREEK ISLANDS AREA BY AREA

The Greek Islands at a Glance

The Greek islands range in size from tiny uninhabited rocks to the substantial islands of Crete and Evvoia. Over the centuries, the sea has brought settlers and invaders and provided the inhabitants with their way of life; it now attracts millions of visitors. Each island has developed its own character through a mix of landscape, climate and cultural heritage. As well as the scattered historical sites, there is enough remote, rugged terrain to satisfy the most discerning walker and, of course, the variety of beaches is extraordinary.

Skópelos
The capital of this rugged island (see pp112–13), Skópelos town, spills dow from the hilltop kástro to the sea.

Corfu
The most visited of the Ionians, Corfu (see pp72–83) is a green, fertile island. Corfu town, its capital, contains a maze of narrow streets overlooked by two Venetian fortresses.

KEY

☐ The Ionian Islands pp68–91

☐ The Argo-Saronic Islands pp92–103

☐ The Sporades and Evvoia pp104–23

☐ The Northeast Aegean Islands pp124–157

☐ The Dodecanese pp158–203

☐ The Cyclades pp204–43

☐ Crete pp244–81

Athens

Aígina
Home to the spectacular and well-preserved ancient Temple of Aphaia, Aígina (see pp96–9) has a rich history due to its proximity to Athens.

Crete
The largest Greek island, Crete (see pp244–81) encompasses historic cities, ancient Minoan palaces, such as Knosós, and dramatic landscapes, including the Samaria Gorge (right).

◁ **Windmills in the village of Olympos, Kárpathos**

Delos
This tiny island (see pp218–19) is scattered with the ruins of an important ancient city. From its beginnings as a centre for the worship of Apollo in 1000 BC until its sacking in the 1st century AD, Delos was a thriving cultural and religious centre.

Chíos
The Byzantine monastery of Néa Moní in the centre of the island (see pp146–53) contains beautiful mosaics, which survived a severe earthquake in 1881. The mastic villages in the south of the island prospered from the wealth generated by the medieval trade in mastic gum.

Pátmos
The "holy island" of Pátmos (see pp162–5) is where St John the Divine wrote the book of Revelation. Pilgrims still visit the Monastery of St John, a fortified complex of churches and courtyards.

0 kilometres 100
0 miles 50

Rhodes
Rhodes town is dominated by its walled medieval citadel founded by the crusading Knights of St John. The island has many fine beaches and, inland, some unspoilt villages and remote monasteries (see pp180–97).

For additional map symbols *see back flap*

THE IONIAN ISLANDS

CORFU · PAXOS · LEFKADA · ITHACA · KEFALLONIA · ZAKYNTHOS

The Ionian Islands are the greenest and most fertile of all the island groups, characterized by olive groves and cypresses. Lying off the west coast of mainland Greece, these islands have been greatly influenced by Western Europe, in part because the Turks never managed to gain control here, except on the island of Lefkáda.

Famous as the homeland of Homer's Odysseus, these islands were colonized by the Corinthians in the 8th century BC and flourished as a wealthy trading post. In the 5th century BC Corfu defeated Corinth and joined the Athenians, instigating the Peloponnesian War. The Ionians first became a holiday destination during the Roman era.

Gorgon pediment in Corfu town's Archaeological Museum

The islands were not politically grouped together until Byzantine times. They were later occupied by the Venetians whose rule began in 1363 and lasted until 1797. After a brief period of French rule the British took over in 1814. The islands were finally ceded to the Greek state in 1864.

Evidence of the various periods of occupation can be seen throughout the islands, especially in Corfu town which contains a mixture of Italian, French and British architecture.

Each island has its own distinct character, from tiny Paxós covered in olive groves, to rocky Ithaca, the rugged beauty of Kefalloniá and mountainous Corfu. The group historically includes Kýthira, but in this guide it is included under the Argo-Saronic Islands due to easier transport connections.

The islands lie on a fault line, which runs south down Greece's west coast, and have been subjected to much earthquake damage. Kefalloniá and Zákynthos in particular suffered massive destruction in the summer of 1953.

Summers are hot and dry but for the rest of the year the islands have a mild climate; the above-average rainfall supports the lush greenery. There is a huge variety of beaches throughout the Ionians, from resorts providing lively nightlife to quieter stretches, virtually untouched by tourism.

Watching from the shade as a ship comes into Sámi town, Kefalloniá

◁ The islet of Vlachérna with its small convent, reached by a short causeway from Corfu island

Exploring the Ionian Islands

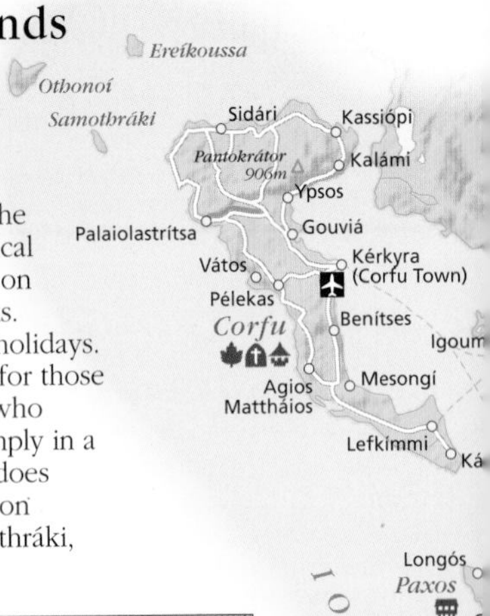

The widely scattered Ionian islands are not particularly well connected with each other, though most are easily reached from the mainland. Corfu is the best base for the northern islands and Kefalloniá for the southern islands. There are few archaeological remains, and museums tend to concentrate on folklore, culture and historical European links. Today's Europeans come mostly for beach holidays. The main islands are large enough to cater for those who like bars and discos, as well as those who prefer a quieter stay, in a family resort or simply in a small fishing village. Traditional Greek life does exist here, inland on the larger islands and on islands such as Meganísi off Lefkáda, or Mathráki, Othonoí and Erikoúsa off northern Corfu.

Ereíkoussa

Othonoí

Samothráki

Sidári · Kassiópi

Pantokrátor 906m · Kalámi

Ypsos

Palaiolastrítsa · Gouviá

Vátos · Kérkyra (Corfu Town)

Pélekas

Corfu · Benítses

Igoum

Agios Matthaíos · Mesongí

Lefkímmi · Ká

Longós

Paxos

IONIAN

Antíp

SEA

ISLANDS AT A GLANCE

Corfu pp72–83
Ithaca pp86–7
Kefalloniá pp88–9
Lefkáda p85
Paxós p84
Zákynthos pp90–91

SEE ALSO

• *Where to Stay* pp302–4
• *Where to Eat* pp330–1
• *Travel Information* pp366–9

Looking down on Plateía Dimarcheíou in Corfu town with the Town Hall on the left

A typical house by the roadside in Stavrós village on Ithaca

KEY

═══ Motorway
──── Main road
──── Minor road
──── Scenic route
----- High-season, direct ferry route
△ Summit

0 kilometres 25

0 miles 25

The mountain landscape of Lefkáda

LOCATOR MAP

GETTING AROUND

Aside from Paxós, all the main Ionians can be reached by air. Préveza airport serves Lefkáda, which is also connected to the mainland by a road bridge. A sea plane service runs from Corfu to Pátra, Iaoninna, Paxos and Lefkada, and there are plans to introduce flights to Kefalloniá and Ithaca. Larger ferries often travel via the mainland but smaller boats offer direct connections between the islands. Islands often have several ports so check specific destinations. Buses in the capitals provide services around the islands. Car and bike hire is widespread but road standards vary, as do local road maps.

An islander working on his boat in Gáios harbour on Paxós

Holiday apartments at Fiskárdo on Kefalloniá

Corfu
Κέρκυρα

Detail from Corfu Town Hall

Corfu is a green island offering the diverse attractions of secluded coves, stretches of wild coast, bands of coast given over totally to resorts and traditional hill-villages. In 229 BC it became a colony of the Roman Empire, remaining so until AD 337. Byzantine rule then began, intermittently broken by the Goths, the Normans and Angevin rule. Situated between Italy and the Greek mainland, its strategic importance continued under Venetian rule (1386–1797). French rule (1807–14) saw the Greek language restored and the founding of the Ionian Academy, set up for the development of the arts. A period of British rule (1814–64) was followed by unification with Greece.

Angelókastro is a ruined 13th-century fortress, which stands across the bay from Palaiokastrítsa *(see p81).*

Myrtiótissa is one of Corfu's finest beaches *(see p82).*

Sidári
Unusual rock formations, produced by the effect of sea on sandstone, give the resort of Sidári its appeal. Legend has it that any couple swimming through the Canal d'Amour will stay together forever **5**

Vátos
This traditional Greek hill-village is set above the fertile Ropa plain **7**

Korisíon Lagoon
This lake is a haven for wildlife and is separated from the Ionian Sea only by some beautiful beaches **8**

KEY

For key to map see back flap

0 kilometres	5
0 miles	3

Palaiokastrítsa
Three main coves cluster around a thickly wooded headland at Palaiokastrítsa. It is now one of the most popular spots on the island and is an ideal base for families, with watersports available and a friendly atmosphere **6**

STAR SIGHTS

★ Corfu Town

Kassiópi

The unspoilt bay at Kassiópi is overlooked by an attractive quayside lined with tavernas, shops and bars ❹

VISITORS' CHECKLIST

🏘 100,000. ✈ 3 km (2 miles) S of Corfu town. ⚓ Xenofóntos Stratigoú, Corfu town. 🚌
ℹ Corfu town (26610 37520, eotcorfu@otenet.gr). 🎭 For cultural events see **www.kerkyra.gr** and **www.corfu.gr**

Mount Pantokrátor

This is the highest point on Corfu and offers excellent views over the island and, on a clear day, as far as Italy ❸

Kalámi

Made famous by the author Lawrence Durrell, this remains an attractive coastal village ❷

★ Corfu Town

Corfu town is a delightful blend of European influences. The Liston, focus of café life, was built during the brief French rule. It overlooks the Esplanade that dates to Venetian rule in the town ❶

Igoumenítsa, Paxos, Pátra

Achílleion Palace

The Empress Elizabeth of Austria built this palace (1890–91) ❿

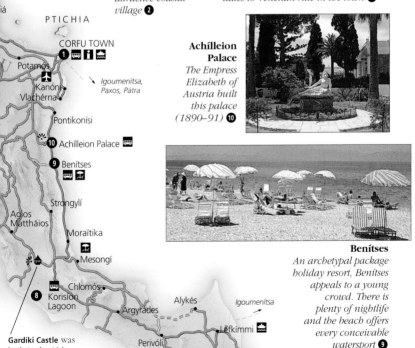

Benítses

An archetypal package holiday resort, Benítses appeals to a young crowd. There is plenty of nightlife and the beach offers every conceivable watersport ❾

Gardíki Castle was built in the 13th century on the site of Paleolithic remains *(see p82)*.

Map labels: PTICHIA, CORFU TOWN, Potamós, Kanóni, Vlachérna, Pontikonísi, Achílleion Palace, Benítses, Strongylí, Aghíos Matthaíos, Moraïtika, Mesongí, Chlomós, Korisíon Lagoon, Argyrádes, Alykés, Igoumenítsa, Léfkimmi, Perivóli, Kávos, Dragótina, Kassiópi, Avláki, Peritheia, Koulóura, Mount Pantokrátor, Nisáki, Kalámi

Street-by-Street: Corfu Old Town ❶

Πόλη της Κέρκυρας

The 21st century has not spoiled Corfu town, and it continues to be a delightful blend of European influences. The Venetians ruled here for over four centuries, and elegant Italianate buildings, with balconies and shutters, can be seen above French-style colonnades. British rule left a wealth of monuments, public buildings, and the cricket pitch, which is part of the Esplanade, or Spianáda (*see pp76–7*). This park is a focus for both locals and tourists, with park games and good walks. On its eastern side is the Old Fortress (*see p78*) standing guard over the town, a reminder that Corfu was never conquered by the Turks. The town was designated a UNESCO World Heritage site in 2007.

New Fortress *(see p78)*

View of the Old Fortress from Corfu old town

FILELLINON

THERMISTOKLEOUS

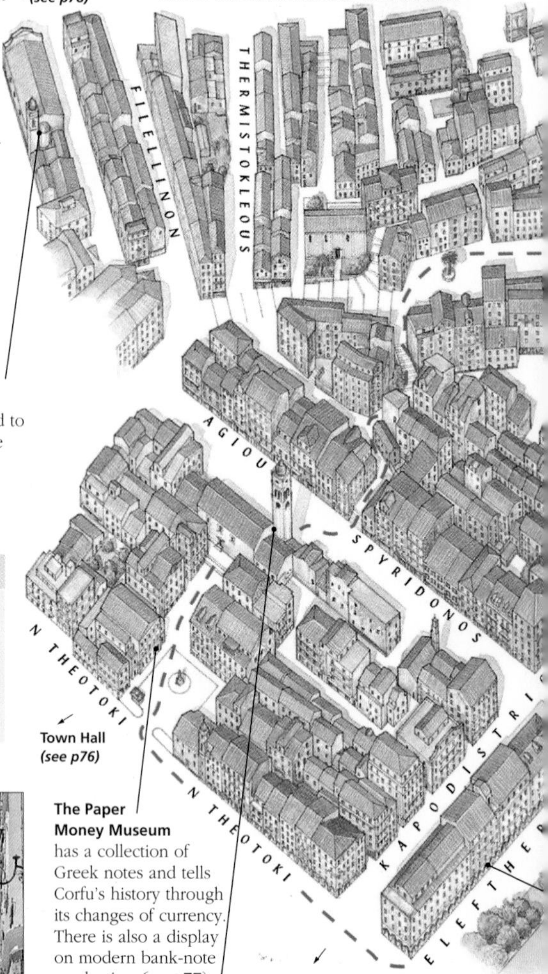

The Mitrópoli was built in 1577, and became Corfu's Orthodox cathedral in 1841. It is dedicated to St Theodora, whose remains are housed here along with some impressive gold icons.

AGIOU SPYRIDONOS

STAR SIGHTS

- ★ Palace of St Michael and St George
- ★ The Liston
- ★ Agios Spyrídon

N THEOTOKI

Town Hall *(see p76)*

KAPODISTRI

N THEOTOKI

The Paper Money Museum has a collection of Greek notes and tells Corfu's history through its changes of currency. There is also a display on modern bank-note production *(see p77)*.

ELEFTHER

Archaeological Museum *(see pp78–9)*

★ Agios Spyrídon

The red-domed belfry of this church is the tallest on Corfu. It was built in 1589 and dedicated to the island's patron saint, whose sarcophagus is just to the right of the altar (see p76).

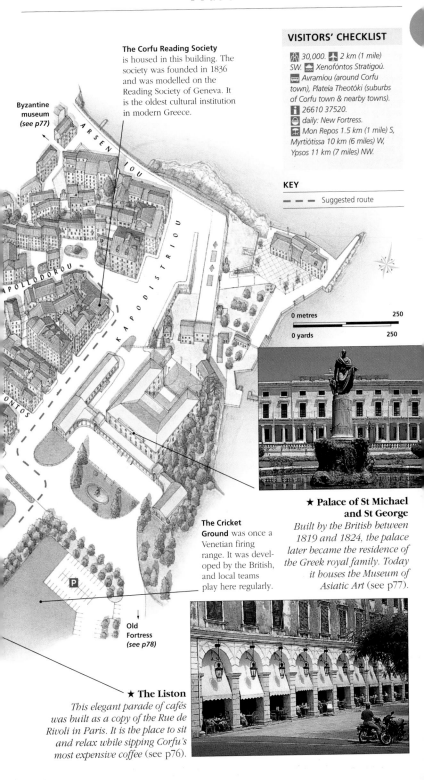

The Corfu Reading Society is housed in this building. The society was founded in 1836 and was modelled on the Reading Society of Geneva. It is the oldest cultural institution in modern Greece.

Byzantine museum (see p77)

VISITORS' CHECKLIST

30,000. 2 km (1 mile) SW. Xenofóntos Stratigoú. Avramíou (around Corfu town), Plateía Theotóki (suburbs of Corfu town & nearby towns). 26610 37520. daily: New Fortress. Mon Repos 1.5 km (1 mile) S, Myrtiótissa 10 km (6 miles) W, Ypsos 11 km (7 miles) NW.

KEY

- - - Suggested route

| 0 metres | 250 |
| 0 yards | 250 |

The Cricket Ground was once a Venetian firing range. It was developed by the British, and local teams play here regularly.

Old Fortress (see p78)

★ Palace of St Michael and St George

Built by the British between 1819 and 1824, the palace later became the residence of the Greek royal family. Today it houses the Museum of Asiatic Art (see p77).

★ The Liston

This elegant parade of cafés was built as a copy of the Rue de Rivoli in Paris. It is the place to sit and relax while sipping Corfu's most expensive coffee (see p76).

Exploring Corfu Town

In midsummer the narrow streets of Corfu's old town may be packed with visitors, but there are always quiet places to be found down alleyways and shady cobbled squares. The Corfiot housewives string washing across the streets from their balconies and, below, silversmiths and woodcarvers' shops are hidden away in the maze of alleys. On Nikifórou Theotóki, the southern boundary of the old town, there are several elegant arcaded sections. Built by the French, they are now home to souvenir shops, chapels and churches. Parts of the surrounding new town are quite modern but many of the buildings date back to French and British rule.

Corfu town by horse and trap

🔒 Agios Spyrídon

Agíou Spyrídonos. *Tel 26610 33059.*
☐ daily.
The distinctive red-domed tower of Agios Spyrídon guides the visitor to this church, the holiest place on the island. Inside, in a silver casket, is the mummified body of the revered saint, after whom many Corfiot men are named.

Spyrídon himself was not from Corfu but from Cyprus, where he was raised as a shepherd. Later he entered the church and rose to the rank of bishop. He is believed to have performed many miracles before his death in AD 350, and others since – not least in 1716 when he is said to have helped drive the Turks from the island after a six-week siege. His body was smuggled from Constantinople just before the Turkish occupation of 1453. It was only by chance that it came to Corfu, where the present church was built in 1589 to house his coffin.

The church is also worth seeing for the immense amount of silverware brought by the constant stream of pilgrims. On four occasions each year (Palm Sunday, Easter Saturday, 11 August and the first Sunday in November) the saint's remains are carried aloft through the streets.

🍁 Esplanade

This mixture of park and town square is one of the reasons Corfu town remains such an attractive place. Known as the Esplanade, or Spianáda, it offers relief from the packed streets in summer, either on a park bench or in one of the elegant cafés lining the square on **The Liston**, overlooking the cricket pitch.

The Liston was designed by a Frenchman, Mathieu de Lesseps, who built it in 1807. The name Liston comes from the Venetian practice of having a "List" of noble families in the *Libro d'Oro* or Golden Book – only those on this list were allowed to promenade here.

There are a number of monuments in and around the Esplanade. Near the fountain is the **Enosis Monument**: the word *énosis* means unification, and this celebrates the 1864 union of the Ionian Islands with the rest of Greece, when British rule came to an end. The marble monument has carvings of the symbols of each of the Ionian Islands.

A statue of **Ioánnis Kapodístrias**, modern Greece's first president in 1827 and a native of Corfu, stands at the end of the street that flanks the Esplanade and bears

Agios Spyridon, seen down one of the many small shopping alleyways

For hotels and restaurants in this region see pp302–4 and pp330–31

A game of cricket on the pitch by the Esplanade

his name. He was assassinated in Náfplio in the Peloponnese in 1831 by two Cretans whose uncle he had imprisoned.

Facing this is the **Maitland Rotunda** (1816), a memorial to Sir Thomas Maitland, who became Britain's first Lord High Commissioner to Corfu after the island became a British Protectorate in 1814, though neither he nor his policies were much liked.

🏛 Palace of St Michael and St George

Plateía Spianáda. *Tel 26610 30443.*
◯ *Tue–Sun.* 🔲 *main public hols.*
The Palace of St Michael and St George was built by the British between 1819 and 1824 from Maltese limestone. It served as the residence of Sir Thomas Maitland, the High Commissioner, and as such is the oldest official building in Greece. When the British left Corfu in 1864 the palace was used for a short time by the Greek royal family but it was later abandoned and left to fall into disrepair.

The palace was carefully renovated in the 1950s by Sir Charles Peake, British Ambassador to Greece, and now houses the traffic police, a library and some government offices. Conferences and exhibitions are also held in the palace from time to time.

The Palace of St Michael and St George also houses the **Museum of Asiatic Art**. The core of the museum's collection is the 10,000 items that were collected by a Corfiot diplomat, Grigórios Mános (1850–1929), during his travels overseas. He

offered his vast collection to the state on condition that he could retire and become curator of the museum. Unfortunately he died before he could realize his ambition. The exhibits include statues, screens, armour, silk and ceramics from China, Japan, India and other Asiatic countries.

In front of the building is a statue of **Sir Frederick Adam**, the British High Commissioner to Corfu from 1824–31. He built the Mon Repos Villa *(see p79),* to the south of town and was also responsible for popularizing the west coast resort of Palaiokastrítsa *(see p81),* one of his favourite spots on the island.

Statue of Sir Frederick Adam

📷 Byzantine Museum

Prosfórou 30 & Arseníou. *Tel 26610 38313.* ◯ *8:30am–2:30pm Tue–Sun.* 🔲 *main public hols.* 🎟
The Byzantine Museum opened in 1984 and is housed in the renovated church of Panagía Antivouniótissa, which provided some of the exhibits. The museum contains about 90 icons dating back to the 15th century. It also has work by artists from the Cretan School. Many of these artists worked and lived on Corfu, as it was a convenient stopping-off point on the journey between Crete and Venice from the 13th to the 17th centuries during the period of Venetian rule.

📷 Paper Money Museum

Ionikí Trápeza, Plateía Iróon Kypriakoú Agóna. *Tel 26610 41552.* ◯ *Tue, Thu.* 🔲 *main public hols.*
This complete collection of Greek bank notes traces the way in which the island's currency has altered as Corfu's society and rulers changed. The first bank note on the island was issued in British pounds, while later notes show the German and Italian currency of the war years. Another intriguing display shows the process of producing a note from the artistic design to engraving and printing.

Maitland Rotunda situated in the Esplanade

The Old Fortress towering above the sea on the eastern side of Corfu town

⚓ Old Fortress

Tel 26610 48310. ☐ Apr–Oct: 8am–7pm daily. ⬛ main public hols. 🎫 except Sun. ♿ limited.

The ruined Old Fortress, or Palaió Froúrio, stands on a promontory believed to have been fortified since at least the 7th or 8th century AD; archaeological digs are still underway. The Old Fortress itself was constructed by the Venetians between 1550 and 1559. The very top of the fortress gives magnificent views of the town and along the island's east coast. Lower down is the church of St George, a British garrison church built in 1840. The fortress is also a venue for concerts and musical events, which are held in the summer months. The fortress is linked to the town by an iron bridge.

⚓ New Fortress

Plateía Solomoú. **Tel** 26610 44436. ☐ Apr–Oct: daily. 🎫

The Venetians began building the New Fortress, or Néo Froúrio, in 1576 to further strengthen the town's defences. It was not completed until 1589, 30 years after the Old Fortress, hence their respective names. The fortress is used by the Greek navy as a training base, while the surrounding moat is the setting for the town's market.

🔒 Mitrópoli

Mitropóleos. **Tel** 26610 39409. ☐ daily.

The Greek Orthodox church of the Panagía Spiliótissa, or Virgin Mary of the Cave, was built in 1577. It became Corfu's cathedral in 1841, when the nave was extended. It is

dedicated to St Theodora, a former Byzantine empress whose remains were brought to Corfu at the same time as those of St Spyrídon. Her body is in a silver coffin near the altar.

🏛 Plateía Dimarcheíou

Town Hall Tel 26610 40402. ☐ daily. ⬛ main public hols. ♿ **Agios Iákovos** ☐ daily.

Within this elegant square stands the **Town Hall**. It is a grand Venetian building, which began life in 1663 as a single-storey *loggia* or meeting place for the nobility. It was then converted into the San Giacomo Theatre in 1720, which made it the first modern theatre in Greece. The British added the second floor in 1903 when it became the Town Hall.

Adjacent to it is the Catholic cathedral **Agios Iákovos**, also known by its Italian name of San Giacomo. Built in 1588 and consecrated in 1633, it was badly damaged by bombing in 1943 with only the bell tower surviving intact. Services are held every day, with three Masses on Sundays.

🏛 Archaeological Museum

Vraíla 1. **Tel** 26610 30680. ☐ Tue–Sun. ⬛ main public hols. 🎫 ♿

The Archaeological Museum is situated a pleasant stroll south from the centre of town, along the seafront. The museum's collection is not large but a visit is worthwhile to see the centrepiece, the stunning Gorgon frieze.

The frieze, dating from the 6th century BC, originally formed part of the west pediment of the Temple of Artemis near Mon Repos Villa. The layout ensures that the

The 17th-century Catholic cathedral Agios Iákovos in Plateía Dimarcheíou

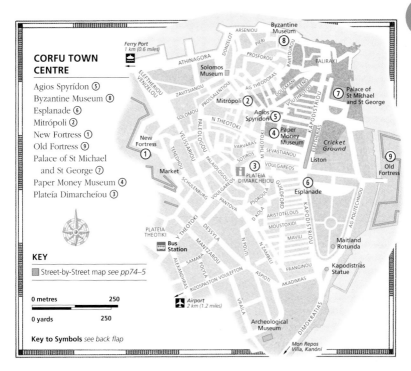

CORFU TOWN CENTRE

Agios Spyrídon ⑤
Byzantine Museum ⑧
Esplanade ⑥
Mitrópoli ②
New Fortress ①
Old Fortress ⑨
Palace of St Michael
   and St George ⑦
Paper Money Museum ④
Plateía Dimarcheíou ③

KEY

▨ Street-by-Street map see pp74–5

0 metres   250

0 yards   250

Key to Symbols see back flap

The Gorgon frieze in Corfu town's Archaeological Museum

frieze, a massive 17 m (56 ft) long, is not seen until the final room. The museum also displays other finds from the Temple of Artemis and the excavations at Mon Repos Villa.

Environs

Garítsa Bay sweeps south of Corfu town, with the suburb of Anemómilos visible on the promontory. Here, in the street named after it, is the 11th-century church of **Agion Iásonos kai Sosipátrou** (saints Jason and Sossipater). These disciples of St Paul brought Christianity to Corfu in the 1st century AD. Inside are faded wall paintings, including an 11th-century fresco.

South of Anemómilos is **Mon Repos Villa**. It was built in 1824 by Sir Frederick Adam, the second High Commissioner of the Ionian state, as a present for his wife, and was

later passed to the Greek royal family. The remains of the **Temple of Artemis** lie nearby. Opposite the villa are the 5th-century ruins of **Agía Kerkýra**, the church of the old city.

An hour's walk or a short bus ride south of Corfu town is **Kanóni**, with the islands of

Vlachérna and **Pontikonísi** just off the coast. Vlachérna, with its tiny white convent, is a famous landmark and can be reached by a causeway. In summer boats go to Pontikonísi, or Mouse Island, said to be Odysseus's ship turned to stone by Poseidon. This caused Odysseus to be shipwrecked on Phaeacia, the island often identified with Corfu in Homer's *Odyssey*.

🏛 **Mon Repos Villa**
Tel 26610 41369. ◯ April–Oct:
Tue–Sun (gardens open daily). 🈳 🎫

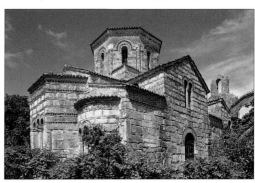

The church of Agion Iásonos kai Sosipátrou

Around Northern Corfu

Northern Corfu, in particular the northeast coast, is emphatically holiday Corfu, with a string of resorts along the main coastal road. These include popular spots such as Kassiópi and Sidári, though there are also quieter villages like Kalámi. In the northwest is one of Corfu's prettiest areas, Palaiokastrítsa, a jigsaw of bays and beaches. Inland stands Mount Pantokrátor, a reminder that there is also a rugged interior to explore.

View looking southwards over the beach at Kalámi Bay

Kalámi ❷
Καλάμι

26 km (16 miles) NE of Corfu town.
👥 18. 🚌 to Kassiópi.

Kalámi village has retained its charm despite its popularity with visitors. A handful of tavernas line its sand and shingle beach, while behind them cypress trees and olive groves climb up to the lower slopes of Mount Pantokrátor. The hills of Albania are a little over 2 km (1 mile) across Kalámi Bay.

Kalámi's obvious appeal attracted the author Lawrence Durrell to the village in 1939.

Only during the day in high season, when visitors from holiday resorts throng his "peaceful fishing village", might Durrell fail to recognize the place. In the evenings and outside the months of July and August, normality returns.

Mount Pantokrátor ❸
Ορος Παντοκράτωρ

29 km (18 miles) N of Corfu town.
🚌 to Petáleia.

Mount Pantokrátor, whose name means "the Almighty", dominates the northeast bulge of Corfu. It rises so steeply that its peak, at 906 m (2,972 ft), is less than 3 km (2 miles) from the beach resorts of Nisáki and Barbati. The easiest approach is from the north, where a rough road goes all the way to the small monastery at the top. The mountain has great appeal to naturalists as well as walkers, but exploring its slopes is not something to be undertaken lightly as Corfu's weather can change suddenly. However, the reward is a view to Albania and Epirus in the east, of Corfu town to the south, and even west to Italy when weather conditions are clear.

Kassiópi ❹
Κασσιόπη

37 km (23 miles) N of Corfu town.
👥 600. 🚌 🚕 Avláki 2 km (1 mile) S.

Kassiópi has developed into one of Corfu's busiest holiday centres without losing either its charm or character. It is set around a harbour that lies between two wooded headlands. Although there is plenty of nightlife to attract younger holiday-makers, there are no high-rise hotels to spoil the setting. The heart of the town is at its harbour, with tavernas and souvenir shops overlooking fishing boats moored alongside motor boats from the many watersports schools.

In the 1st century AD the Emperor Nero is said to have visited a Temple of Jupiter, which was situated on the western side of the harbour, where the church of **Kassiopítissa** now stands. The ruins of a 13th-century castle are a short walk further west.

Fishing boats moored in Kassiópi harbour, east of the castle ruins

For hotels and restaurants in this region see pp302–4 and pp330–31

British High Commissioner, Sir Frederick Adam *(see p77)*, loved to picnic here but did not like the awkward journey from Corfu town, so he had a road built between the two.

On the main headland stands **Moní Theotókou**, which dates from the 17th century, although the first monastery stood here in 1228. The church's ceiling features a fine carving of the *Tree of Life*.

Views from the monastery include **Angelókastro**, the ruined 13th-century fortress of Michaíl Angelos Komninós II, the Byzantine despot of Epirus. Situated above the cliffs west of Palaiokastrítsa, the fortress was never taken and in 1571 it sheltered locals from another failed Turkish attempt to conquer Corfu. The remains include a hilltop chapel and some hermit cells and caves.

Outlying Islands

Corfu has three offshore islands. **Mathráki** offers the simplest Greek island life, with two villages and only a few rooms to rent. **Ereikoússa** is the most popular island, largely because of its glorious sandy beaches. **Othonoí** is the largest island and has the best facilities but lacks the finer beaches.

Caretaker monk at Moní Theotókou, Palaiokastrítsa

Sidári ❺
Σιδάρι

31 km (20 miles) NW of Corfu town.
🚶 *300.* 🚌 🚆 *Róda 6 km (4 miles) E.*

One of Corfu's first settlements, the village of Sidári has pre-Neolithic remains dating back to about 7000 BC. Today it is a bustling holiday centre with the twin attractions of sandy beaches and unusual rock formations. Erosion of the sandstone has created a number of caves and tunnels, the most famous being a channel between two rocks known as the Canal d'Amour *(see p72).*

Palaiokastrítsa ❻
Παλαιοκαστρίτσα

26 km (16 miles) NW of Corfu town.
🚶 *600.* 🚌

Palaiokastrítsa is one of Corfu's most popular spots. Three main coves cluster around a wooded headland, dividing into numerous other beaches which are popular with families because swimming is safe. Watersports are available as well as boat trips out to see the nearby grottoes. Until the early 19th century the place was noted for its beauty but access was difficult. The

WRITERS AND ARTISTS IN CORFU

The poet Dionýsios Solomós lived on Corfu from 1828 until his death in 1857. He is best known for his poem Hymn to Freedom, part of which was adopted as the national anthem after Independence. Other writers have also found inspiration on Corfu, including the British poet and artist Edward Lear, who visited the island in the 19th century, and the Durrell brothers, who both wrote about Corfu. Gerald described his idyllic 1930s childhood in *My Family and Other Animals*, while Lawrence produced *Prospero's Cell* in 1945. He wrote this while staying in Kalámi, where he was visited by Henry Miller, whose 1941 book *The Colossus of Maroussi* is one of the most accurate and endearing books about Greece.

A view from the Benítses road near Gastoúri, by Edward Lear

Around Southern Corfu

Less mountainous but more varied than the north, southern Corfu encompasses Benítses' wild nightlife and the shy wildlife of the Korisíon Lagoon. Much of Corfu's produce grows in the fertile Rópa Plain north of Vátos. To the south lies Myrtiótissa, once described as the world's most beautiful beach. Bus services are good but to explore off the beaten track you will need your own car.

View inland over the freshwater Korision Lagoon

Vátos ⑦

Βάτος

24 km (15 miles) W of Corfu town.
🏠 480. 🚌 🚋 Myrtiótissa 2 km
(1 mile) S, Ermones 2 km (1 mile) W.

In the hillside village of Vátos, the whitewashed houses with flower-bedecked balconies offer a traditional image of Greece. Vátos has two tavernas and a handful of shops and has mostly remained untainted by the impact of tourism. From the village, a steep climb leads up the mountainside to the top of Agios Geórgios (392 m; 1,286 ft). Below lies the fertile Rópa Plain and a beach at Ermones.

Environs
The glorious beach at **Myrtiótissa**, 2 km (1 mile) south of Vátos, is named after the 14th-century monastery behind it dedicated to Panagía Myrtiótissa (Our Lady of the Myrtles). The beach is a long golden sweep of sand backed with cypress and olive trees. Lawrence Durrell was fond of the area and, in his book *Prospero's Cell*, referred to Myrtiótissa as "perhaps the loveliest beach in the world".

South of Vátos lies **Pélekas**, another picturesque and unspoilt hillside village. Its traditional houses tumble down wooded slopes to the small and secluded beach below. Above this is the **Kaiser's Throne**, the hilltop

from which Kaiser Wilhelm II of Germany loved to watch the sunset while staying at the Achílleion Palace.

Korisíon Lagoon ⑧

Λίμνη Κορισσίων

42 km (26 miles) S of Corfu town.
🚋 Gardíki 1 km (0.5 mile) N.

The Korisíon Lagoon is a 5-km (3-mile) stretch of water, separated from the sea by some of the most beautiful dunes and beaches on Corfu. The lake remains a haven for wildlife, despite the Greek love of hunting. At the water's edge are a variety of waders such as sandpipers and avocets, egrets and ibis. Flowers include sea daffodils and Jersey orchids.

Almost 2 km (1 mile) north lies **Gardíki Castle**, built in the 13th century by Michaíl Angelos Komninós II *(see p81)*, with the ruined towers and outer castle walls still standing. The site is also known for a find of Paleolithic remains, now removed.

Benítses ⑨

Μπενίτσες

14 km (9 miles) S of Corfu town.
🏠 1,400. 🚌 🚋 Benítses.

Benítses has become the archetypal package holiday resort. Its appeal is to young people, and not to those seeking peace and quiet or a real flavour of Greece.

The beaches offer every conceivable watersport, and at the height of the season are extremely busy. The nightlife is also very lively: the bars and discos close about the same time as the local fishermen return from their night at sea.

There are few sights of interest in Benítses other than the remains of a Roman bathhouse near the harbour square.

A whitewashed house in the attractive village of Vátos

Achílleion Palace ⑩
Αχίλλειον

19 km (12 miles) SW of Corfu town.
Tel 26610 56210. **Palace &
gardens** daily.

A popular day trip from any
of Corfu's resorts, the Achílleion Palace was built in
1890–91 by the Italian architect Raphael Carita for the
Empress Elizabeth of Austria
(1837–98), formerly Elizabeth
of Bavaria and best known as
Princess Sissy. She used it as
a personal retreat from her
problems at the Hapsburg
court: her health was poor
and her husband, Emperor
Franz Josef, notoriously
unfaithful. After the assassina-
tion of the Empress Elizabeth
by an Italian anarchist in
1898, the palace lay emp-
ty for nearly a decade
until it was bought by
Kaiser Wilhelm II in
1907. It is famous as
the set used for the
casino in the James
Bond film *For
Your Eyes Only.*

The outer entrance to the Achilleion's gardens

The Gardens
The lush green
gardens below
the palace are
terraced on a slope
which drops 150 m
(490 ft) to the coast road. The
views along the rugged coast
both north and south are
spectacular. In the grounds

**A 19th-century painting
of Elizabeth of Bavaria
by Franz Xavier**

the walls are draped with
colourful bougainvillea and a
profusion of palm trees. The
gardens are also dotted
with numerous statues,
especially of Achilles, who
was the empress's hero,
after whom the palace
is named. One moving
bronze of the
Dying Achilles is
by the German
sculptor, Ernst
Herter. The statue
is thought to have
appealed to the
unhappy empress
following the
tragic suicide of her second
son, the Archduke Rudolph, at
Mayerling. Another impressive
statue of the hero Achilles is

the massive 15-m (49-ft) high,
cast-iron figure, which was
commissioned by Kaiser
Wilhelm II.

The Palace
There have been numerous
attempts to describe the
Achílleion's architectural style,
ranging from Neo-Classical to
Teutonic, although Lawrence
Durrell was more forthright,
and declared it "a monstrous
building". The empress was
not particularly pleased with
the finished building, but her
fondness for Corfu made her
decide to stay.

The palace does however
contain a number of interesting
artifacts. Inside, some original
furniture is on display and on
the walls there are
some fine paintings
of Achilles, echoing
the bronze and stone
statues seen in the
gardens. Another
exhibit is the strange
saddle-seat that was
used by Kaiser
Wilhelm II when-
ever he was writing
at his desk.

Visitors requiring
a pick-me-up after
touring the palace
can try the Vasilákis
Tastery, opposite
the entrance, and
sample this local
distiller's many
products, which
include a number of
Corfiot wines, ouzo
and the speciality
kumquat liqueur.

THE LEGEND OF ACHILLES

Shortly after his birth, Achilles was immersed in the River Styx by his
mother Thetis. This left him invulnerable apart from the heel where she
had held him. Achilles' destiny lay at Troy *(see pp56–7)*; Helen, the wife
of King Menelaos of Sparta, was held by Paris at Troy where Menelaos
and his allies laid siege. As the Greeks' mightiest warrior, it was Achilles
who killed the Trojan hero Hector. However, he did not live to see Troy
fall, since he was struck in the heel by a fatal arrow from Paris's bow.

Achilles victoriously dragging the body of Hector around the walls of Troy

Local fishing boats moored at the eastern end of the harbour at Gáïos

Paxós
Παξοί

🏠 2,700. ⛴ Gáïos, Lákka. 🚌 Gáïos.
ℹ Gáïos (26620 32222). 🎣
Mogonísi 3 km (2 miles) SE of Gáïos.

Paxos is green and wooded, with a few farming and fishing villages. The thick groves of olive trees are still a major part of the island's economy. In mythology, Poseidon created Paxós for his mistress, and its small size has saved it from the turbulent history of its larger neighbours. Paxós became part of the Greek state along with the other Ionians in 1864.

Gaios

Gáïos is a lively, if small-scale, holiday town with two harbours: the main port where ferries dock and, a short walk away, the small harbour, lined with 19th-century houses with Venetian-style shutters and balconies. At the waterfront stands Pyropolitís, a statue of Constantínos Kanáris, hero in the Greek Revolution (see pp42–3). The grandest house was once residence of the British High Commissioner of Corfu. Behind it are narrow old streets, bars and tavernas.

Around the Island

One main road goes from the south to the north of the island. There are few cars and the best way to get about is by bicycle or moped. Many pleasant tracks lead through woods to high cliffs or secluded coves. At the end of a deep, almost circular inlet on Paxós's northern coast lies the town of **Lákka**. This pretty coastal town is backed by olive groves and pine-covered hills. Lákka is popular with day-trippers from Corfu, but at night it returns to being a quiet fishing village, with a few rooms to rent and only a scattering of restaurants and cafés.

To the east is the small village of **Pórto Longós**, which is the most attractive of the island's settlements. It has a pebble beach, a handful of houses, a few shops, and tavernas whose tables stand at the water's edge. Pórto Longós is a peaceful place where the arrival of the boat bringing fruit and vegetables every few days is a major event. Paths from the village lead through olive groves to several quiet coves, good for swimming.

Outlying Islands

Around 100 people live on **Antípaxos**, south of Paxós, and mostly in Agrapidiá, although there are a few hamlets inland. The island is unusual in that olive trees are easily outnumbered by vines, which produce Antípaxos's potent and good-quality wine. There is little tourism and no accommodation available, although the sandy beaches do fill up in summer with visitors from Paxós. Offshore from Gáïos lie the two islets of Panagiá and Agios Nikólaos.

Statue of Pyropolitís on the waterfront in Gáïos

View overlooking Lákka to the south

Houses on a hillside near Kalamítsi

Lefkáda
Λευκάδα

🏠 25,000. 🚤 Nydri, Vasilikí. 🚌 Dimitroú Golémi, Lefkáda town. ℹ️ Lefkáda town (26450 29379/ 25295). 🎭 Lefkáda town: daily.

Lefkada offers variety, from mountain villages to beach resorts. It has had a turbulent history, typical of the Ionian Islands, since the Corinthians took control of the island from the Akarnanians in 640 BC, right up until the British left the island in 1864.

Lefkada Town
The town has suffered repeated earthquakes, but there are interesting back streets and views of the beautiful ruins of the 14th-century **Sánta Mávra fortress**. Situated on the mainland opposite, the fortress is connected to Lefkáda by a causeway. The main square, Plateía Agíou Spyrídona is named after the 17th-century church with its rare metal bell towers. Nearby, the **Phonograph Museum** houses a private collection of records and old

phonographs. The small **Folk Museum** has local costumes and old photographs of island life. Above the town, **Moní Faneroménis** was founded in the 17th century, though the present buildings date from the 19th century. Its icon of the Panagía is also 19th century.

🎧 **Phonograph Museum**
Konstantínou Kalkáni 10.
⬜ daily. ⬤ main public hols. ♿

🏛️ **Folk Museum**
Theódoro Strátou. **Tel** 26450 33443. ⬜ call for opening times. ⬤ main public hols.

A bell at Moní Faneroménis

Around the Island
The best way to see the island is to hire a moped or bike, although bus services operate from Lefkáda town. **Agios Nikítas** is a traditional small resort with a harbour and beach. To the south, **Kalamítsi** is a typical Lefkáda mountain village. In the south, the main hill-village is **Agios Pétros**, still a rural community despite the nearby resort of **Vasilikí**, a windsurfer's paradise. **Nydrí** is the main resort on the east coast, with splendid views of the offshore islands.

Outlying Islands
Meganísi has retained its rural lifestyle. Most boats from Nydrí stop at Vathý, the main port. Uphill, the small village of Katoméri has the island's only hotel. **Skorpios** is a private island owned by Aristotle and now, Athina, Onassis.

LEFKADA TOWN
Moní Faneroménis
Agios Nikítas
Kalamítsi
Eláti Stavrotá
1,157m
3,795ft
Nydrí
SPARTI
SKORPIOS
Agios Pétros
Vathý
Vasilikí
Spartochóri
MEGANISI
Ithaca,
Kefalloniá
Kefalloniá

KEY
For key to map see back flap

0 kilometres 5
0 miles 3

Sailing boats off the white-sand beach at Vasilikí

The pebble beach of Pólis Bay on the northwest coast of Ithaca

Ithaca
Ιθάκη

🏛 *4,000.* ⛴ *Vathý.* 🚌 ℹ *Vathý*
(26740 33733). 🚤 *Pólis Bay 20 km*
(12 miles) NW of Vathý.

Small and rugged, Ithaca is
famous, according to
Homer's epic

the *Odyssey*, as the home of
Odysseus. Finds on Ithaca date
back as far as 4000–3000 BC,
and by Mycenaean times it had
developed into the capital of
a kingdom that included its
larger neighbour, Kefalloniá.

Vathy
The capital, also known as
Ithaca town, is an attractive
port, its brown-roofed houses
huddled around an indented
bay. The surrounding hills were
the site for the first settlement,
but the harbour itself was
settled in the medieval period,

and Vathý became the capital
in the 17th century. Destroyed
by an earthquake in 1953, it
was reconstructed and declared
a traditional settlement, which
requires all new buildings to
match existing styles.

The **Archaeological
Museum** contains a collection
mainly of vases and votives
from the Mycenaean period. In
the church of **Taxiárchis** is a
17th-century icon of Christ,
believed to have been painted
by El Greco *(see p268)*.

🏛 **Archaeological Museum**
Behind OTE office. *Tel 26740
32200.* ◗ *Tue–Sun.* ● *main
public hols.* ♿ ⤵

Around the Island
With just one main town, high
hills, a few pebble beaches and
little development, Ithaca is a
pleasant island to explore. A
twice-daily bus (four in season)
links Vathý to villages in the
north and there are some taxis.

Stavrós, the largest village in
northern Ithaca, has only 300
inhabitants but is a thriving hill
community and market centre.
Nearby Pólis Bay is thought to
have been the old port of
ancient Ithaca, and site of an
important cave sanctuary to the
Nymphs. **Odysseus' Palace**
may have stood above Stavrós
on the hill known as Pilikáta.
To find it ask for directions at
the one-room **Archaeological
Museum**, whose curator gives
guided tours in several
languages. Among the varied
local finds is a piece of a
terracotta mask from Pólis
cave bearing the inscription
"Dedicated to Odysseus".

🏛 **Archaeological Museum**
Stavrós. *Tel 26740 23955.* ◗ *variable.*
● *Mon, main public hols.* ♿

Exogi
Platreithiás
Kefalloniá
Pilikáta
Frikes   *Lefkáda*
Stavrós
Pólis
Bay   Kióni
Léfki   Anogí
          Astakós
Agios Ioánnis
          *Kefalloniá
          Pátras*

0 kilometres   5
0 miles   2

VATHY
Píso Aetós
Perachóri
Kefalloniá
Taxiárchis   Filiatró

KEY

For key to map see back flap

The red-domed roof of a church in Stavrós

The Legend of Odysseus' Return to Ithaca

Odysseus, the king of Ithaca, had been unwilling to leave his wife Penelope and infant son Telemachos and join Agamemnon's expedition against Troy *(see pp56–7)*. But once there his skills as warrior and speaker, and his cunning, ensured he played a vital role. However, his journey home was fraught with such perils as the monstrous one-eyed Cyclops, the witch Circe, and the seductive Calypso. His blinding of the Cyclops angered the god Poseidon who ensured that, despite the goddess Athena's support, Odysseus lost all his companions, before the kindly Phaeacians brought him home, ten years after he left Troy. On Ithaca, Odysseus found Penelope besieged by suitors. Disguising himself as a beggar, and aided by his loyal swineherd Eumaios and his son, he killed them all and returned to his marriage bed and to power.

Odysseus' homecoming *is depicted in this 15th-century painting attributed to Coracelli. Odysseus had been washed ashore on Phaeacia (Corfu), where King Alkinoös took pity and ferried him back to Ithaca.*

Penelope *wove a shroud for Odysseus' father Laertes, shown in this 1920 illustration by A F Gorguet. She refused to remarry until the shroud was finished; each night she would unpick the day's weaving.*

Eumaios, *Odysseus' faithful swineherd, gave his disguised master food and shelter for the night on his arrival in Ithaca. Eumaios then demonstrated his loyalty by praising his absent king while describing the situation on Ithaca to Odysseus. Their meeting is shown on this 5th-century BC Athenian vase.*

Argus, *Odysseus' aged dog, recognized his master without prompting, a feat matched only by Odysseus' old nurse, Eurykleia. Immediately after their meeting Argus died.*

Telemachos *had challenged Penelope's suitors to string Odysseus' bow and thereby to win his mother's hand in marriage. The suitors all failed the test. Odysseus locked them in the palace hall, strung the bow, and revealed his identity before slaughtering them.*

Kefalloniá
Κεφαλλονιά

Archaeological finds date Kefalloniá's first inhabitants to about 50,000 BC. In Mycenaean times the island flourished and remained Greek until the 2nd century BC when it was captured by the Romans. It was squabbled over by many powers but from 1500 to 1700 it shared the Ionians' history of Venetian occupation. Kefalloniá's attractions range from busy beach resorts to Mount Aínos National Park, which surrounds the Ionians' highest peak.

A church tower in the countryside between Argostóli and Kástro

Argostóli
A big, busy town with lush surrounding countryside, Kefalloniá's capital is situated by a bay with narrow streets rising up the headland on which it stands. Its traditional appearance is deceptive as Argostóli was destroyed in the 1953 earthquake and rebuilt with donations from emigrants. The destruction and re-building is shown in a photographic collection at the **Historical and Folk Museum**. Other exhibits range from rustic farming implements to traditional folk costumes.

The nearby **Archaeological Museum** includes finds from the Sanctuary of Pan, based at the Melissáni Cave-Lake and an impressive 3rd-century AD bronze head of a man found at Sámi. From the waterfront you can see the **Drápanos Bridge**, built during British rule in 1813.

⌂ Historical and Folk Museum
Ilía Zervoú 12. **Tel** 26710 28835.
🕘 9am–2pm Mon–Sat. ⬤ Dec or Jan for cleaning, main public hols. 🐾

⌂ Archaeological Museum
Rókkou Vergotí. **Tel** 26710 28300.
🕘 8:30am–3pm Tue–Sun.
⬤ main public hols.

Around The Island
It takes time to travel around Kefalloniá, the largest of the Ionian Islands. Despite this, driving is rewarding, with some beautiful spots to discover. The island's liveliest places are **Lássi** and the south-coast resorts; elsewhere there are quiet villages and the scenery is stunning. A bus service links Argostóli with most parts of the island.

Capital of Kefalloniá until 1757, the whitewashed village of **Kástro** still flourishes outside the Byzantine fortress of Agios Geórgios. The Venetians renovated the fortress in 1504 but it was damaged by earthquakes in 1636 and 1637, and the 1953 earthquake finally ruined it. The large and overgrown interior is a haven for swallowtail butterflies.

In 1264 there was a convent on the site of **Moní Agíou Andréa**. The original church was

Lefkáda
Fiskárdo
Ithaca (Píso Aetós)

Asos
Agios Spyrídon
Mýrtou Bay
Zóla
Sinióri
Kardakáta
Agía Efthimía
Agía Thékla
Agios Dimítrios
Melissáni Cave-Lake
Ithaca (Vathý)
Ithaca (Píso Aetós)
Fársa
Lixoúri
Dilináta
Agriliá
Ithac (Vath
Drogkaráti Cave
Sámi
Pátra
ARGOSTÓLI
Fragkáta
Lássi
Kástro
Kyllíni
Peratáta
Forest Station
Póros
Miniés
Moní Agíou Andréa
Vlacháta
Mount Aínos
Pèssáda
1,630 m
5,350 ft
Pástra
Zákynthos
Atsoupádes
Markópoulo
Néa Skála

0 kilometres 10
0 miles 5

KEY
For key to map see back flap

Visitors to the blue waters of the subterranean Melissáni Cave-Lake

VISITORS' CHECKLIST

🏠 31,000. ✈ 9 km (5.5 miles) S of Argostóli. ⛴ Argostóli, Fiskárdo, Agía Efthimía, Sámi, Póros, Pessáda. 🚌 Ioánnou Metaxá, Argostóli. 🛈 Waterfront, Argostóli (26710 22248). 🛍 daily, Argostóli. 🎭 Panagía or Snake Festival at Markópoulo: 15 Aug; Wine Festival at Fragáta: 1st Sat after 15 Aug.

damaged in 1953, but has been restored as a museum to house icons and frescoes made homeless by the earthquake. The new church houses the monastery's holiest relic, supposedly the foot of the apostle Andrew.

There was once a sanctuary to Aenios Zeus at the summit of **Mount Aínos**, which is 1,630 m (5,350 ft) high. Wild horses live in the Mount Aínos National Park, and the slopes of the mountain are covered with the native

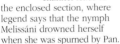

Apostle Andrew from the Moní Agíou Andréa

fir tree, *Abies cephalonica*. A road leads up towards the mountain's summit, but soon becomes a very rough track.

On the east coast, **Sámi** has ferry services to the Peloponnese and Ithaca. Nearby are two caves, Drogkaráti Cave, 3.5 km (2 miles) southwest and the Melissáni Cave-Lake, 2 km (1 mile) to the north. **Drogkaráti** drips with stalactites. It is the size of a large concert hall and is sometimes used as such due to its fine acoustics. The subterranean **Melissáni Cave-Lake** was a sanctuary of Pan in Mycenaean times. Part of its limestone ceiling has collapsed creating a haunting place with deep, blue water. A channel leads to the enclosed section, where legend says that the nymph Melissáni drowned herself when she was spurned by Pan.

Fiskárdo is Kefalloniá's prettiest village, undamaged by the 1953 earthquake. Its pastel-painted 18th-century Venetian houses cluster by the harbour, which is a popular berth for yachts. It is also busy in the summer with daily ferry services and day trips from elsewhere on Kefalloniá. Despite the crowds and gift shops Fiskárdo retains its charm.

Asos is an unspoilt village on Kefalloniá's west coast. The surrounding hilly terrain is noted for its stone terracing, which once covered the island. On the peninsula across the isthmus from Asos is a ruined Venetian fortress, built in 1595, which has seen occupation by Venetians, and stays by the French and Russians in the 19th century. Now Asos sees mostly day-trippers, as there is little accommodation in the village. South of Asos is **Mýrtou Bay**, a lovely cove with the most beautiful beach on the island.

🏛 **Moni Agíou Andréa**
Peratáta village. **Tel** 26710 69700.
◻ daily. 🎫 museum only.

A view overlooking Asos in the northwest of the island

Zákynthos
Ζάκυνθος

Zákynthos was inhabited by Achaians until Athens took control in the 5th century BC. They were followed by a succession of rulers, including the Spartans, Macedonians, Romans and Byzantines. The Venetians ruled from 1484 until 1797 and Zákynthos finally joined the rest of Greece in 1864. An attractive and green island, there are mountain villages, monasteries, fertile plains and beautiful views to reward exploration.

Statue of the poet Solomós in the main square, Zákynthos town

Zákynthos Town
Completely destroyed in the 1953 earthquake that hit the Ionian Islands, Zákynthos town has now been rebuilt with efforts to recapture its former grace. The traditional arcaded streets run parallel to the waterfront, where fishing boats arrive each morning to sell their catch. Further down the waterfront the ferry boats dock alongside grand Mediterranean cruise ships.

At the southern end of the harbour is the impressive church of **Agios Dionýsios**, the island's patron saint (1547–1622). The church, which houses the body of St Dionýsios in a silver coffin, was built in 1925 and survived the earthquake. The **Byzantine Museum** has a scale model of the pre-earthquake town, an elegant city built by the Venetians. It also houses a breathtaking collection of icons and frescoes rescued from the island's destroyed churches and monasteries.

North of here is the **Solomós Museum**, which contains the tomb of the poet Dionýsios Solomós (1798–1857), author of the Greek national anthem. The collection details lives of prominent Zákynthiot citizens.

A short walk north from the town centre, **Stráni hill** offers good views, while the Venetian kástro, above the town, has even more impressive views of the mainland. The ruined walls contain remnants of several churches and an abundance of plants and wildlife.

🏛 **Byzantine Museum**
Tel 26950 42714. ⬜ *Tue–Sun.*
⬤ *main public hols.* 🎫 ♿

🏛 **Solomós Museum**
Tel 26950 48982. ⬜ *daily.*
⬤ *main public hols.* 🎫

Around the Island
Outside the main resorts there is little tourist development on Zákynthos. It is possible to drive around the island in a day as most of the roads are in good condition. Hiring a

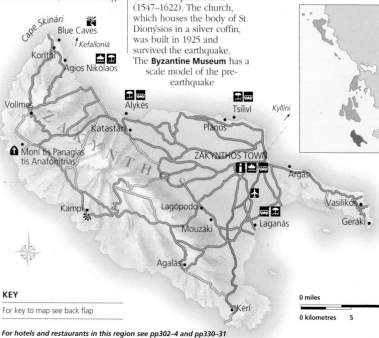

LOGGERHEAD TURTLES

The Mediterranean green loggerhead turtle *(Caretta caretta)* has been migrating from Africa to Laganás Bay, its principal nesting site, for many thousands of years. These giant sea creatures can weigh up to 180 kg (400 lb). They lay their eggs in the sand, said to be the softest in Greece, at night. However, disco and hotel lights disorientate the turtles' navigation and few now nest successfully. Of the eggs that are eventually laid, many are destroyed by vehicles or by

the poles of beach umbrellas. The work of environmentalists has led to some protection for the turtles, with stretches of beach now off-limits, in an attempt to give the turtles a chance to at least stabilize their numbers.

VISITORS' CHECKLIST

🏯 30,000. ✈ 4 km (2 miles) S of Zákynthos town. ⛴ Zákynthos town; Agios Nikólaos. 🚌 Zákynthos town. 🛈 Tzouláti 2, Zákynthos town (26950 42064). 🎭 Zákynthos town Festival: Jul.

car or a powerful motorbike is the best idea, though buses from Zákynthos town are frequent to resorts such as Alykés, Tsiliví and Laganás.

The growth of tourism on Zákynthos has been heavily concentrated in **Laganás** and its 14-km (9-mile) sweep of soft sand. This unrestricted development has decimated the population of loggerhead turtles that nests here – only an estimated 800 remain. Efforts are now being made to protect the turtles and to ensure their future survival. Visitors may take trips out into the bay in glass-bottomed boats to see the turtles, and all sorts of turtle souvenirs fill the large number of trinket shops. An equally large

number of bars and discos ensure the nightlife here continues till dawn.

Head to the north coast for the busy beach resorts of **Tsiliví** and **Alykés**, the latter being especially good for windsurfing.

The 16th-century **Moní tis Panagías tis Anafonítrias** in the northwest has special appeal for locals as it was here the island's patron saint, Dionýsios, spent the last years of his life as an abbot. During his time here, it is said that Dionýsios heard a murderer's confession; the murderer received the saint's

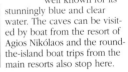

Coat of arms at Moní tis Panagías tis Anafonitrias

forgiveness, never knowing that his victim was the abbot's brother. When questioned by the authorities, Dionýsios denied seeing the man, which was the only lie he ever told. Dionýsios lived in a cell here which still stands and contains many of the saint's revered possessions. The three-aisled church and the tiny chapel alongside are rare in that they survived the 1953 earthquake. At the northernmost tip of the island are the unusual **Blue Caves**, formed by the relentless action of the sea on the coastline. The principal cave, the Blue Grotto, lies directly underneath the lighthouse on Cape Skinári. It was discovered in 1897 and has become well known for its stunningly blue and clear water. The caves can be visited by boat from the resort of Agios Nikólaos and the round-the-island boat trips from the main resorts also stop here.

The Blue Caves of Zákynthos on the northern tip of the island

THE ARGO-SARONIC ISLANDS

SALAMINA · AIGINA · POROS · YDRA · SPETSES · KYTHIRA

*A*lthough still supporting fishing and farming communities, the Argo-Saronic Islands have succumbed to a degree of tourism. Many Athenians visit the islands at weekends, when the beaches can become very busy. Kýthira, off the tip of the Peloponnese, shares its history of Venetian and British rule with the Ionians, but is today administered with the Argo-Saronics.

The islands' location close to Athens has given them a rich history. Aígina was very prosperous in the 7th century BC as a maritime state that minted its own coins and built the magnificent temple of Aphaia. Salamína is famed as the site of the Battle of Salamis (480 BC), when the Greek fleet defeated the Persians. Wealth gained from maritime trading also assured the Argo-Saronics' cultural and social development, seen today in the architectural beauty of Ydra and in the grand houses and public buildings of Aígina. Ydra and Spétses were important in the War of Independence *(see pp42–3)*, both islands producing brave fighters, including the notorious Laskarína Bouboulína and Admiral Andréas Miaoúlis.

Terracotta ornament

Salamína and Aígina are so easy to reach from the capital that they are often thought of as island suburbs of Athens. Póros hardly seems like an island at all, divided from the Peloponnese by a narrow channel. However, despite modern colonization peaceful spots can still be found. Póros and Spétses are lush and green, covered with pine forests and olive groves, in contrast to the other more barren and mountainous islands. Scenically, Kýthira's rugged coastline has more in common with the Ionians than the Argo-Saronics. The island's position on ancient shipping routes has led to some major finds, such as the bronze *Youth of Antikýthira*, now in the National Archaeological Museum *(see p286)*.

The chapel of Agios Nikólaos on Aígina

Póros town with the mountains of the Peloponnese in the background

Exploring the Argo-Saronic Islands

Close proximity to Athens makes the Argo-Saronic Islands suitable for short visits as well as longer stays. The islands have a lush landscape, with pine forests and crystal-clear waters in secluded bays. Aígina is an ideal base and, like the other islands, has picturesque ports with cobbled streets and Neo-Classical buildings. Packed with smart bars and shops, the cosmopolitan atmosphere of the Argo-Saronics is tempered by harbourside caïques selling vegetables and horse-drawn carriages driving along the seafront. Horse power is particularly evident in Póros, Ydra and Spétses where no cars are allowed. Kýthira remains a well-kept secret. This large island has beautiful villages and deserted beaches to explore.

The harbour in Aígina town

ISLANDS AT A GLANCE

Aígina *pp96–9*
Kýthira *pp102–3*
Póros *p100*
Salamína *p96*
Spétses *p101*
Ydra *pp100–101*

SEE ALSO

• **Where to Stay** pp304–6

• **Where to Eat** pp331–2

• **Travel Information** pp366–9

KEY

▬	Main road
▭	Minor road
▬	Scenic route
---	High season, direct ferry route
△	Summit

The rugged scenery of Palaióchora on Kýthira

0 kilometres   20

0 miles   10

Ermió
Kranídi
Portochéli
Kósta
Spétses Town
Agia Paraskeví
Agioi Anárgyroi
Spétses
Spetsopou

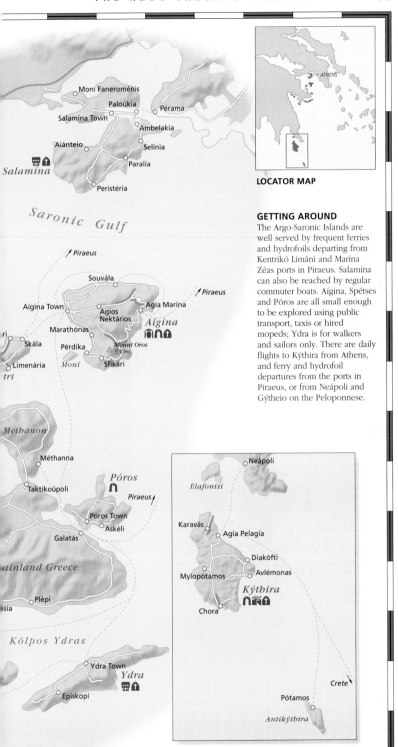

LOCATOR MAP

GETTING AROUND

The Argo-Saronic Islands are well served by frequent ferries and hydrofoils departing from Kentrikó Limáni and Marína Zéas ports in Piraeus. Salamína can also be reached by regular commuter boats. Aígina, Spétses and Póros are all small enough to be explored using public transport, taxis or hired mopeds; Ydra is for walkers and sailors only. There are daily flights to Kýthira from Athens, and ferry and hydrofoil departures from the ports in Piraeus, or from Neápoli and Gýtheio on the Peloponnese.

Salamína
Σαλαμίνα

🏛 23,000. 🚢 Paloúkia & Selínia.
🚌 Salamína town. 🛍 Thu at
Salamína town, Sat at Aiánteio.

Salamína is the largest of
the Saronic Gulf islands, and
so close to Athens that most
Greeks consider it part of the
mainland. The island
is famed as the site of the
decisive Battle of Salamis
in 480 BC, when the Greeks
defeated the Persians. The
king of Persia, Xerxes, watched
the humiliating sight of his
cumbersome ships being
destroyed in Salamis Bay,
trapped by the faster triremes
of a smaller Greek fleet under
Themistokles. The island
today is a cheerful medley of
holiday homes, immaculately
whitewashed churches and
cheap tavernas, although its
east coast is lined with a
string of marine scrapyards
and naval bases.

The west coast capital of
Salamína town is a charmless
place, straddling an isthmus of
flat land filled with vineyards.
Both the town and the island
are known as Koúlouri, nick-
named after a biscuit that
resembles the island's shape.

East of Salamína town
Agios Nikólaos has far
more character, with 19th-
century mansions lining
the quayside and small
caïques off-loading
their catch of fish. A
road from Paloúkia
meanders across the
south of the island to
the villages of Selínia,
Aiánteio and Peristéria.

In the northwest of the
Salamína, the 17th-century
Moní Faneroménis looks
across a narrow gulf to
Ancient Eleusis on the
Attic coast. The monas-
tery was used during
the War of Indepen-
dence *(see pp42–3)*
as a hiding place for
Greek freedom fighters. Its
Byzantine church was restored
by the Venetians, and has fine
18th-century frescoes vividly
depicting the *Last Judgment*.
Today nuns welcome visitors,
and tend the gardens, home to
a number of peacocks.

**Shrine
opposite Moní
Faneroménis**

Fishing boats sailing into Aígina harbour

Aígina
Αίγινα

🏛 14,000. 🚌 Aígina town.
ℹ Leonárdou Ladá, Aígina town
(22970 27777).

Only 20 km (12 miles)
southwest of the port of
Piraeus, Aígina has been inhabited for over 4,000 years,
and has remained an important settlement throughout
that time. According to Greek
mythology, the island's name
was changed from Oinóni to
Aígina, who was the daughter
of the river god Asopós,
after Zeus installed her on
the island as his mistress.
By the 7th century
BC the second-largest
Saronic island was the
first place in Europe
to mint its own silver
coins, which became
accepted currency
throughout the Greek-
speaking world. Plying
the Mediterranean and
the Black Sea, the people
of Aígina controlled most
foreign trade in Greece.
However, their legendary
nautical skills and vast
wealth finally incurred
the wrath of neigh-
bouring Athens, who
settled the long-term
rivalry by conquering
the island in 456 BC. Aígina's
most famous site is the well-
preserved **Temple of Aphaia**
(see pp98–9), built in about
490 BC, prior to Athenian
control. Later, the island declined during the centuries
of alternating Turkish and
Venetian rule and the constant
plague of piracy. However,
Aígina enjoyed fame again for
a brief period in 1828 when
Ioánnis Kapodístrias (1776–
1857) declared it the first
capital of modern Greece.

**The ruinous Venetian Pýrgos
Markéllou in Aígina town**

Aigina Town
This picturesque island town
is home to many churches,
including the pretty 19th-
century **Agía Triáda**, next to
the fish market overlooking
the harbour. At the quayside,
horse-drawn carriages take
visitors through narrow streets
of Neo-Classical mansions to
the Venetian tower **Pýrgos
Markéllou** near the cathedral.
Agios Nektários cathedral,
inaugurated in 1994, is said to
be the second-biggest Greek
Orthodox church after Agía
Sofia in Istanbul. Octopuses
are hung out to dry at tavernas in the street leading to
the fish market. To the north-

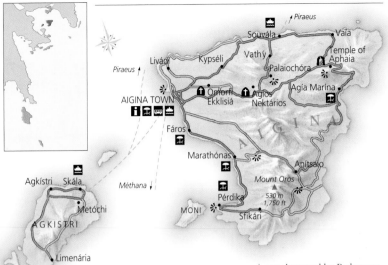

Agios Nektários cathedral in Aígina town

west, past shops selling pistachio nuts and earthenware jugs, are the remains of the 6th-century BC **Temple of Apollo**. The 6th-century Sphinx of Aígina, now in the **Aígina Museum**, was discovered here.

Aígina Museum
Kolóna 8. **Tel** 22970 22637.
Tue–Sun. main public hols.

Environs
North of Aígina town, in Livádi, a plaque marks the house where Níkos Kazantzákis wrote *Zorba the Greek (see p276)*.

Around the Island
Aígina, at only 8 km (5 miles) across, is easy to explore by bicycle. Just off the main road east from Aígina town is the

KEY

For key to map see back flap

13th-century Byzantine church, **Omorfi Ekklisiá**, which has some fine frescoes. Pilgrims take this road to pay homage at **Agios Nektários**. Archbishop Nektários (1846–1920) was the first man to be canonized in modern times (1961) by the Orthodox Church. Visitors can see his quarters and the chapel where he rests.

On the opposite hillside are the remains of the deserted town of **Palaiochóra**. Populated since Byzantine times, it

The scattered ruins of Byzantine chapels around the deserted town of Palaiochóra

was destroyed by Barbarossa, the general of Sultan Suleiman I, in 1537. The area around the town was abandoned in 1826.

South from Aígina town, the road hugs the shore, beneath the shadow of **Mount Oros** at 530 m (1,750 ft). Passing the pistachio orchards and the fishing harbour of **Fáros**, this scenic route ends at **Pérdika** at the southwestern tip of the island. Overlooking the harbour, this small, picturesque fishing village has some excellent fish tavernas that are packed at weekends with Athenians over for a day trip.

Outlying Islands
Just 15 minutes by caïque from Pérdika is the island of **Moni**, popular for its emerald-green waters, secluded coves and hidden caves.

Agkístri is easily accessible by caïque from Aígina town or by ferry from Piraeus. Originally settled by Albanians, today this island is colonized by Germans who have bought most of the houses in the village of Metóchi, just above Skála port. Although many hotels, apartments and bars have been built in Skála and Mílos, its other main port, the rest of this hilly, pine-clad island remains largely unspoilt. Limenária, in the south of the island, is a more traditional, peaceful community of farmers and fishermen.

Aigina: Temple of Aphaia
Ναός της Αφαίας

Surrounded by pine trees, on a hilltop above the busy resort of Agía Marína, the Temple of Aphaia is one of the best-preserved Doric temples in Greece *(see pp60–61)*. The present temple dates from around 490 BC, but the site is known to have been a place of worship from the 13th century BC. In 1901 the German archaeologist Adolf Furtwängler found an inscription to the goddess Aphaia, disproving theories that the temple was dedicated to Athena. Although smaller, the building is similar to the temple of Zeus at Olympia, built 30 years later.

The east pediment sculptures, with Athena at the centre, were replacements for an earlier set. The west pediment sculptures are Archaic in style.

Inner Walls
The inner wall was built with a thickened base and a minimal capital to correspond with the capitals of the colonnade.

Aerial view of the site from the south

Triglyph

Metope

Architrave

Corner Columns
These columns were made thicker for emphasis and to counteract the appearance of thinness in a column that was seen against the sky.

Ramp from altar to temple

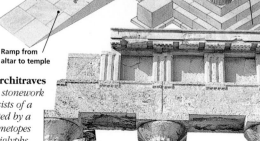

Corner Architraves
Still in good condition, the stonework above the capitals consists of a plain architrave surmounted by a narrow band of plain metopes alternating with ornate triglyphs.

Inner Columns

The cella is enclosed by two storeys of Doric columns, one on top of the other. The taper of the upper columns is continuous with that of the lower.

The roof was made of terracotta tiles with Parian marble tiles at the edges.

Opisthodomos, or rear porch

VISITORS' CHECKLIST

12 km (7 miles) E of Aígina town.
Tel 22970 32398. **Site** ◯ Apr–
Oct: 8:15am–7pm daily;
Nov–Mar: 8:15am–5pm daily.
Museum ◯ 9am–1pm
Tue–Sun (entry at 9am, 11am,
noon, 1pm only). ● 1 Jan, 25
Mar, Good Fri am, Easter Sun, 1
May, 25, 26 Dec. 🖾 ◙ 🖵

View of the Cella

The cella was the inner room of the temple, and the home of the cult statue. Some temples had more than one, the back cella being reserved for the priestess alone.

RECONSTRUCTION OF THE TEMPLE OF APHAIA

Viewed from the northwest, this reconstruction shows the temple as it would have been in c.490 BC. Built of local limestone covered in stucco and painted, it was highly colourful.

Cult statue of the goddess Aphaia

The pool of olive oil was a collection of the many libations (offerings) made to the goddess.

TEMPLE PEDIMENTS

The famous sculptures from the pediments of the temple of Aphaia were discovered by a group of British and German architects and artists, including John Foster, C R Cockerell, and Baron Haller von Hallerstein, in April 1811. They were later sold to the Crown Prince of Bavaria at auction and are now housed in the Glyptothek in Munich. They portray the struggles of various mythological heroes. The sculptures from the west pediment date from around 490 BC and are in the late Archaic style. Those from the east, with their more fluid movements and serious expressions, date from approximately 480 BC and foreshadow the Classical style.

Reconstruction of the *Warriors* sculpture from the west pediment

Póros
Πόρος

🏛 4,500. 🚢 🚌 *Póros town.*
ℹ️ *Póros town (22980 22462).*
🎪 *Fri (am) at Paidikí Chará.*

Póros takes its name from the
400-m (1,300-ft) passage
(póros) separating it from the
mainland at Galatás. Póros is
in fact two islands, joined by
a causeway: pine-swathed
Kalávria to the north, and the
smaller volcanic islet of Sfairía
in the south over which
Póros town is built. In spite of
much tourist development,
the town is an appealing
place, extending along the
narrow straits, busy with
shipping. Its 19th-century
houses climb in tiers to its
apex at a clock tower.

The **National Naval
Academy**, northwest of the
causeway and Póros town,
was set up in 1849. An old
battleship is usually at anchor
there for training naval cadets.

The attractive 18th-century
Moní Zoödóchou Pigís can be
found on Kalávria, built
around the island's only spring.
There are the ruins of a 6th-
century hilltop **Temple of
Poseidon** near the centre of
Kalávria, next to which the
orator Demosthenes poisoned
himself in 323 BC rather than
surrender to the Macedonians.
In antiquity the site was
linked to ancient Troezen in
the Peloponnese. The temple
has unlimited access.

The busy waterfront on Ydra

Ydra
Ύδρα

🏛 2,700. 🚢 *Ydra town.*
ℹ️ *Ydra town (22980 52205).*
🏖 *Mandráki 1.5 km (1 mile) NE
of Ydra town; Vlychós 2 km (1 mile)
SW of Ydra town.*

Along, narrow mass of
barren rock, Ydra had
little history before the
16th century when it
was settled by Ortho-
dox Albanians, who
then turned to the sea
for a living. Ydra town
was built in a brief
period of prosperity in
the late 18th and early
19th centuries, boosted
by blockade-running
during the Napoleonic
wars. After Indepen-
dence, Ydra lapsed
into obscurity again,
until foreigners

**Bell tower of
Ydra's Panagía
church**

rediscovered it after World
War II. By the 1960s, the trickle
had become a flood of outsid-
ers who set about restoring the
old houses, transforming Ydra
into one of the most exclusive
resorts in Greece. Yet the
island has retained its charm,
thanks to an architectural
preservation order which has
kept the town's appear-
ance as it was in the
1820s, along with a
ban on motor vehicles.
Donkey caravans
perform all haulage
on steep stair streets.

Ydra Town
More than a dozen
three- or four-storeyed
mansions *(archontiká)*
survive around the
port, though none are
regularly open to the
public. Made from
local stone, they were

Póros town, its houses clustered on the hillside of Sfairía

For hotels and restaurants in this region see pp304–6 and pp331–2

built by itinerant craftsmen between 1780 and 1820. On the east side of the harbour the **Tsamadoú mansion** is now the National Merchant Marine Academy. On the west, the **Tompázi mansion** is a School of Fine Arts. Just behind the centre of the marble-paved quay is the monastic church of the **Panagía**, built between 1760 and 1770 using masonry from Póros's Temple of Poseidon. The marble belfry is thought to have been erected by a master stonemason from Tínos.

Around the Island

Visitors must walk virtually everywhere on Ydra, or hire water taxis to go along the coast. **Kamíni**, 15 minutes' walk southwest along the shore track, has been Ydra's main fishing port since the 16th century. The farm hamlet of **Episkopí**, in the far southwest of the island, used to be a summer refuge and a hunting resort for the upper classes. An hour's steep hike above the town is the convent of **Agía Efpraxía**, which still houses nuns who are keen to sell you handicrafts. The adjacent 19th-century **Profítis Ilías** functions as a monastery. In the island's eastern half, visible from Profítis Ilías, are three uninhabited monasteries, dating from the 18th and 19th centuries. They mark the arduous 3-hour-long route to **Moní Panagía**, situated out near Cape Zoúrvas to the northeast of the island.

Spétses
Σπέτσες

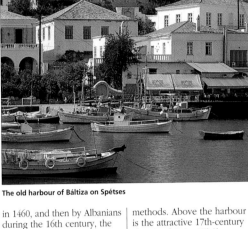

The old harbour of Báltiza on Spétses

🏠 3,900. 🚢 🚌 Spétses town.
🛈 Spétses town (22980 73100).
🛥 Wed at Kokinária.

Spétses is a corruption of Pityoússa, or "Piney", the ancient name for this round, green island. Occupied by the Venetians in 1220, by the Turks

in 1460, and then by Albanians during the 16th century, the island developed as a naval power, and supplied a fleet for the Greek revolutionary effort.

Possibly the most famous Spetsiot was Laskarína Bouboulína, the admiral who menaced the Turks from her flagship *Agamemnon* and reputedly seduced men at gunpoint. She was shot in 1825 by the father of a girl her son had eloped with. During the 1920s and 30s, Spétses was a fashionable resort for British expatriates and anglophile Greeks. The ban on vehicles is not total: mopeds and horsecabs can be hired in town, and there are buses to the beaches.

Statue of Bouboulína in Spétses town

Spétses Town
Spétses town runs along the coast for 2 km (1 mile). Its centre lies at Ntápia quay, fringed by cafés. The *archontiká* of Chatzi-Giánnis Méxis, dating from 1795, is now the **Chatzi-Giánnis Méxis Museum**. Bouboulína's coffin is on display as well as figureheads from her ship. Her former home is now the privately run **Bouboulína Museum**. Southeast from here lies the old harbour at **Báltiza** inlet, where wooden boats are still built using the traditional

methods. Above the harbour is the attractive 17th-century church of **Agios Nikólaos**, which has some fine pebble mosaics and a belfry made by craftsmen from Tínos.

🏛 **Chatzi-Giánnis Méxis Museum**
300 m (985 ft) from the port.
Tel 22980 72994. ⬤ Tue–Sun.
⬤ main public hols. 🎫 📷

🏛 **Bouboulína Museum**
Behind Plateía Ntápia. *Tel* 22980 72416. ⬤ 25 Mar–28 Oct: 9:30am–9pm daily. 🎫 📷 🎥

Around the Island

A track, only partly concreted, runs all the way round the island, and the best way to get around is by bicycle or moped. East of town stands the Anargýreios and Korgialéneios College, which is now closed. British novelist John Fowles taught there briefly in the early 1950s. He later used Spétses as the setting for *The Magus*. The pebble beaches on Spétses are the best in the Argo-Saronic group, including **Ligonéri**, **Vréllas**, and **Agía Paraskeví**. **Agioi Anárgyroi** is the only sandy one.

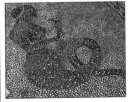

Pebble mosaic from the church of Agios Nikólaos, Spétses town

Kýthira
Κύθηρα

Called Tserigo by the Venetians, Kýthira is one of the legendary birthplaces of Aphrodite. Historically, the island shared Venetian and British rule with the Ionian islands; today it is governed from Piraeus with the other Argo-Saronics. Clumps of eucalyptus seem emblematic of the Island's modern alias of "Kangaroo Island"; return visits from 60,000 Australian Kythirans are central to Kythiran life. The island is also popular with Athenians seeking unspoilt beaches and holiday homes, many of which are the typical mix of Aegean and Venetian architecture.

Kapsáli harbour seen from Chóra

KYTHIRA

0 kilometres 5

0 miles 3

KEY

For key to map see back flap

Chóra

Chóra has been Kýthira's capital only since the destruction of Palaióchora in 1537. Its magnificent **kástro** was built in two phases during the 13th and 15th centuries. A multidomed cistern lies intact near the bottom of the castle; at the summit, old cannons surround the church of **Panagía Myrtidiótissa**. The steepness of the drop to the sea below and Avgó islet, thought to be the birthplace of Aphrodite, is unrivalled throughout the Greek islands. A magnet for wealthy Athenians, the appealing lower town with its solidly built, flat-roofed mansions dates from the 17th to 19th centuries. The **Archaeological Museum** just outside Chóra has finds from Mycenaean and Minoan sites, plus gravestones dating from the British occupation of 1809–64.

🏛 Archaelogical Museum
Tel 27360 31739. ◯ Jul–Sep:
10am–8pm daily. ● main public
hols. 📷

Environs
Yachts, hydrofoils and large ferries drop anchor at the harbour of Kapsáli, just east of Chóra. The beach is mediocre, but most foreigners stay here. In the cliff above the pine wood is the 16th-century Moní Agios Ioánnis sto Gkremó, built on to the cliff edge. The nearest good beaches are pebbly Fyrí Ammos, 8 km (5 miles) northeast via Kálamos, with sea caves at its south end; and sandy Chalkós, 7 km (4 miles) south of Kálamos.

The houses of Chóra clustered on the hillside at dusk

For hotels and restaurants in this region see pp304–6 and pp331–2

Whitewashed house in Mylopótamos

3,400. 22 km (14 miles) NE of Chóra. Agía Pelagía & Kapsáli. runs between Agía Pelagía & Kapsáli and between Diakófti & Kapsáli. 27360 33222. Sun at Potamós.

been much altered, and the Baroque relief plaque over the door is a rarity in Greece.

To the north, the main port, **Agía Pelagía**, has a handful of hotels. **Karavás**, 5 km (3 miles) northwest is, in contrast, an attractive oasis village, with clusters of houses overhanging the steep banks of a stream valley.

Agía Sofía Cave
Mylopótamos. *Tel* 27360 33754.
Tue–Sun. Nov–Mar.
Jul & Aug.

Outlying Islands
Directly north of Kýthira, the barren islet of **Elafonísi** is visited mostly by Greeks for its fantastic desert-island beaches. The better of the two is **Símos** on the east side of a peninsula 5 km (3 miles) southeast of the port town. The remote island of **Antikýthira**, southeast of Kýthira, has a tiny population and no beaches.

Around the Island
Like many Greek islands, the best way to get around Kýthira is by car, particularly as it is quite mountainous. A bus runs to the main towns once a day during the summer from Agía Pelagía to Kapsáli. **Avlémonas**, with its vaulted warehouses and double harbour, forms an attractive fishing port at the east end of a stretch of rocky coast. Just offshore the *Mentor*, carrying many of the Elgin Marbles, sank in 1802. Excellent beaches extend to either side of Kastrí point. The 6th-century hilltop church of **Agios Geórgios**, which has a mosaic floor, sits high above Avlémonas.

On the other side of the island is **Mylopótamos**. From here a track leads west to the small Fónissa waterfall, downstream from which is a millhouse, and a tiny stone bridge.

In its blufftop situation with steep drops to the north and west, and a clutch of locked chapels, the Venetian kástro at **Káto Chóra** superficially resembles Palaióchora. It was not a military stronghold but a refuge prepared in 1565 for the peasantry in unsettled times. The Lion of St Mark presides over the entrance; nearby an English-built school of 1825 is being restored.

Agía Sofía Cave, 2.5 km (1.5 miles) from Káto Chóra and 150 m (490 ft) above the sea, has formed in black limestone strata. At the entrance, a frescoed shrine, painted by a 13th-century hermit, depicts Holy Wisdom and three attendant virtues. Palaióchora, the Byzantine "capital" of Kýthira after 1248, was sited so as to be nearly invisible from the sea, but the pirate Barbarossa detected and destroyed it in 1537. The ruins of the town perch on top of a sheer 200-m (655-ft) bluff. Among six churches in Palaióchora, the most striking and best preserved is the 14th-century **Agía Varvára**.

To the south, **Moní Agíou Theodórou** is the seat of Kýthira's bishop. The church, originally 12th century, has

Roadside shrine on Kýthira

View to the east across a gorge from Palaióchora

THE SPORADES AND EVVOIA

SKIATHOS · SKOPELOS · ALONNISOS
SKYROS · EVVOIA

The lush landscape of Evvoia and the Sporades comes as a surprise after barren and arid islands such as the Cyclades. Since ancient times, settlers and pirates alike have been lured by the pine-clad mountains, abundant springs and rivers, endless beaches and hidden coves that are found throughout these islands.

Being close to the mainland, the Sporades and Evvoia have been easily conquered throughout history. They were colonized in the prehistoric era by nearby Iolkos (Vólos), and also by the Minoans, who introduced vine and olive cultivation. More than any other island, Evvoia reveals its diverse history in the large number of buildings remaining from the long periods of Venetian and Turkish occupation. Susceptible to pirate raids, the inhabitants of the Sporades lived in the safety of fortified towns until as late as the 19th century. Even in Evvoia, when life proved too difficult in coastal villages such as Límni, the residents simply migrated to Skiáthos for a few generations. The islanders

Mariner statue in Kárystos, Evvoia

have a rich heritage of maritime trading around the Aegean and are still noted today as sailors. The islands' patchworked interiors of fertile fields and orchards, watered by ample springs and rivers, also encouraged agricultural self-sufficiency and wealth. Particularly on remote and rugged Skýros, such insularity has nurtured some unique folk art and colourful traditions. Its inaccessible coastline enables it to remain relatively unaffected by the numerous tourist hotel complexes that have sprung up on Skiáthos and Skópelos.

The size of Evvoia also means it is one of the few places in the Greek islands where life carries on during the summer, undeterred by the annual invasion of holiday-makers.

Castel Rosso near Kárystos on Evvoia

◁ **The harbour of Agnóntas on Skópelos in the evening sun**

Exploring the Sporades and Evvoia

The rich and famous first flocked in their yachts to the deserted beaches of Skiáthos, Skópelos and Alónnisos in the 1960s and 1970s. Although no longer so exclusive, the beautiful coastlines of these islands still lure Greek and foreign holiday-makers alike. There are facilities for windsurfing and boats for hire on most beaches. Skópelos and Skiáthos have a sophisticated array of nightclubs and bars. Quieter Skýros and Evvoia, offering a varied culture and landscape, are perfect for rambling holidays, punctuated by visits to local folk art museums and lingering days on the fine beaches.

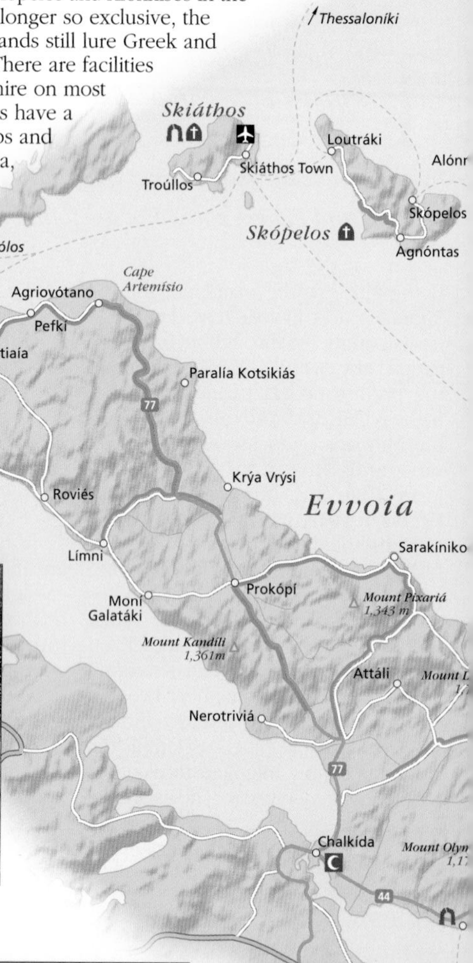

Thessaloníki

Skiáthos
Loutráki
Alónr
Skiáthos Town
Troúllos
Skópelos
Skópelos
Agnóntas

Vólos

Cape Artemísio
Agriovótano
Pefkí
Istiaía
Glýfa
Paralía Kotsikiás
Agiókampos
Loutrá Aidipsoú
Agios Geórgios
Krýa Vrýsi
Roviés
Evvoia
Límni
Sarakíniko
Moní Galatáki
Prokópi
Mount Pixariá 1,343 m
Mount Kandíli 1,361m
Attáli
Mount L
Nerotriviá
Chalkída
Mount Olyn 1,17
Skála Oropoú
1 E75

A house in Stení on Evvoia, with Mount Dírfys in the background

The harbour of Skópelos town

SEE ALSO

- *Where to Stay* pp306–7
- *Where to Eat* p332
- *Travel Information* pp366–9

For additional map symbols *see back flap*

Gioúra

Pipéri

Kyrá Panagía

Gérakas

Alónnisos

eni Vála

Peristéra

Sporades Marine Park

Skantzoúra

ISLANDS AT A GLANCE

Alónnisos *pp114–15*
Evvoia *pp118–23*
Skiáthos *pp108–9*
Skópelos *pp112–13*
Skýros *pp116–17*

LOCATOR MAP

ATHENS

*A E G E A N
S E A*

Valáxa

Atsítsa

Mólos

Skýros Town

Skýros

Linariá

*Kochýlas
792 m*

Sarakiní

Sýros, Andros, Tínos

Kými

Paralía Kýmis

Paralía

Ochthoniá

Alivéri

Lépoura

44

GETTING AROUND

Skýros and Skiáthos are both connected
with Athens by internal flights. Skiáthos's
international airport also caters for
charter flights. Island-hopping is
easy in the summer season, with
frequent ferries and Flying
Dolphin hydrofoils plying
between the Sporades,
Evvoia and the main-
land. It is also possible
to connect by ferry
with the Cyclades and
Thessaloníki. Kárystos
is the best base for touring the south of Evvoia;
stay at Kými for the east
coast, and Límni or Loutrá
Aidipsoú for a tour of the
north. There are good roads
around Evvoia and a frequent,
reliable bus service.

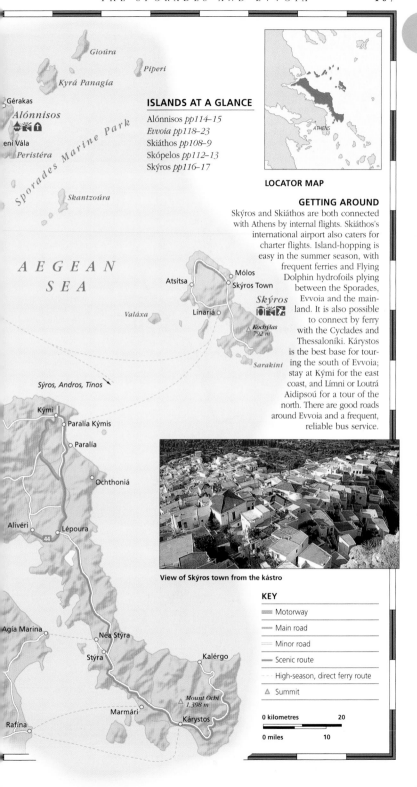

View of Skýros town from the kástro

Agía Marina

Néa Stýra

Stýra

Kalérgo

*Mount Óchi
1,398 m*

Marmári

Kárystos

Rafína

KEY

▬▬	Motorway
▬▬	Main road
▭▭	Minor road
▬▬	Scenic route
- - -	High-season, direct ferry route
△	Summit

0 kilometres   20

0 miles   10

Skiáthos
Σκιάθος

Skiáthos has always been an unashamedly
hedonistic island from its early tourist
development in the 1960s, when it attracted the
rich and famous with its legendary beaches, to
its current role as bucket-and-spade paradise for
family package tours. Although the introduction
of direct package flights has diminished
Skiáthos's exclusive status, the luxury yachts are
still in evidence off Koukounariés beach. In spite
of the tourism, the island retains its scenic beauty
and a scattering of atmospheric
churches and monasteries.

The sweeping bay of Koukounariés

[Map of Skiáthos showing locations including: Kástro, Laláre, Panagía Kardási, Moní Agíou Charalámpou, Moní Evangelismoú, Agios Apóstolos, Taxiárchis, Profítis Ilías, Kechria, Agios Ioánnis, Asélinos, Moní Panagías Kounístras, Mandráki, Koukounariés, Máratha, Troúllos, Kalamáki, Ftelá, SKIÁTHOS TOWN, and sea routes to Thessaloniki/Skópelos, Loutráki, ARGOS, Andros, Sýros/Tínos, Vólos, TSOUGKRIA, TSOUGRIAKI]

KEY

For key to map see back flap

0 kilometres 2
0 miles 1

Skiáthos Town

Still picturesque, the island
town is a charming place with
its red-tiled roofs and maze
of cobbled back streets. It is
built on two small hills,
dominated by the large 19th-
century churches of **Trión
Ierarchón** and **Panagía
Limniá**, which offer excellent
views of the bustling harbour
below. The main street winds
up between the two hills to
the old quarter of Limniá, a
quiet neighbourhood of
restored sea captains' houses,
covered with trailing
bougainvillea and trellised
vines. The town is excellent

for shopping, full of aromatic
bakeries, smart boutiques and
antique shops, some of which
specialize in genuine folk artifacts, including ceramics, icons,
jewellery and embroidery.

The town has twin harbours,
separated by **Bourtzi** islet,
which is reached by a narrow

causeway. The pine-covered
islet, once a fortress, is now a
cultural centre and hosts the
annual Aegean festival of
dance, theatre and concert
performances each summer.
Bourtzi is dominated by a
handsome Neo-Classical
building, with a statue of
the famous Greek
novelist Aléxandros
Papadiamántis
standing guard.
Life in Skiáthos
town centres
on the long,
sweeping quay-
sides lined with
numerous *kafeneía*,
specializing in
loukoumádes (small
honeyed fritters). In
the evenings the
waterside attracts
many people for a stroll
in the cool night air. During
the day there is the spectacle
of arriving and departing
flotilla yachts, ferries and
hydrofoils. The western end
of the quay has a good fish
market, and an *ouzerí* frequented by locals. It is also
where small boats and caïques
depart for day trips to some
of the island's famous beaches,

View of Skiáthos town from the church of Profítis Ilías

For hotels and restaurants in this region see pp306–7 and p332

An ornate fresco in the Christós sto Kástro church

VISITORS' CHECKLIST

🏠 5,000. ✈ 2 km (1 mile) NE of Skiáthos town. 🚢 🚌 Harbour-front, Skiáthos town. ℹ 24270 23172. 🎭 Aegean Festival of Dance, Skiáthos town: Jul.

such as Koukounariés and Laláría, or to the nearby islands of Tsougkriá and Argos.

Behind the harbour is the **Papadiamántis Museum**, former home of the locally born novelist, whose name it takes. Although tiny, the museum shows the simplicity of local island life prior to the invasion of tourism.

🏛 **Papadiamántis Museum**
Tel 24270 23843. ⬜ May–Oct. 🎫

Moní Agíou Charalámpou, set in the hills above Skiáthos town

Around the Island

The interior of the northern side of the island, with its verdant landscape of pine and olive trees, reveals deserted monasteries and churches, springs and plenty of birdlife. This is in contrast to the overdeveloped southern coast. It is still possible to find deserted beaches and coves scattered along the northern coast. Many of these, such as **Kechriá** and **Mandráki**, can only be visited when the excursions stop for a few hours on their day trips around the island.

The main road south from Skiáthos town passes Fteliá and branches to the west just before Troúllos for Asélinos beach and **Moní Panagías Kounístras**. The monk who founded this 17th-century monastery, originally called Panagía Eikonístria, discovered a miraculous icon in a nearby tree. The icon is kept in Trión Ierarchón in Skiáthos town.

The path north from here leads to **Agios Ioánnis**, where it is customary to stop and ring the church bell after completing the steep walk through pine trees.

Further north still is the tiny 19th-century chapel of **Panagía Kechriás**, with its blue ceiling covered in stars, which perches high above **Kástro**. Abandoned in 1829, remains of the 300 houses are still visible in this deserted town and three churches have been restored. The 17th-century **Christós** church has a fine iconostasis.

On the road heading northwest out of Skiáthos town lies the barrel-vaulted 20th-century church of **Profítis Ilías**, which has a good taverna nearby with stunning views over the town. Continuing north, past rich farms and the 20th-century **Agios Apóstolos** church, the track descends through sage and bracken to **Moní Agíou Charalámpou**, built in 1809. Aléxandros Moraïtidis, the writer, spent his last days here as a monk in the early 1920s. Just south of here is **Moní Evangelismoú**. Founded in 1775 by monks from Mount Athos, it played a crucial role in the War of Independence (*see pp42–3*), hiding many freedom fighters.

To the south of Moní Agíou Charalámpou, on the way back to Skiáthos town, is the beautiful church of **Taxiárchis**. It is covered in plates in the shape of a cross, and the best mineral spring water on the island flows out of a tap that is by the church.

ALEXANDROS PAPADIAMANTIS

The island's most famous native is one of Greece's outstanding literary figures. Aléxandros Papadiamántis spent his early childhood on the island, with five brothers and sisters, before leaving to study in Athens where he began his career in journalism. He wrote more than 100 novellas and short stories, all set against the back-drop of island life. Among his best-known works are *The Gypsy*, *The Murderess*, a compulsive psychological drama, and *The Man Who Went to Another Country*. In 1908 he returned to Skiáthos where he died a few years later in 1911 at the age of 60.

Kalamáki beach, Skiáthos ▷

Skópelos
Σκόπελος

Surprisingly, given its close proximity to Skiáthos, Skópelos has not totally succumbed to tourism. It is known to have been colonized by the Minoans as far back as 1600 BC and was used as a place of exile by the Byzantines. The Venetians held power for about 300 years after 1204. Famed for its wine in ancient times, Skópelos is still renowned for its fruit today. It offers many good beaches, and has a beautiful pine-covered interior.

The way up to Panagía tou Pýrgou above Skópelos town

KEY

For key to map see back flap

the Ghisi family in the 13th century, the castle stands on the site of the 5th-century BC acropolis of ancient Skópelos. The church nearest the castle is **Agios Athanásios**. It was built in the 11th century, but the foundations date from the 9th century. There are some fine 16th-century frescoes inside. The **Folk Art Museum** is situated behind the harbourfront in a 19th-century mansion. Pieces of furniture and examples of traditional local costumes and embroidery are on display.

⌂ Folk Art Museum
Chatzistamáti. **Tel** 24240 23494.
◯ Nov–May: 8:30am–2:30pm daily; Jun–Oct: 10:30am–2:30 pm & 6–10:30pm daily. 🎫 📷

Environs
In the hills above Skópelos town there are numerous impressive monas-teries. Reached by the road going east out of the town, they all

Skópelos Town

This charming town proudly reveals its rich pedigree with 123 churches, many fine mansion houses and myriad shops selling local delicacies such as honey, prunes and various delicious sweets. The cobbled streets wind up from the waterfront, and are covered with intricate designs made from sea pebbles and shells. There are numerous classic examples of the old Sporadhan town house, with its wooden balcony and fish-scale, slate-tiled roof.

In the upper town the cruciform church of **Panagía Papameletíou** is particularly splendid. Built in 1662, it is also known as Koímisis tis Theotókou. It has a well-kept interior, with an interesting display case of ecclesiastical *objets d'art* and a carved iconostasis by the Cretan

craftsman Antónios Agorastós. Perched on a clifftop above the town, the landmark church of **Panagía tou Pýrgou**, with its shining fish-scale roof, overlooks the harbour.

The old quarter of Skópelos town, the Kástro, sits above the modern town and is topped by the remains of the Venetian **castle**. Built by

The attractive bay of Skópelos town, viewed from the Kástro

For hotels and restaurants in this region see pp306–7 and p332

Fish tavernas around the bay at Agnóntas in the late afternoon

VISITORS' CHECKLIST

3,000. Skópelos town.
Harbourfront (24240 23221).
Panagía: 15 Aug.

have immaculate churches with carved iconostases and icons. **Moní Evangelistrías** (also known as Evangelismós) was built in 1712 and is one of the largest on the island. The nuns sell their handicrafts, including weavings, embroidery and food. Further up the road is **Metamórfosis tou Sotíros**, one of the oldest monasteries on Skópelos. It was built in the 16th century and is now inhabited by a solitary monk.

Moní Timíou Prodrómou, north of Moní Metamórfosis tou Sotíros, was restored in 1721. It has been inhabited by nuns, who also sell crafts, since the 1920s, and has a commanding view of Skópelos. From here a rough track leads up to **Mount Paloúki**. The deserted **Moní Taxiarchón** is reached by a track from Mount Paloúki that hugs the *sares*, the local name for the steep cliffs facing Alónissos.

Around the Island

The island is easy to explore, with its main road traversing the developed southern coast, and continuing as far as Glóssa to the northwest. It has a beautiful interior, full of plum orchards, pine forests and *kalývia* (farmhouses), but beware of the lack of signposts when travelling inland.

A steep road leads down to the popular beaches south of Skópelos town, Stáfylos and Velóna. **Agnóntas**, which serves as a port for ferries in rough weather, is quieter than Skópelos town. It is popular with locals who come for the

fish tavernas beside its pebble beach. Nearby **Limnonári**, with its stunning pebble beach and azure-coloured water, is reached by boat or along the narrow clifftop road.

Whitewashed houses in Glóssa with colourful doors and shutters

Before reaching the modern village of Elios, there are two thriving resorts at Miliá and Pánormos. For a quieter location, the tiny beach of **Adrína** nearby is often deserted. Sitting oposite the beach is wooded Dasiá island, named after a female pirate who drowned there long ago.

Glóssa is the other major settlement on the island, and sits directly opposite Skiáthos. Reminders of the Venetian occupation of Skópelos are evident in the picturesque remains of Venetian towers and houses. The small port of **Loutráki** below Glóssa is a sleepy place with little charm, but most ferries stop here as well as at Skópelos town.

On the north coast, caïques shuttle every half-hour between the pebbled **Glystéri** beach and Skópelos town. From Glystéri a winding road leads inland to the wooded region just east of the island's highest peak, **Mount Délfi**. A short walk through the enchanting pine forest leads to four mysterious niches, signposted as *sentoúkia*, literally "crates", that are carved in the rocks. Believed to be Neolithic sarcophagal tombs, their position offers fine views over the island.

KALÝVIA

Skópelos's interior is covered with an unusual array of beautiful *kalývia* (farmhouses). Some of these traditional stone buildings are still occupied all year round, others are only used during important seasonal harvests or for celebratory feasts on local saints' days. They all have distinctive outdoor prune ovens – a legacy from the days when Skópelos was renowned for its prunes. They provide a rare insight into the rural life that has virtually disappeared on neighbouring islands.

A traditional *kalívi* among olive and cypress trees

Two of the old houses in Palaiá Alónnisos in the process of restoration

Alónnisos

Αλόννησος

2,800. 🚢 Patitíri.
🚌 Kokkinókastro 6 km (4 miles)
N of Patitíri.

Sharing a history of attacks
by the pirate Barbarossa with
the other Sporades and
having endured earthquake
damage in 1965, Alónnisos
has suffered much over the
years. However, the island is
relatively unspoilt by tourism,
and most of the development
is centred in the main towns
of Patitíri and Palaiá Alónnisos.

Patitiri

The port of Patitíri is a centre
of bustling activity. Boats are
available for day trips to the
neighbouring islands, and
there is excellent swimming
off the rocks, northeast of the
port. The picturesque backstreets display typical Greek
pride in the home, evident in
the immaculate whitewashed
courtyards and pots of flowers.

Fishing vessels and cargo boats
moored in Patitíri harbour

Rousoúm Gialós and Vótsi,
3–4 km (1–2 miles) north of
Patitíri, are quieter alternatives
with their natural cliff-faced
harbours and tavernas.

Around the Island

This quiet island has a surfeit
of beaches and coves and
the interior is crisscrossed
by dirt tracks accessible only
to intrepid shepherds and
motorbikes. The old capital
of **Palaiá Alónnisos**, west of
Patitíri, perches precariously
on a clifftop. There are ruins
of a 15th-century Venetian
castle and a beautiful small
chapel, Tou Christoú, that has
a fish-scale roof. The town
was seriously damaged by the
earthquake in 1965, and the
inhabitants were
forced to leave their
homes. They were
rehoused initially in
makeshift concrete
homes at Patitíri.
Today, the houses
of Palaiá Alónnisos
have been bought
and restored by
German and British
families, and the
town retains all the
architectural beauty
of a traditional
Sporadhan village.

The road across
the island, northeast
from Patitíri, reveals
a surprisingly fertile
land of pine, olive
and arbutus trees.
At **Kokkinókastro**,
a popular pebble

beach edged by red cliffs and
pines, there are scant remains
of the site of ancient Ikos – the
old name of the island.

Further north lies the seaside
village of **Stení Vála**. From
here, a road snakes towards
Gérakas, at the wild northern
tip of the island. In summer,
the beach here is home to
the research centre for the
HSSPMS (Hellenic Society for
the Study and Protection of the
Monk Seal). The organisation's
main premises are located in
the harbour area of Patitíri.

🐋 **Hellenic Society for the
Study and Protection of the
Monk Seal (HSSPMS)**
Patitíri. *Tel* 24240 66350. ⬜ Apr–
Oct: daily; Nov–Mar: on request. ♿

Taverna at Stení Vála

Two endangered Mediterranean monk seals

Sporades Marine Park
Θαλάσσιο Πάρκο

🚢 from Skiáthos, Skópelos, Alónnisos.

Founded in 1992, the National Marine Park of Alónnisos and the Northern Sporades, to give it its full name, is an area of great environmental importance. It is the only such park in the Aegean, and includes not just Alónnisos but also its uninhabited outlying islands of Peristéra, Skantzoúra and Gioúra. Day trips by boat are possible but access is limited.

The park was created to protect an important breeding colony of the endangered Mediterranean monk seal and a fragile marine ecosystem of other rare wildlife, flora and fauna. Thanks to the pioneering efforts of marine biologists from the University of Athens, who first formed the Hellenic Society for the Study and Protection of the Monk Seal in 1988, Greece's largest population of the elusive Mediterranean monk seal is now scientifically monitored. Fewer than 500 of these seals exist worldwide, making it one of the world's most endangered species. There is an estimated population of 300 seals around the Aegean, with about 50 in the marine park. A recent campaign to promote awareness of the endangered status of the seals and restrictions on fishing in the area seems to be paying off.

Sightings of seals are not always guaranteed and there is no longer access for the public to view the wild goats on Gioúra, Audouin's gull or Eleonora's falcons on the islet of Skantzoúra: only scientists are now permitted.

The marine park is also an important route and staging post for many migrant birds during the spring and autumn. Land birds, ranging in size from tiny warblers through to elegant pallid harriers, pass through the region in large numbers to and from their breeding grounds in northeast Europe.

MARINE WILDLIFE IN THE SPORADES

Visitors can observe a wide range of other wildlife in the Sporades while watching out for monk seals. Grey herons and kingfishers are both birds of the coast here, a surprise for many birdwatchers from northern Europe who usually associate them with freshwater habitats. Spring and autumn in particular are good times for seeing several species of gulls and terns and, when venturing close to sea cliffs, keep an eye out for the Eleonora's falcons which nest on the inaccessible ledges; in the air, they are breathtakingly acrobatic birds.

Further out to sea, look for jellyfish in the water and the occasional group of common dolphins which may accompany the boat for a while. Cory's shearwaters fly with rigid wings close to the waves and head towards the shore in high winds and as dusk approaches. If you are at sea after dark, you are likely to see a glowing bioluminescence on the surface of the waves, caused by microscopic marine animals.

Cory's shearwaters glide low over the water. They are a common sight around Alónnisos.

Jellyfish flourish in the seas off the Sporadic islands. This is a Pelagia noctiluca.

Mediterranean gulls are easily recognized by the pure white wings and black hood that characterize their summer plumage.

Common dolphins can sometimes be seen in small groups diving in and out of the waves around the boat's wake or swimming alongside.

Skýros
Σκύρος

Skýrian pony

Renowned in myth as the hiding place of Achilles *(see p83)* and the home-in-exile of the hero Theseus, Skýros has always played an important role in Greek history. A rich Athenian colony from 476 BC, it later became a place of exile for the wealthy from Byzantine Constantinople. Currently one of the homes of the Greek Navy and Air Force, its unique heritage, landscape and architecture bear more resemblance to the Dodecanese than the Sporades.

An example of traditional Skýrian embroidery in the Faltáits Museum

Skýros Town
The main town is architecturally unusual in the Aegean; it has a fascinating mixture of cube-shaped houses, Byzantine churches and spacious squares. Although its main street has been spoilt by loud tavernas and bars, many backstreets give glimpses into Skýrian homes. Traditional ceramics, wood carving, copper and embroidery are always proudly on display.

Topping the kástro of the old town with its impressive mansion houses are the remains of the **Castle of Lykomedes**, site of both an ancient acropolis and later a Venetian fortress. It is reached through a tunnel underneath the whitewashed **Moní Agíou Georgíou**, which contains a fine painting of St George killing the dragon. The views from the kástro of the bay below are quite breathtaking. Nearby are the remains of two Byzantine churches, and three tiny chapels, with colourful pastel

Immortal Poetry in Plateía Rupert Brooke

pink and blue interiors. The town has two good museums. The **Archaeological Museum** displays some bracelets and pottery that were discovered during excavations of minor Neolithic and Mycenaean sites around the island. The museum also presents a traditional Skýrian town house that has been accurately recreated with local furnishings.

Housed in an old mansion owned by the Faltáits family, the excellent **Faltáits Museum** was opened in 1964 by one of their descendants, Manos Faltáits. It has a diverse collection of folk art including rare books and manuscripts, photographs and paintings, which reveal much about Skýrian history and culture. It not only shows how craftsmen absorbed influences from the Byzantine, Venetian and Ottoman occupations, but also

how the development of a wealthy aristocracy actively helped transform the island's woodcarving, embroidery, ceramics and copperware into highly sophisticated artforms.

One place to learn some of these crafts is the **Skýros Centre**, a unique holiday centre which also has courses in such wide-ranging subjects as yoga, reflexology, creative writing and windsurfing. The main branch is in Skýros town, with another branch at Atsítsa, on the west coast of the island.

Plateía Rupert Brooke, above the town, is famous for its controversial statue of a naked man by M Tómpros. Erected in 1930 in memory of the British poet Rupert Brooke who died on the island, the statue is known as *Immortal Poetry*.

Archaeological Museum
Plateía Brooke. *Tel* 22220 91327.
☐ Tue–Sun. ● main public hols. ⬡

Faltáits Museum
Palaiópyrgos. *Tel* 22220 91232.
☐ daily. ⬡ ⬡ ⬡

Skýros Centre
Tel 020-7267 4424 (contact London office for bookings, www.skyros.co.uk).
☐ Apr–Oct.

Environs
Beneath Skýros town are the resorts of **Mólos** and **Magaziá**. Around these two resorts there are plenty of decent hotels, tavernas and rooms to rent. Further along the coast from Magaziá, there is another sandy stretch of beach at **Pouriá**, which offers excellent spear-fishing and snorkelling. At **Cape Pouriá** itself, the chapel of Agios Nikólaos is built into a cave. Just off the coast are the islets of Vrikolakonísia where the incurably ill were sent during the 17th century.

The Castle of Lykomedes towering above Skýros town

For hotels and restaurants in this region see pp306–7 and p332

KEY

For key to map see back flap

VISITORS' CHECKLIST

3,000. 18 km (11 miles) NW of Skýros town. Linariá. Skýros town. 22220 91600. Carnival around island: end Feb–early Mar.

Access to Vounó, the mountainous southern part of the island, is through a narrow fertile valley south of **Ormos Achíli** between the island's two halves. The road continues south to **Kalamítsa** bay, and beyond to **Treís Mpoúkes**, a natural deep-water harbour used by pirates in the past and the Greek Navy today. Reached by dirt-track road, this is also the site of poet Rupert Brooke's simple marble grave, set in an olive grove. Brooke (1887–1915) died on a hospital ship that was about to set sail to fight at Gallipoli.

Around the Island

The island divides into two distinct halves bisected by the road from Skýros town to the port of Linariá. Meroí, the northern part of the island, is where most people live and farm on the fertile plains of Kámpos and Trachý.

Skýros is famous for its indigenous ponies, thought by some to be the same breed as the horses that appear on the Parthenon frieze (see p290). It is certainly known that the animals have been bred exclusively on Skýros since ancient times and can still be seen in the wild on the island today, particularly in the south, near the grave of Rupert Brooke.

The road running north from Skýros town leads first to the airport and then west around the island through pine forests to **Kalogriá** and **Kyrá Panagiá**,

two leeward beaches sheltered from the *meltémi* (north wind). From here, the road leads to the small village and pine-fringed beach of **Atsítsa**, where there are rooms to rent and a good taverna. As noted above, Atsítsa is also home to the other branch of the Skýros Centre, the island retreat offering alternative holidays. A little way south are the two beaches of **Agios Fokás** and **Péfkos**. The road loops back from Péfkos to the port of **Linariá**. Caïques depart from here to the inaccessible sea caves at Pentekáli and Diatrýpti on the east coast.

The azure waters and tree-lined sand of Péfkos beach

THE SKÝROS GOAT DANCE

This famous goat dance is one of Greece's few rites that have their roots in pagan festivals. It forms the centrepiece of the pre-Lenten festivities in Skýros, celebrated with dancing and feasting. Groups of masquerading men parade noisily around the narrow streets of Skýros town. Each group is led by three central characters, the *géros* (old man), wearing a traditional shepherd's outfit and a goatskin mask and weighed down with noisy bells, the *koréla*, a young man in Skýrian women's clothing, and the *frángos*, or foreigner, a comic figure wearing dishevelled clothes.

The *géros* in full costume

Evvoia
Εύβοια

After Crete, Evvoia is Greece's largest island. It is
generally unspoilt by tourism, and its diverse
landscape and history make it a microcosm of the
whole country. From Macedonian rule in 338 BC, to
Turkish government until 1833, the island has suffered
many occupations. Traces of Evvoia's mixed
history are widely evident, from the
range of religious cultures in
Chalkída to the descendants
of 15th-century Albanian
immigrants who still
speak their own
dialect of Arvanitika.

Cape Artemísio
*This is the site of the
Battle of Artemisium
which took place
in 480 BC* **8**

Istiaía is the main town in the northern part
of the island. It is a pretty market town with
sleepy squares *(see p123)*.

Límni
*This picturesque fishing
town is full of narrow streets
lined with white houses, and
colourful flowers that pour
out on to the pavement* **10**

★ **Loutrá Aidipsoú**
*Old-fashioned, this charming
resort has attracted visitors for
centuries with its warm spa
waters. Local fishermen still
continue their trade in the
wide bay* **9**

Prokópi
*The large Kandíli estate, belonging to the
English Noel-Baker family, sits just outside
the quiet village of Prokópi* **7**

STAR SIGHTS

★ Loutrá Aidipsoú

★ Kárystos

Chalkída
*A modern town, Chalkída
is the capital of the island,
and has a mixed populace of
Muslims, Jews and Orthodox
Greeks. By the waterfront is a
flourishing market* **1**

Stení

Nestling in the green hills of Mount Dírfys, Stení's cool climate makes it a pleasant escape from the summer heat and a popular place for a day trip 6

Kými

A wealthy port in the 1880s, Kými is quieter today, with a fine Folk Museum displaying traditional crafts such as this embroidered picture frame 5

Ochthoniá

The wild and exposed beaches surrounding Ochthoniá are quiet and often deserted, offering a relaxing break from the busy village 4

Mount Dírfys, the highest point on Evvoia, is a trekker's paradise *(see p118)*.

★ Kárystos

The traditional seaside and port town of Kárystos is overlooked by the dramatic slopes of Mount Ochi 3

Lake Dýstos is a large swampy area on the road to Néa Stýra *(see p121).*

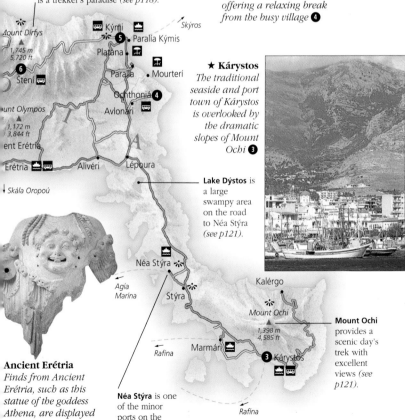

Mount Ochi provides a scenic day's trek with excellent views *(see p121).*

Ancient Erétria

Finds from Ancient Erétria, such as this statue of the goddess Athena, are displayed in the modern town's Archaeological Museum 2

Néa Stýra is one of the minor ports on the island for ferries to the mainland *(see p121).*

KEY

For key to map see back flap

0 kilometres 15

0 miles 5

Chalkída ❶

Χαλκίδα

Ancient Chalkis was one of the major independent city-states until it was taken by Athens in 506 BC, and it remained an Athenian ally until 411 BC. Briefly Macedonian, the town was under Roman rule by 200 BC. There followed the same history of Byzantine, Frankish and Venetian rule that exists in the Sporades. A bridge has spanned the fast-flo wing Evripos channel since the 6th century BC. According to legend, Aristotle was so frustrated at his inability to understand the ever-changing currents that he threw himself into the water.

VISITORS' CHECKLIST

🏠 69,000. 🚢 🚆 Athinón.
🚌 corner of Athanasíou Diákou
& Frízi. 🛈 22210 77777.
🛥 Mon–Sat. 🎭 Agía Paraskeví
celebrations: 26 Jul–1 Aug.

Chalkída's waterfront market

Exploring Chalkída

Although much of modern Chalkída is dominated by commercial activity, there are two areas of the town that are worth a visit: the waterfront which overlooks the Evripos channel, and the old Kástro quarter, on the slopes overlooking the seafront.

The Waterfront

Lined with old-fashioned hotels, cafés and restaurants, Chalkída's waterfront also has a bustling enclosed market where farmers from the neighbouring villages sell their produce. This often leads to chaotic traffic jams i n the surrounding narrow streets, an area still known by its Turkish name of Pazári, where there are interesting shops devoted to beekeeping (No. 6 Neofýtou) and other rural activities.

Kástro

In the old Kástro quarter, southeast of the Evripos bridge, the deserted streets reveal a fascinating architec-tural history. Many houses still bear the traces of their Venetian and Turkish ancestry, with timbered façades or marble heraldic

carving. Now inhabited by Thracian Muslims who settled here in the 1980s, and the surviving members of the oldest Jewish community in Greece, the Kástro also has an imposing variety of religious buildings. Three examples of these include the 19th-century **synagogue** on Kótsou, the beautiful 15th-century mosque, **Emir Zade**, in the square marking the entrance to the Kástro, and the church of **Agía Paraskeví**. The mosque is usually closed, but outside is an interesting marble fountain with an Arabic inscription.

Agía Paraskeví, situated near the Folk Museum, reveals the diverse history of Evvoia more than any other building in Chalkída. This huge 13th-century basilica is built on the site of a much earlier Byzantine church. Its exterior

Roman horse head in Archaeo-logical Museum

resembles a Gothic cathedral but the interior is a patch-work of different styles, a result of years of modification by invading peoples, including the Franks and the Turks. It has a marble iconostasis, a carved wooden pulpit, brown stone walls and a lofty wooden ceiling. Opposite the church on a house lintel is a carving of St Mark's winged lion, the symbol of Venice.

Housed in the vaults of the old Venetian fortress at the top of the Kástro quarter, the **Folk Museum** presents a jumble of local costumes, engravings and a bizarre set of uniforms from a brass band, suspended with their instruments from the ceiling. The **Archaeological Museum** is a more organized collection of finds from ancient Evvoian sites such as Kárystos. Ex-hibits include some 5th-century BC grave-stones and vases.

🏛 **Folk Museum**
Skalkóta 4. **Tel** 22210
21817. ⏰ 10am–1pm
Wed–Sun. 🎫 📷 💷

🏛 **Archaeological Museum**
Venizélou 13. **Tel** 22210 76131. ⏰
Tue–Sun. ⏰ main public hols. 🎫 📷

The 15th-century mosque in the Kástro, home to some Byzantine relics

Around Evvoia

The forests of Pine and Chestnut trees, rivers and deserted beaches in the fertile north contrast dramatically with the dry and scrubby south. Separated by the central mountains, the south becomes rough and dusty with sheep grazing in flinty fields, snaking roads along cliff tops and the scree slopes of Mount Ochi.

Picturesque Kárystos harbour, with Mount Ochi in the background

Ancient Erétria ❷

Αρχαία Ερέτρια

22 km (14 miles) SE of Chalkída.

Excavations begun in the 1890s in the town of Néa Psará have revealed the sophistication of the ancient city-state of Erétria, which was destroyed by the Persians in 490 BC and the Romans in AD 198. At the height of its power it had colonies in both Italy and Asia Minor. The ancient harbour is silted up, but evidence of its maritime wealth can be seen in the ruined agora, temples, gymnasium, theatre and sanctuary, which still remain around the modern town.

Artifacts from the ancient city are housed in the **Archaeological Museum**. The tomb finds include some bronze cauldrons and funerary urns. There are votive offerings from the Temple of Apollo, gold jewellery and a terracotta gorgon's head, which was found in a 4th-century BC Macedonian villa.

Archaeologists have also restored the **House with Mosaics** (ask for the key at the museum). Its floor mosaics are of lions attacking horses, sphinxes and panthers.

📷 **Archaeological Museum**
On the road from Chalkída to Alivéri. **Tel** 22290 62206.
8:30am–3pm Tue–Sun.

Environs

Past **Alivéri**, with its medieval castle and ugly power station, the road divides at the village of **Lépoura**. Venetian towers can be seen on the hill-side here, and also around the Dýstos plain north-wards to Kými and south to Kárystos. A road twists through tiny villages such as **Stýra**, with their surrounding wheat fields. Below lie the seaside resorts of Néa Stýra and Marmári, both of which provide ferry services to the mainland port of Rafína.

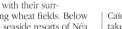

Gorgon's head, Archaeological Museum, Erétria

Kárystos ❸

Κάρυστος

130 km (80 miles) SE of Chalkída.
👥 5,000.

Kárystos, overlooked by the imposing Castel Rosso and the village of Mýloi where plane trees surround the *kafeneía*, is a picturesque town. The modern part of the town dates from the 19th century, and was built during the reign of King Otto. Kárystos has five Neo-Classical municipal buildings, excellent waterfront fish tavernas close to its Venetian Bourtzi fortress and a **Folk Museum**. Set up as a typical Karystian house, the museum contains examples of rural life – copper pots and pans, oil amphorae and ornate 19th-century furniture and embroidery. Kárystos is also famed for its green and white marble and green slate roof and floor tiles.

📷 **Folk Museum**
50 m (165 ft) from the town square. **Tel** 22240 22452.
8am–10pm Tue & Thu. main public hols.

Environs

Southeast of Kárystos, remote villages, such as Platonistós and Amigdaliá, hug the slopes of Mount Ochi. Caïques from these villages take passengers on boat trips to visit nearby coves where there are prehistoric archaeological sites.

DRAGON HOUSES

Off the main road at Stýra, a signpost points the way to the enigmatic dragon houses, known locally as *drakóspita*. Red arrows mark the trail that leads to these low structures. Constructed with huge slabs of stone, they take their name from the only creatures thought capable of carrying the heavy slabs. There are many theories about the *drakóspita*, but the most plausible links them to two other similar sites, on the summits of Mount Ochi and Mount Ymittós in Attica. All three are near marble quarries, and it is believed that Carian slaves from Asia Minor (where there are similiar structures) built them as temples in around the 6th century BC.

Scenic road running through olive groves between Ochthoniá and Avlonári

Ochthoniá ❹

Οχθωνιά

90 km (56 miles) E of Chalkída.
🏘 *1,140.*

Both Ochthoniá and its neighbouring village of Avlonári, with their Neo-Classical houses clustered around ruined Venetian towers, are reminiscent of protected Umbrian hill-towns.

A Frankish castle overlooks the village of Ochthoniá, and just west of Avlonári is the distinctive 14th-century basilica of Agios Dimítrios, which is the largest Byzantine church in Evvoia. Beyond the fertile fields that surround these villages, wild beaches, such as those at Agios Merkoúris and Mourterí, stretch out towards the forbidding cliffs of Cape Ochthoniá.

Kými ❺

Κύμη

90 km (56 miles) NE of Chalkída.
🏘 *4,000.* 🚌 Sat.
Platána 7 km (4.5 miles) S.

Four km (2 miles) above Paralía Kýmis, lies the thriving town of Kými. With a commanding view of the sea, this remote settlement had surprisingly rich resources, derived from silk production and maritime trading, in the 19th century. In the 1880s,

45 ships from Kými plied the Aegean sea routes. The narrow streets of elegant Neo-Classical houses testify to its past wealth. It is known today mainly for the medicinal spring water from nearby Choneftikó, and a statue in the main square of Dr Geórgios Papanikoláou, Kými's most famous son and inventor of the cervical smear "Pap test". An extensive and well-organized **Folk Museum** contains many exhibits from Kymian life, such as a fine collection of unique cocoon embroideries and costumes. On the road north of Kými, the 17th-century **Moní Metamórfosis tou Sotíra**, now inhabited by nuns, perches on the cliff edge.

Dr Papanikoláou (1883–1962)

🏛 **Folk Museum**
Tel 22220 22011. 🕐 *by appointment only.* 🔴 *main public hols.*

Stení ❻

Στενή

31 km (20 miles) NE of Chalkída.
🏘 *1,250.*

This mountain resort is much loved by Greeks who come for the cool climate and fine scenery. Stení is also popular with hikers setting their sights on Mount Dírfys, the island's highest peak at 1,745 m (5,720 ft), with spectacular views from the summit. A brisk walk followed by a lazy lunch of classic mountain cuisine – grilled meats and oven-baked beans – make for a pleasant day. The main square is also good for shops selling local specialities, such as wild herbs and mountain tea.

The road from Stení to the northern coast snakes up the mountain. It passes through spectacular scenery of narrow gorges filled with waterfalls and pine trees, and cornfields that stretch down to the sea.

Moní Sotíra in the mountains near Kými overlooking the sea

Prokópi ●
Προκόπι

52 km (32 miles) NW of Chalkída.
🏠 1,200. 🚌 🚗 Sun. 🚉 Krýa Vrýsi
15 km (9 miles) N.

Sleepy at most hours, Prokópi only wakes when the tourist buses arrive with pilgrims coming to worship the remains of St John the Russian (Agios Ioánnis o Rósos), housed in the modern church of Agíou Ioánnou tou Rósou. Souvenir shops and hotels around the village square cater fully for the visiting pilgrims. In reality a Ukranian, John was captured in the 18th century by the Turks and taken to Prokópi (present-day Ürgüp) in central Turkey. After his death, his miracle-working remains were brought over to Evvoia by the Greeks during the exodus from Asia Minor in 1923.

Prokópi is also famous for the English Noel-Baker family, who own the nearby Kandíli estate. Although the family have done much for the region, local feeling is mixed about the once-feudal status of this estate. Many locals, however, now accept the important role Kandíli plays in its latest incarnation as a specialist holiday centre, by bringing money into the local economy.

Environs
The road between Prokópi and Mantoúdi runs by the river Kiréa, and a path leads to one of the oldest trees in Greece, said to be over 2,000 years old. This huge plane tree has a circumference of over 4.5 m (15 ft). Sadly, it is sinking into the sludge created by a nearby mine.

Façade of the mansion on the Noel Baker Kandíli estate, Prokópi

View across the beach at Cape Artemísio

Cape Artemísio ●
Ακρωτήριο Αρτεμίσιο

105 km (65 miles) NW of Chalkída.
🚌 to Agriovótano. 🛈 Agriovótano
(22260 41720) 🚉 Psaropoúli 15 km
(9 miles) SE.

Below the Picturesque village of Agriovótano sits Cape Artemísio, site of the Battle of Artemisium. Here the Persians, led by King Xerxes, defeated the Greeks in 480 BC. In 1928, local fishermen hauled the famous bronze statue of Poseidon out of the sea at the cape. It is now on show in the National Archaeological Museum in Athens (see p286).

Old Mercedes truck delivering produce

Environs
About 20 km (12 miles) east lies Istiaía, a pleasant market town with sleepy squares, white chapels and ochre-coloured houses.

Loutrá Aidipsoú ●
Λουτρά Αιδηψού

100 km (62 miles) NW of Chalkída.
🏠 3,000. 🚌 🛈 22260 22456.
🚗 Mon–Sat 🚉 Giáltra 15 km
(9 miles) SW.

Loutrá Aidipsoú is Greece's largest spa town, popular since antiquity for its cure-all sulphurous waters. These waters bubble up all over the town and many hotels are built directly over hot springs to provide a supply to their treatment rooms. In the rock pools of the public baths by the sea, the steam rises in winter scalding the red rocks.

The old hotel Thérmai Sýlla has a rickety lift and a marble staircase down to its splendid basement treatment rooms. These luxuries are reminders of the days when the rich and famous came to take the cure. Other faded Neo-Classical hotels along the seafront also recall the town's days of glory in the late 19th century. The town has a relaxed atmosphere and in summer the beach is popular with Greek families.

Environs
In the summer a ferry service goes across the bay to Loutrá Giáltron where warm spring water mixes with the shallows of a quiet beach edged with tavernas.

Límni ●
Λίμνη

87 km (54 miles) NW of Chalkída.
🏠 2,100. 🚌 🛈 22270 32111.

Once a wealthy 19th-century seafaring power, the pleasant town of Límni has elegant houses, cobbled streets and a charming seafront. Just south of the town is the magnificent Byzantine Moní Galatáki, the oldest monastery on Evvoia, etched into the cliffs of Mount Kandíli. Inhabited by nuns since the 1940s, its church is covered with beautiful frescoes. The Last Judgment is shown in particularly gory detail, with some souls frantically climbing the ladder to heaven, while others are dragged mercilessly into the leviathan's jaws.

THE NORTHEAST AEGEAN ISLANDS

THASOS · SAMOTHRAKI · LIMNOS · LESVOS
CHIOS · IKARIA · SAMOS

*M*ore than any other archipelago in Greece, the seven major
islands of the Northeast Aegean defy easy categorization.
Though they are neighbours, sharing a common history of
rule by the Genoese and lively fishing industries, the islands are
culturally distinct, encompassing a range of landscapes and lifestyles.

Although Sámos and Chíos were prominent in ancient times, few traces of that former glory remain. Chíos offers the region's most compelling medieval monuments, including the Byzantine monastery of Néa Moní and the mastic villages, while Sámos has a fascinating museum of artifacts from the long-venerated Heraion shrine. In Límnos's capital, Mýrina, you encounter evidence of the Genoese and Ottoman occupations, in the form of its castle and domestic architecture.

Assumption of the Virgin by
Theófilos (1873–1934), Mytilini's
Byzantine Museum, Lésvos

Lésvos shares the fortifications and volcanic origin of Límnos, though the former's monuments are grander and its topography more dramatic. To the south, the islands of Sámos, Chíos and Ikaría have mountainous profiles and are forested with pine, olive and cypress trees. Most of the pines of Thásos were devastated by forest fires in the 1980s, though Samothráki remains unspoilt; its numerous hot springs and waterfalls, as well as the brooding summit of Mount Fengári, are a counterpoint to the long-hallowed Sanctuary of the Great Gods.

Beaches come in all sizes and consistencies, from the finest sand to melon-sized volcanic shingle. Apart from Thásos, Sámos and Lésvos, package tourism is scarce in the north where summers are shorter. Wild Ikaría, historically a backwater, will appeal mostly to spa-plungers and beachcombers, while its tiny dependency, Foúrnoi, is an ideal do-nothing retreat owing to its convenient beaches and abundant seafood.

Mólyvos harbour, Lésvos, overlooked by the town's 14th-century Genoese castle

◁ The broad, sandy beach near the village of Kámpos, Ikaría

Exploring the Northeast Aegean Islands

For its beaches and ancient ruins, both composed of white marble, Thásos is hard to fault, while Samothráki has long been a destination for hardy nature lovers. Less energetic visitors will find Límnos ideal, with picturesque villages and beaches close to the main town. Olive-rich Lésvos offers the greatest variety of scenery but requires time and effort to tour. For first-time visitors to the eastern isles, Sámos is the best touring base, though the cooler climate of Chíos is more attractive, and its main town offers good shopping. Connoisseurs of relatively unspoiled islands will want to sample a slower pace of life on Ikaría, Psará or Foúrnoi.

SEE ALSO

- *Where to Stay* pp307–9
- *Where to Eat* pp333–4
- *Travel Information* pp366–9

Fishing boat in Mólyvos harbour, Lésvos

ISLANDS AT A GLANCE

Byzantine monastery of Néa Moní, Chios, seen from the southwest

KEY

▬▬	Motorway
▬	Main road
═	Minor road
▬	Scenic route
---	High-season, direct ferry route
▬▬	International border
△	Summit

For additional map symbols see back flap

Volcanic landscape near Kontiás, Límnos

LOCATOR MAP

GETTING AROUND

Thásos and Samothráki have
no airports, but are served by
ferries from Alexandroúpoli and
Kavála on the mainland, while
Límnos and Lésvos have air and
ferry links with Athens and
Thessaloníki. Bus services vary
from virtually non-existent on
Límnos and Samothráki, or
Lésvos's functional schedules,
to Thásos's frequent coaches.
Chíos, Ikaría and Sámos are
served by flights from Athens,
and are connected by ferry.
Chíos has an adequate bus
service but is best explored by
car; Sámos has more frequent
buses, and is small enough to
be toured by motorbike; Ikaría
has skeletal public transport
and steep roads requiring
sturdy vehicles.

Sandy Messaktí beach, Ikaría

Thásos
Θάσος

Thásos has been inhabited since the Stone Age, with settlers from Páros colonizing the east coast during the 7th century BC. Spurred by revenues from gold deposits near modern Thásos town, Ancient Thásos became the seat of a seafaring empire, though its autonomy was lost to the Athenians in 462 BC. The town thrived in Roman times, but lapsed into medieval obscurity. Today, the island's last source of mineral wealth is delicate white marble, cut from quarries whose scars are prominent on the hillsides south of Thásos town.

Exterior of the Archaeological Museum, Thásos

Thásos town harbour, viewed from the agora

Thásos Town ❶
Λιμένας

🏠 3,130. 🚍 🚌 ℹ 25930 23111. 🚢 daily. 🚌 Pachýs 9 km (6 miles) W.

Modern Liménas, also known as Thásos town, is an undistinguished resort on the coastal plain which has been settled for nearly three millennia. Interest lies in the vestiges of the ancient city and the manner in which they blend into the modern town. Foundations of a Byzantine basilica take up part of the central square, while the road to Panagiá cuts across a vast shrine of Herakles before passing a monumental gateway.

🏛 Ancient Thásos
Site & Museum *Tel* 25930 22180.
☐ Tue–Sun. ● main public hols.
📷 📹 ♿

Founded in the 7th century BC, Ancient Thásos is a complex series of buildings, only the remains of which can be seen today. French archaeologists have conducted excavations here since 1911; digs were recently resumed at a number of locations in Thásos town. The **Archaeological Museum**, next to the agora, houses treasures from the site.

Well defined by the ruins of four stoas, the Hellenistic and Roman **agora** covers a vast area behind the ancient military harbour, today the picturesque Limanáki, or fishing port. Though only

SIGHTS AT A GLANCE

Alykí ❸
Kástro ❻
Megálo Kazavíti ❽
  Moní Archangélou Michaïl ❹
Potamiá ❷
Sotíras ❼
Thásos Town ❶
Theológos ❺

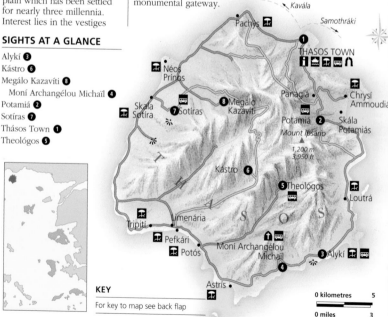

KEY
For key to map see back flap

0 kilometres   5

0 miles   3

PLAN OF ANCIENT THASOS

KEY TO PLAN

1. Archaeological Museum
2. Agora
3. Temple of Dionysos
4. Theatre
5. Citadel
6. Walls
7. Temple to Athena Poliouchos
8. Shrine to Pan
9. Gate of Parmenon

0 kilometres 5

0 miles 3

VISITORS' CHECKLIST

16,000. Thásos town.
Panagiá: 15 Aug.

The Gate of Parmenon in the south wall of Ancient Thásos

a few columns have been re-erected, it is easy to trace the essentials of ancient civic life, including several temples to gods and deified Roman emperors, foundations of heroes' monuments and the extensive drainage system.

Foundations of a **Temple of Dionysos**, where a 3rd-century BC marble head of the god was found, mark the start of the path up to the acropolis. Partly overgrown by oaks, the Hellenistic **theatre** has spectacular views out to sea. The Romans adapted the stage area for their bloody spectacles; it is now being excavated with the intent of complete restoration.

The ancient **citadel**, once the location of an Apollo temple, was rebuilt during the 13th century by the Venetians and Byzantines. It was then ceded by Emperor Manuel II Palaiológos to the Genoese Gatelluzi clan in 1414, who enlarged and occupied it until 1455. Recycled ancient masonry is conspicuous at the south gateway. By the late 5th century BC, substantial walls of more than 4 km (2 miles) surrounded the city, the sections

by the sea having been mostly wrecked on the orders of victorious besiegers in 492 and 462 BC.

Foundations of a **Temple to Athena Poliouchos** (Patroness of the City), dated to the early 5th century BC, are just below the acropolis summit; massive retaining walls support the site terrace. A cavity hewn in the rocky outcrop beyond served

as a **shrine to Pan** in the 3rd century BC; he is depicted in faint relief playing his pipes.

Behind the summit point, a steep 6th-century BC stairway descends to the **Gate of Parmenon** in the city wall. The gate retains its lintel and takes its name from an inscription "Parmenon Made Me" (denoting its mason), on a nearby wall slab.

Columns of the agora, with the town church in the background

Around Thásos Island

Sculpture at the
Vágis Museum

Thásos is just small enough to explore by motorbike, though the bus service along the coastal ring road is good and daily hydrofoils link Thásos town with the western resorts. The best beaches are in the south and east, though the coastal settlements are mostly modern annexes of inland villages, built after the suppression of piracy in the 19th century.

Boats in the peaceful harbour of Skála Potamiás

Potamiá ❷
Ποταμιά

9 km (6 miles) S of Thásos town.
🏘 1,000. 🚌 🚢 daily. 🚢 Loutrá
12 km (7 miles) S; Chrysi Ammoudiá
5 km (3 miles) E.

Potamiá is a small village, with one of the most popular paths leading to the 1,200-m (3,950-ft) summit of Mount Ipsário. Following bulldozer tracks upstream brings you to the trailhead for the ascent, which is a 7-hour excursion; although the path is way-marked by the Greek Alpine Club, it is in poor condition.

The sculptor and painter Polýgnotos Vágis (1894–1965) was a native of the town, although he emigrated to America at an early age. Before his death, the artist bequeathed most

Blue-washed house
in Panagiá

of his works to the Greek state and they are now on display at the small **Vágis Museum**, situated in the village centre. His work has a mythic, dreamlike quality; the most compelling sculptures are representations of birds, fish, turtles and ghostly faces which he carved on to boulders or smaller stones.

🔲 Vágis Museum
Tel 25930 61400. ⬜ 10am–1pm,
6–9pm Tue–Sun. 📷 📷

Environs
Many visitors stay and enjoy the traditional Greek food at **Skála Potamiás**, 3 km (2 miles) east of Potamiá, though **Panagiá**, 2 km (1 mile) north, is the most visited of the inland villages. It is superbly situated above a sandy bay,

has a lively square and many of its 19th-century houses have been preserved or restored.

Alykí ❸
Αλυκή

29 km (18 miles) S of Thásos town.
🚌 🚢 Astris 12 km (7 miles) W.

Perhaps the most scenic spot on the Thasian shore, the headland at Alykí is tethered to the body of the island by a slender spit, with beaches to either side. The westerly cove is fringed by the hamlet of Alykí, which has well-preserved 19th-century vernacular architecture due to its official classification as an archaeological zone. A Doric temple stands over the eastern bay, while behind it, on the headland, are two fine Christian basilicas, dating from the 5th century, with a few of their columns re-erected.

Local marble was highly prized in ancient times; now all that is left of Alykí's quarries are overgrown depressions on the headland. At sea level, "bathtubs" (trenches scooped out of the rock strata) were once used as evaporators for salt-harvesting.

Moní Archangélou Michaïl,
perched on its clifftop

Moní Archangélou Michaïl ❹
Μονή Αρχαγγέλου Μιχαήλ

34 km (21 miles) S of Thásos town.
Tel 25930 31500. 🚌 ⬜ daily.

Overhanging the sea 3 km (2 miles) west of Alykí, Moní Archangélou Michaïl was founded early in the 12th century by a hermit called Luke, on the spot where a spring had appeared at the behest of the Archangel. Now a dependency of Moní Filothéou on Mount Athos in northern Greece, its most treasured relic is a Holy Nail from the Cross. Nuns have occupied the grounds since 1974.

Slate-roofed house with characteristically large chimney pots, Theológos

Theológos ❺
Θεολόγος

50 km (31 miles) S of Thásos town. ⛺ 900. ▦ ▢ daily. ✈ Potós 10 km (6 miles) SW.

Well inland, secure from attack by pirates, Theológos was the Ottoman-era capital of Thásos. Tiered houses still exhibit their typically large chimneys and slate roofs. Generous gardens and courtyards give the village a green and open aspect. A ruined tower and low walls on the hillside opposite are evidence of Theológos's original 16th-century foundation by Greek refugees from Constantinople.

Kástro ❻
Κάστρο

45 km (28 miles) SW of Thásos town. ⛺ 6. ✈ Tripiti 13 km (8 miles) W of Limenária.

At the centre of Thásos, 500 m (1,640 ft) up in the mountains, the village of Kástro was even more secure than Theológos. Founded in 1403 by Byzantine Emperor Manuel II Palaiológos, it became a stronghold of the Genoese, who fortified the local hill which is now the cemetery. Kástro was slowly abandoned after 1850, when a German mining concession created jobs at Limenária, on the coast below.

This inland hamlet has now been reinhabited on a seasonal basis by sheep farmers. The *kafeneío*, on the ground floor of the former school, beside the church, shelters the single telephone; there is no mains electricity.

Sotíras ❼
Σωτήρας

23 km (14 miles) SW of Thásos town. ⛺ 12. ✈ Skála Sotíra 3 km (2 miles) E.

Facing the sunset, Sotíras has the most alluring site of all the inland villages – a fact not lost on the dozens of foreigners who have made their homes here. Under gigantic plane trees watered by a triple fountain, the tables of a small taverna fill the relaxed balcony-like square. The ruin above the church was a lodge for German miners, whose exploratory shafts still yawn on the ridge opposite.

Traditional stone houses with timber balconies, Megálo Kazavíti

Megálo Kazavíti ❽
Μεγάλο Καζαβίτι

22 km (14 miles) SW of Thásos town. ⛺ 1,650. ▦ ▢ daily. ✈ Néos Prínos 6 km (4 miles) NE.

Greenery-shrouded Megálo Kazavíti (officially Ano Prínos) surrounds a central square, which is a rarity on Thásos. There is no better place to find examples of traditional domestic Thasian architecture with its characteristic mainland Macedonian influence: original house features include narrow-arched doorways, balconies and overhanging upper storeys, with traces of the indigo, magenta and ochre plaster pigment that was once commonly used across the Balkans.

Taverna overhung by plane trees in Sotíras village

Samothráki
Σαμοθράκη

🏛 2,700. ⛴ 🚌 Kamariótissa.
🚌 Pachiá Ammos 15 km (9 miles)
SW of Kamariótissa.

With virtually no level terrain,
except for the western cape,
Samothráki is synonymous
with the bulk of Mount
Fengári. In the Bronze Age
the island was occupied by
settlers from Thrace. Their
religion of the Great Gods
was incorporated into the
culture of the Greek colonists
in 700 BC, and survived under
Roman patronage until the
4th century AD. The rawness
of the weather seems to go
hand in hand with the brooding landscape, making it easy
to see how belief in the Great
Gods endured.

Chóra
Lying 5 km (3 miles)
east of Kamariótissa,
the main port of the
island, Chóra is the
capital of Samothráki.
The town almost fills
a pine-flecked hollow
which renders it in-
visible from the sea.
  With its labyrinthine
bazaar, and cobbled
streets threading past
sturdy, tile-roofed
houses, Chóra is the
most handsome
village on the island.
A broad central
square with two
tavernas provides an
elegant vantage point,
looking out to sea beyond the
Genoese **castle**. Adapted from
an earlier Byzantine fort, little
other than the castle's gate-

The town of Chóra with the remains of its
Genoese castle in the background

way remains, though more
substantial fortifications can
be found downhill at Chóra's
predecessor, **Palaiópoli**; here

Sanctuary of the Great Gods
Ερείπια του Ιερού των Μεγάλων Θεών

The sanctuary of the Great Gods
on Samothráki was, for almost a
millennium, the major religious
centre of ancient Aeolia, Thrace
and Macedonia. There were
similar shrines on Límnos and
Ténedos, but neither commanded
the following or observed the same
rites as the one here. Its position in a
canyon at the base of savage,
plunging crags on the northeast slope
of Mount Fengári was perhaps
calculated to inspire awe; today,
though thickly overgrown, it is
scarcely less impressive. The sanctuary was expanded and improved
in Hellenistic times by Alexander's
descendants, and most of the ruins
visible today date from that period.

Nike Fountain
*A marble centrepiece,
the Winged Victory of
Samothráki, once
decorated the
fountain. It was
discovered by the
French in 1863 and is
now on display in the
Louvre, Paris.*

The stoa is 90 m
(295 ft) long and
dates to the early
3rd century BC.

**Hall for votive
offerings**

The theatre held
performances of
sacred dramas in
July, during the
annual festival.

Hieron
*The second stage of initiation, epopteia, took place
here. In a foreshadowing of Christianity, this involved
confession and absolution followed by baptism in
the blood of a sacrificed bull or ram. Rites took
place in an old Thracian dialect until 200 BC.*

three Gatelluzi *(see p138)* towers of 1431 protrude above the extensive walls of the ancient town.

Around the Island

Easy to get around by bike or on foot, Samothráki has several villages worth visiting on its southwest flank, lost in olive groves or poplars. The north coast is moister, with plane, chestnut and oak trees lining the banks of several rivers. Springs are abundant, and waterfalls meet the sea at Kremastá Nerá to the south. Stormy conditions compound the lack of adequate harbours.

Thérma has been the island's premier resort since the Roman era, due to its hot springs and lush greenery. You

Three Gatelluzi towers at ancient Palaiópoli

can choose among two rustic outdoor pools of about 34° C (93° F), under wooden shelters; an extremely hot tub of 48° C (118° F) in a cottage, only for groups; and the rather sterile modern bathhouse at 39° C (102° C). Cold-plunge fans will find rock pools and low waterfalls 1.5 km (1 mile) east at **Krýa Váthra**. These

are not as impressive or cold as the ones only 45 minutes' walk up the Foniás canyon, 5 km (3 miles) east of Thérma.

The highest summit in the Aegean, at 1,600 m (5,250 ft), is the granite mass of **Mount Fengári**. Although often covered with cloud, it serves year round as a sea-faring landmark and the views from the top are superb. In legend, the god Poseidon watched the Trojan War from this mountain. The peak is usually climbed from Thérma as a 6-hour round trip, though there is a longer and easier route up from Profítis Ilías village on its southwest flank.

Arsinoeion

At over 20 m (66 ft) across, this rotunda is the largest circular building known to have been built by the Greeks. It was dedicated to the Great Gods in the 3rd century BC.

VISITORS' CHECKLIST

6 km (4 miles) NE of Kamariótissa. 🚌 *to Palaiópoli.* **Site & Museum Tel** 25510 41474. ⬜ *8am–7:30pm (site till 8:30pm May–Sep) Tue–Sun.* 📷 ⬜

Sanctuary of Anaktoron

This building was where myesis, *the first stage of initiation into the cult, took place. This involved contact with the* kabiri *mediated by prior initiates.*

The Temenos is a rectangular space where feasts were probably held.

Small theatre

DEITIES AND MYSTERIES OF SAMOTHRAKI

When Samothráki was colonized by Greeks in 700 BC, the settlers combined later Olympian deities with those they found here. The principal deity of Thrace was Axieros, the Great Mother, an earth goddess whom the Greeks identified with Demeter, Aphrodite and Hekate. Her consort was the fertility god Kadmilos and their twin offspring were the *kabiri* – a Semitic word meaning "Great Ones" which soon came to mean the entire divine family. These two deities were later recognized as the *dioskouri* Castor and Pollux, whose emblems were snakes and a star. The cult was open to all comers of any age or gender, free or slave, Greek or barbarian. Details of the mysteries are unknown as adherents honoured a vow of silence.

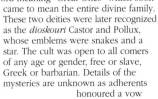

The twin *kabiri*, Castor and Pollux

The Propylon (monumental gate) was dedicated by Ptolemy II of Egypt in 288 BC.

useum

Límnos
Λήμνος

The mythological landing place of Hephaistos, the god of metalworking cast out of Olympos by Zeus, Límnos is appropriately volcanic; the lava soil crumbles into broad beaches and grows excellent wine and herbal honey. Controlling the approaches to the Dardanelles, the island was an important outpost to both the Byzantines and the Turks, under whom it prospered as a trading station. The Greek military still controls much of the island, but otherwise it is hard to imagine a more peaceful place.

VISITORS' CHECKLIST

🏘 18,000. ✈ 22 km (14 miles) NE of Mýrina. ⛴ Mýrina. 🚌 Plateía Kída, Mýrina. 🛈 Town Hall, on the waterfront Mýrina (22540 22996). 🎭 15 Aug.

SERGITSI

Pláka

Kabeirio

Ifaisteía

Propoúli

Katálakko

Mount Skopiá
430 m
1,410 ft

Dáfni   Atsikí

Sardés

Karpási

Város

Kondopoúli

Kalliópi

Repanídi

Romanó

Kavála

Kornós

Livadochóri

Roussopoúli

Kάspakas

Avlónas

Portianó

Moúdros

Poliόchni

Thessaloniki

Kontiás

Mount Paradeísi
260 m
850 ft

MYRINA

Agios Pávlos

Platí   Tháños

Fisíni

Mount Fakós
265 m
865 ft

Skandáli

↓
Rafina

Lésvos
➝ *Agios Efstrátios*

KEY

For key to map see back flap

0 kilometres   5

0 miles   3

Mýrina

Successor to ancient Mýrina Límnos's second town in antiquity, modern Mýrina sprawls between two sandy bays at the foot of a rocky promontory. Not especially touristed, it is one of the more pleasant island capitals in the North Aegean, with cobbled streets, an unpretentious bazaar and

imposing, late-Ottoman houses. The most ornate of these cluster behind the northerly beach, Romeíkos Gialós, which is also the centre of the town's nightlife. The south beach, Toúrkikos Gialós, extends beyond the compact fishing port with its half-dozen quayside tavernas. The only explicitly Turkish relic is a fountain on Kída, inscribed with Turkish calligraphy, from which delicious potable water can still be drawn.

Housed in an imposing 19th-century mansion behind Romeíkos Gialós, the recently redesigned **Archaeological Museum** is exemplary in its display of artifacts belonging to the four main ancient cities of Límnos. The most

prestigious items have been sent to Athens, however, leaving a collection dominated by pottery shards which may only interest a specialist. The most compelling ceramic exhibits are a pair of votive lamps in the form of sirens from the temple at Ifaisteía, while metalwork from Poliόchni is represented by bronze tools and a number of decorative articles.

Spread across the headland, and overshadowing Mýrina, the **kástro** boasts the most dramatic position of any North Aegean stronghold. Like others in the region, it was in turn an ancient acropolis and a Byzantine fort, fought over and refurbished by Venetians and Genoese until the Ottomans took the island in 1478. Though dilapidated, the kástro makes a rewarding evening climb for views over western Límnos.

Mýrina harbour, overlooked by the kástro in the background

The volcanic landscape of Límnos, viewed from the village of Kontiás

Around the Island

Though buses run from Mýrina to most villages in summer, the best way to travel around Límnos is by car or motorbike; both can be hired at Mýrina. Southeast from Mýrina, the road leads to **Kontiás**, the third-largest settlement on Límnos, sited between two volcanic outcrops supporting the only pine woods on the island. Sturdily constructed, red-tiled houses, including some fine *belle époque* mansions, combine with the landscape to make this the island's most appealing inland village.

The bay of **Moúdros** was Commonwealth headquarters during the ill-fated 1915 Gallipoli campaign. Many casualties were evacuated to hospital here; the unlucky ones were laid to rest a short walk east of Moúdros town on the road to Roussopoúli. With 887 graves, this ranks as the largest Commonwealth cemetery from either world war in the Greek islands; 348 more English-speaking servicemen lie in another graveyard across the bay at **Portianoú**.

Founded just before 3000 BC, occupying a clifftop site near the village of Kamínia, the fortified town of **Polióchni** predates Troy on the coast of Asia Minor just across the water. Like Troy, which may have been a colony, it was levelled in 2100 BC by an earthquake. It was never resettled. The suddenness of

the catastrophe gave many people no time to escape – skeletons were unearthed among the ruins. Polióchni was noted for its metalsmiths, who refined and worked raw ore from Black Sea deposits, and shipped the finished objects to the Cyclades and Crete. A hoard of gold jewellery, now displayed in Athens, was found in one of the houses. Italian archaeologists continue the excavations every summer, and have penetrated four distinct layers since 1930.

The patron deity of Límnos was honoured at **Ifaisteía**, situated on the shores of Tigáni Bay. This was the largest city on the island until the Byzantine era. Most of the site has yet to be completely revealed. Currently, all that is visible are outlines of the Roman theatre, parts of a necropolis and foundations of Hephaistos's temple.

Looking down on the remains of a Roman theatre, Ifaisteía

Rich grave offerings and pottery found on the site can be seen in the Mýrina Archaeological Museum.

The ancient site of the **Kabeirio** (Kavírio in modern Greek) lies across Tigáni Bay from Ifaisteía and has been more thoroughly excavated. The Kabeirioi, or Great Gods, were worshipped on Límnos in the same manner as on Samothráki *(see pp132–3)*, though at this sanctuary little remains of the former shrine and its adjacent stoa other than a number of column stumps and bases.

Below the sanctuary ruins, steps lead down to a sea grotto known as the Cave of Philoctetes. It takes its name from the wounded Homeric warrior who was supposedly abandoned here by his comrades on their way to Troy until his infected leg injuries had healed.

Outlying Islands

Certainly the loneliest outpost of the North Aegean, tiny, oak-covered **Agios Efstrátios** (named after the saint who was exiled and died here) has scarcely a handful of tourists in any summer. The single port town was damaged by an earthquake in 1967, with dozens of islanders killed; some pre-quake buildings survive above the ferry jetty. Deserted beaches can be found an hour's walk to either side of the port.

Lésvos
Λέσβος

Once a favoured setting for Roman holidays, Lésvos, with its thick southern forests and idyllic orchards, was known as the "Garden of the Aegean" to the Ottomans. Following conquest by them, in 1462, much of the Greek population was enslaved or deported to Constantinople, and most physical traces of Genoese or Byzantine rule were obliterated by both the Turks and the earthquakes the island is prone to. Lésvos has been the birthplace of a number of artists, its most famous child being the great 7th-century BC lyric poet Sappho.

Ouzo from Plomári

★ Mólyvos
The tourist capital of the island, Mólyvos has a harbour overlooked by a Genoese castle with fine views of Turkey 6

Pétra
This popular resort takes its name from the huge perpendicular rock at its heart. Steps in the rock lead to an 18th-century church on the summit 7

Kalloní
Known mainly for the sardines caught off the coast of nearby Skála Kallonís, this is a cross-roads for most of the island's bus routes 8

Kámpos

Moní Ypsiloú 10

Sigri 11

Antissa 9

Moní Perivolís

Skalochóri

Vatoússa

Chídira

Kallon

Skála Kalloní

Moní Leimónos

Mólyv

Pétra

Anax

Eresós

Skála Eresoú 12

Mesótopos

Κólpos Kallon

Antissa
Situated just below a pine grove, this is the largest village in the area. It has several excellent kafeneía *in its central square, over-shadowed by huge plane trees* 9

Moní Ypsiloú
Straddling the summit of an extinct volcano on the edge of a fossilized forest, 12th-century Ypsiloú has a museum of ecclesiastical treasures 10

L

Vater

Sígri
Near the westernmost point of the island, this small chapel stands at the waterfront on the edge of the remote village of Sígri 11

Skála Eresoú
One of the largest resorts on the island, the beach at Eresós lies only a short walk from the birthplace of the poet Sappho 12

Mantamádos

This attractive village is famous for both its pottery and the "black" icon at the enormous Moní ton-Taxiarchón ❹

Sykaminiá

The harbour below the hill-town of Sykaminiá, birthplace of modern novelist Strátis Myrivílis, is one of the most picturesque in Greece ❺

★ Mytilíni

Just outside Mytilíni is a museum devoted to the work of the painter Theófilos Chatzimichaíl ❶

0 kilometres	10
0 miles	5

Plomári

This large coastal resort, with its Varvagiánnis distillery, is the ouzo capital of Lésvos ❷

Agiásos

Widely regarded as the most beautiful hill-town of the island, Agiásos's main church has an icon supposedly painted by St Luke ❸

STAR SIGHTS

★ **Mólyvos**

★ **Mytilíni**

KEY

For key to map see back flap

Mytilíni ❶
Μυτιλήνη

Ottoman inscription above the castle gate

Modern Mytilíni has assumed both the name and site of the ancient town. It stands on a slope descending to an isthmus bracketed by a pair of harbours. An examination of Ermoú reveals the heart of a lively bazaar. Its south end is home to a fish market selling species rarely seen elsewhere, while at the north end the roofless shell of the Gení Tzamí marks the edge of the former Turkish quarter. The Turks ruled from 1462 to 1912 and Ottoman houses still line the narrow lanes between Ermoú and the castle rise. The silhouettes of such *belle époque* churches as Agioi Theódoroi and Agios Therápon pierce the tile-roofed skyline.

The dome of Agios Therápon

♣ Kástro Mytilónis
Tel 22510 27970. ◻ Tue–Sun.
◼ main public hols. 🌐 📷
Surrounded by pine groves, this Byzantine foundation of Emperor Justinian (527–65) still impresses with its huge curtain walls, but it was even larger during the Genoese era. Many ramparts and towers were destroyed during the Ottoman siege of 1462 – an Ottoman Turkish inscription can be seen at the south gate. Over the inner gate the initials of María Palaiologína and her husband Francesco Gatelluzi – a Genoan who helped John Palaiológos regain the Byzantine throne – complete the resumé of the castle's various occupants. The ruins include those of the Gatelluzi palace, a Turkish *medresse* (theological school) and a dervish cell; a Byzantine cistern stands by the north gate.

📷 Archaeological Museum
Argýris Eftaliótis. **New wing:**
Corner of 8 Noemvríou & Melínas Merkoúri. *Tel* 22510 28032. ◻
Tue–Sun. ◼ main public hols. 🌐
♿ (new wing only).
Lésvos's archaeological collection occupies a *belle époque* mansion and a small annexe in its back garden. The most famous exhibits are Roman villa mosaics. Neolithic finds from the 1929–33 British excavations at Thermí, just north of town, can also be seen. A new building nearby displays additional finds.

📷 Byzantine Museum
Agios Therápon. *Tel* 22510 28916.
◻ mid-May–mid-Oct: 9am–1pm Mon–Sat. 🌐
This ecclesiastical museum is devoted almost entirely to exhibiting icons. The collection ranges from the 13th to the 18th century and also includes a more recent, folk-style icon by Theófilos Chatzimichaïl.

Environs
The **Theófilos Museum**, 3 km (2 miles) south, in Vareiá village, offers four rooms of canvases by Theófilos Chatzi-michaïl (1873–1934), the Mytilíni-born artist. All were commissioned by his patron Tériade in 1927 and created over the last seven years of the painter's life. Theófilos detailed the fishermen, bakers and harvesters of rural Lésvos and executed creditable portraits of personalities he met on his travels. For his depictions of historical episodes or landscapes beyond his experience, Theófilos relied on his imagination. The only traces of our age are occasional aeroplanes or steamboats in the background of his landscapes.

Just along the road is the **Tériade Museum**, housing the collection of Stratís Eleftheriádis – a local who emigrated to Paris in the early 20th century, adopting the name Tériade. He became a publisher of avant-garde art and literature. Miró, Chagall, Picasso, Léger and Villon were some of the artists who took part in his projects.

📷 Theófilos Museum
Vareiá. *Tel* 22510 41644. ◻ 10am–4pm Tue–Sun. 🌐 ♿

📷 Tériade Museum
Vareiá. *Tel* 22510 23372. ◻ 9am–2pm, 5–8pm daily. 🌐 ♿

VISITORS' CHECKLIST

🏛 27,000. ✈ 8 km (5 miles) S.
⛴ 🚌 Pávlou Kountourióti.
ℹ Aristárchou 6 (22510 42512).
🎭 Agios Ermogénis 12 km
(7 miles) S; Charamída 14 km
(9 miles) S. 🎪 15 Jul–15 Aug.

Daphnis and Chloe, by Marc Chagall (1887–1984), in the Tériade Museum

Olive Growing in Greece

The Cretan Minoans are thought to have been the first people to have cultivated the olive tree, around 3800 BC. The magnificent olive groves of modern Greece date back to 700 BC, when olive oil became a valuable export commodity. According to Greek legend, Athena, goddess of peace as well as war, planted the first olive tree in the Athenian Acropolis – the olive has thus become a Greek

Branch of ripening olives

symbol for peace. The 11 million or so olive trees on Lésvos are reputed to be the most productive oil-bearing trees in the Greek islands; Crete produces more and better-quality oil, but no other island is so dominated by olive monoculture. The fruits can be cured for eating throughout the year, or pressed to provide a nutritious and versatile oil; further crushing yields oil for soap and lanterns, and the pulp is a good fertilizer.

In myth, *the olive is a virgin tree, sacred to Athena, tended only by virgin males. Its abundant harvest has been celebrated in verse, song and art since antiquity. This vase shows three men shaking olives from a tree, while a fourth gathers the harvest into a basket.*

Olive groves *on Lésvos largely date from after a killing frost in 1851. The best olives come from the hillside plantations between Plomári and Agiásos, founded in the 18th century by local farmers desiring land relatively inaccessible to Turkish tax collectors.*

Greek olive oil, *greenish-yellow after pressing, is believed by the Greeks to be of a higher quality than its Spanish and Italian counterparts, owing to hotter, drier summers which promote low acid levels in olive fruit.*

The olive harvest *on Lésvos takes place from late November to late December. Each batch is brought to the local elaiotriveío (olive mill), ideally within 24 hours of being picked, pressed separately and tested for quality.*

TYPES OF OLIVE

From the mild fruits of the Ionians to the small, rich olives of Crete, the Greek islands are a paradise for olive lovers.

Kalamáta, the most famous Greek olive, is glossy-black, almond-shaped and cured in red-wine vinegar.

Elítses are small, sweetly flavoured olives from the island of Crete.

Tsakistés are picked young and lightly cracked before curing in brine.

Throúmpes are a true taste of the countryside, very good as a simple *mezés* with olive-oil bread.

Thásos olives are salt-cured and have a strong flavour that goes well with cheese.

Ionian greens are mild, mellow-flavoured olives, lightly brine-cured.

Around Eastern Lésvos

Eastern Lésvos is dominated by the two peaks of Lepétymnos in the north and Olympos in the south, both reaching the same height of 968 m (3,176 ft). Most of the island's pine forests and olive groves are found here, as well as the two major resort areas and the most populous villages after the port and capital. There are also several thermal spas, the most enjoyable being at Loutrá Eftaloús, near Mólyvos.

Miraculous icon of Agiásos

With an early start from Mytilíni, which provides bus connections to all main towns and villages, the east of the island can be toured in a single day.

Plomári ❷
Πλωμάρι

42 km (26 miles) SW of Mytilíni.
3,400. Mon–Sat.
Agios Isidoros, 3 km (2 miles) NE; Melinta, 6 km (4 miles) NW.

Plomári's attractive houses spill off the slope above its harbour and stretch to the banks of the usually dry Sedoúntas river which runs through the central commercial district. The houses date mostly from the 19th century, when Plomári became wealthy through its role as a major shipbuilding centre. Today, Plomári is known as the island's "ouzo capital", with five distilleries in operation, the most famous being Varvagiánnis.

Agiásos ❸
Αγιάσος

28 km (17 miles) W of Mytilíni.
3,100. Mon–Sat.
Vaterá, 31 km (19 miles) S.

Hidden in a forested ravine beneath Mount Olympos, Agiásos is possibly the most

beautiful hill-town on Lésvos. It began life in the 12th century as a dependency of the central monastic church of the **Panagía Vrefokratoússa** which was constructed to enshrine a miraculous icon reputed to have been painted by St Luke.

After exemption from taxes by the Sultan during the 18th century, Agiásos swelled rapidly with Greeks fleeing hardship elsewhere on the island. The town's tiled houses and narrow, cobbled lanes have changed little in recent years, except for stalls of locally crafted souvenirs which line the way to the church with its belfry and surrounding bazaar. The presence of shops built into the church's foundations, with rents going towards its upkeep, is an ancient arrangement. It echoes the country-fair element of the traditional religious *panigýria* (festivals), where pilgrims once came to buy and sell as well as perform devotions. Agiásos

musicians are hailed as the best on Lésvos – they are out in force during the 15 August festival of the Assumption of the Virgin, considered one of the liveliest in Greece. The pre-Lenten carnival is also celebrated with verve at Agiásos; there is a special club devoted to organizing it.

Mantamádos ❹
Μανταμάδος

36 km (22 miles) NW of Mytilíni.
1,500. Mon–Sat.
Tsónia, 12 km (7 miles) N.

The attractive village of Mantamádos is famous for its pottery industry and the adjacent **Moní Taxiarchón**. The existing monastery dates from the 17th century and houses a black icon of the Archangel Michael, reputedly made from mud and the blood of monks slaughtered in an Ottoman raid. A bull is publicly sacrificed here on the third Sunday after Easter and its meat eaten in a communal stew, the first of several such rites on the island's summer festival calendar. Mantamádos ceramics come in a wide

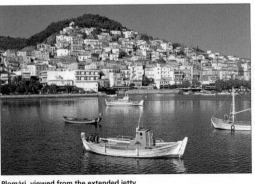

Plomári, viewed from the extended jetty

For hotels and restaurants in this region see pp307–9 and pp333–4

Fishing boats at Mólyvos harbour with the castle in the background

range of sizes and colours, from giant *pythária* (olive oil containers) to smaller *koumária* (ceramic water jugs).

Sykaminiá ❺
Συκαμινιά

46 km (29 miles) NW of Mytilíni.
🚶 *300.* 🚌 🚢 *Mon–Sat.* 🚕 *Kágia 4 km (2 miles) E; Skála Sykaminiás, 2 km (1 mile) N.*

Flanked by a deep valley and overlooking the straits to the Asia Minor coast, Sykaminiá has the most spectacular position of any village on Mount Lepétymnos, which stands at a height of 968 m (3,176 ft). Novelist Efstrátios Stamatópoulos (1892–1969), known as Strátis Myrivílis, was born close to the atmospheric central square. The jetty church, which featured in his novel *The Mermaid Madonna,* can be seen down in Skála Sykaminiás. One of Skála's tavernas is named after the *mouriá* or mulberry tree in which Myrivílis slept on hot summer nights.

Mólyvos (Míthymna) ❻
Μόλυβος (Μήθυμνα)

61 km (38 miles) NW of Mytilíni.
🚶 *1,500.* 🚌 ℹ *22530 71347.* 🚢 *Mon–Sat.*

Situated in a region celebrated in antiquity for its vineyards, Mólyvos is the most popular and picturesque town on Lésvos. It was the birthplace of Arion, the 7th-century BC

poet, and the site of the grave of Palamedes, the Achaian warrior buried by Achilles. According to legend, Achilles besieged the city until the king's daughter fell in love with him and opened the gates – though he killed her for her treachery. There is little left of the ancient town apart from the tombs excavated near the tourist office, but its ancient name, Míthymna, has been revived and is used as an alternative to Mólyvos (a Hellenization of the Turkish "Molova"). Artifacts from Ancient Míthymna are on display in the Archaeological Museum in Mytilíni town *(see p138)*.

Before 1923 over a third of the population was Muslim, forming a landed gentry who built many sumptuous three-storey town houses and graced Mólyvos with a dozen street fountains, some of which retain original ornate inscriptions. The mansions, or *archontiká*, are clearly influenced by eastern architecture *(see p22)*; the living spaces are arranged on the top floor around a central stairwell, or *chagiáti* – a design which had symbolic, cosmological meaning in the original Turkish mansions from which it was taken. The picturesque harbour and cobbled lanes of tiered stone houses are all protected by law; any new development must conform architecturally with the rest of the town.

Overlooking the town, and affording splendid views of the Turkish coast, stands a sizeable Byzantine **kástro**. The castle was modified by the Genoese adventurer Francesco Gatelluzi *(see p138)* in 1373, though it fell into Turkish hands during the campaign of Mohammed the Conqueror in 1462. Restored in 1995, the castle still retains its wood and iron medieval door and a Turkish inscription over the lintel. During summer, the interior often serves as a venue for concerts and plays.

A boatyard operates at the fishing harbour, a reminder of the days when Mólyvos was one of the island's major commercial ports.

🏰 **Kástro**
Tel 22530 71347. 🕐 *May–Oct: 8:30am– 3pm Tue–Sun; Jun–Sep: 8am– 7pm Tue–Sun.* 🔴 *main public hols.* 📷 📷

Colourfully restored Ottoman-style houses in Mólyvos

Tiered stone houses rising above the picturesque harbour of Mólyvos ▷

Around Western Lésvos

Though mostly treeless and craggy, western Lésvos has a severe natural beauty, broken by inland villages, beach resorts and three of Lésvos's most important monasteries. Many of the island's famous horses are bred in this region, and where the streams draining the valleys meet the sea, reedy oases form behind the sand providing a haven for bird-watchers during spring. Bus schedules are too infrequent for touring the area, but cars can be hired at Mólyvos.

Tiered houses of the village of Skalochóri

Pétra **7**
Πέτρα

55 km (34 miles) NW of Mytilíni.
🏠 3,700. 🚌 🚕 Anaxos 3 km (2 miles) W.

The village of Pétra takes its name (meaning "rock") from the volcanic monolith at its centre. By its base is the 16th-century basilica of **Agios Nikólaos**, still with its original frescoes, while a flight of 103 steps climbs to the 18th-century church of **Panagía Glykofiloúsa** church. The **Archontikó Vareltzídainas**, one of the last of the Ottoman dwellings once widespread on Lésvos (*see p141*), is also 18th century.

🏛 **Archontikó Vareltzídainas**
Sapphous. *Tel 22530 41510.*
◻ Tue–Sun. ⦿ *main public hols.*

Kalloní **8**
Καλλονή

40 km (25 miles) NW of Mytilíni.
🏠 1,600. 🚌 🚕 Mon–Sat.
🚕 Skála Kallonís 2 km (1 mile) S.

An important crossroads and market town, Kalloní lies 2 km (1 mile) inland from its name-sake gulf. Sardines are netted at the beach of **Skála Kallonís**.

Environs
In 1527, the abbot Ignatios founded **Moní Leimónos**, the second most important monastery on Lésvos. You can still view his cell, maintained as a shrine. A carved wood ceiling, interior arcades and a holy spring distinguish the central church. Moní Leimónos also has various homes for the infirm, a mini-zoo and two museums: one ecclesiastical and one of folkloric miscellany.

🏛 **Moní Leimónos**
5 km (3 miles) NW of Kalloní.
Tel 22530 22289. **Ecclesiastical Museum** ◻ *daily.* **Folk Museum** ◻ *on request.*

Antissa **9**
Αντισσα

76 km (47 miles) NW of Mytilíni.
🏠 1,410. 🚌 🚕 *daily.*
🚕 *Kámpos 4 km (2.5 miles) S.*

The largest village of this part of Lésvos, Antissa merits a halt for its fine central square alone, in which a number of cafés and tavernas stand overshadowed by three huge plane trees. The ruins of the eponymous ancient city, destroyed by the Romans in 168 BC, lie 8 km (5 miles) below by road, near the remains of the Genoese **Ovriókastro**. This castle stands on the shore, east of the tiny fishing port of Gavathás and the long sandy beach of Kámpos.

Environs
Although, unlike Antissa, there is no view of the sea, **Vatoússa**, 10 km (6 miles) east, is the area's most attractive village. Tiered **Skalochóri**, another 3 km (2 miles) north, does overlook the north coast and – like most local villages – has a ruined mosque dating to the days before the 1923 Treaty of Lausanne (*see p43*).

Hidden in a lush river valley, 3 km (2 miles) east of Antissa, stands the 16th-century **Moní Perivólis**, situated in the middle of a riverside orchard. The narthex features three 16th-century frescoes, restored in the 1960s: the apocalyptic the *Earth and Sea Yield Up Their Dead*, the *Penitent Thief of Calvary* and the *Virgin* (flanked by Abraham). The interior is lit by daylight only, so it is advisable to visit the monastery well before dusk.

Frescoes adorning the narthex of Moní Perivólis

Moní Ypsiloú ⑩
Μονή Υψηλού

62 km (39 miles) NW of Mytilíni.
🚌 *Tel* 22530 56259. ☐ *daily.*

Spread across the 511-m
(1,676-ft) summit of Mount
Ordymnos, an extinct volcano,
Moní Ypsiloú was founded in
the 12th century and is now
home to just four monks. It has
a handsome double gate, and
a fine wood-lattice ceiling in its
katholikón (main church) beside which a rich exhibition of
ecclesiastical treasures can be
found. In the courtyard outside
stand a number of fragments of
petrified trees. The patron saint
of the monastery is John the
Divine (author of the book of
Revelation), a typical dedica-
tion for religious communities
located in such wild, forbidding scenery.

Triple bell tower of Moní Ypsiloú

Environs
The main entry to Lésvos's
petrified forest is just west of
Ypsiloú. Some 15 to 20 million
years ago, Mount Ordymnos
erupted, beginning the process
whereby huge stands of sequ-
oias, buried in the volcanic ash,
were transformed into stone.

Sígri ⑪
Σίγρι

93 km (58 miles) NW of Mytilíni.
🏠 *400.* 🚌

An 18th-century Ottoman
castle and the church of **Agía
Triáda** dominate this sleepy
port, protected from severe
weather by long, narrow Nisópi
island. Sígri's continuing
status as a naval base has
discouraged tourist development, though it has a couple
of small beaches; emptier ones
are only a short drive away.

The peaceful harbour of Sígri

Skála Eresoú ⑫
Σκάλα Ερεσού

89 km (55 miles) W of Mytilíni town.
🏠 *1,500.* 🚌

Extended beneath the acropolis of ancient Eresós, the
wonderful, long beach at Skála
Eresoú supports the island's
third-largest resort. By climbing
the acropolis hill, you can spot
the ancient jetty submerged in
the modern fishing anchorage.
Little remains at the summit, but
the Byzantine era is represented
in the ancient centre by the
foundations of the basilica of
Agios Andreás; its 5th-century
mosaics await restoration.

Environs
The village of **Eresós**, 11 km
(7 miles) inland, grew up as a
refuge from medieval pirate
raids; a vast, fertile plain
extends between the two
settlements. Two of Eresós's
most famous natives were the
philosopher Theophrastos, a
pupil of Aristotle *(see p59)*, and
Sappho, one of the greatest
poets of the ancient world.

SAPPHO, THE POET OF LESVOS

One of the finest lyric poets of any era, Sappho (c.615–562
BC) was born, probably at Eresós, into an aristocratic family
and a society that gave women substantial freedom. In her
own day, Sappho's poems were known across the
Mediterranean, though Sappho's poetry was to be
suppressed by the church in late antiquity and now survives only in short quotations and on papyrus scraps. Many
of her poems were also addressed to women, which has
prompted speculation about Sappho's sexual orientation.
Much of her work was inspired by female companions:
discreet homosexuality was unremarkable in her time. Even

less certain is the manner of
her death; legend asserts that
she fell in love with a younger
man whom she pursued as
far as the isle of Lefkáda.
Assured that unrequited
love could be cured by
leaping from a cliff, she did
so and drowned in the sea:
an unlikely, and unfortunate,
end for a poet reputed to be
the first literary lesbian.

Chíos
Χíος

Although Chíos has been prosperous since antiquity, today's island is largely a product of the Middle Ages. Under the Genoese, who controlled the highly profitable trade in gum mastic *(see pp148–9)*, the island became one of the richest in the Mediterranean. It continued to flourish under the Ottomans until March 1822, when the Chians became the victims of one of the worst massacres *(see p151)* of the Independence uprising. Chíos had only partly recovered when an earthquake in 1881 caused severe damage, particularly in the south.

Chíos Town ❶
Χíος

🏠 *25,000.* ✈ 🚌 *Polytechníou (around island), Dimokratías (environs).* ℹ *Kanári 18 (22710 44389).* 🛒 *Mon–Sat.* 🚢 *Karfás 7 km (4 miles) S.*

Chíos town, like the island, was settled in the Bronze Age and was colonized by the Ionians from Asia Minor by the 9th century BC. The site was chosen for its convenient position for travelling to the Turkish mainland opposite, rather than good anchorage: a series of rulers have been obliged to construct long breakwaters as a consequence. Though it is a modernized island capital (few buildings predate the earthquake of 1881), there are a number of museums and other scattered relics from the town's eventful past. Besides the kástro,

Shopfront in Chíos town bazaar

the most interesting sights are the lively bazaar at the top of Roḯdou, and an ornate Ottoman fountain dating to 1768 at the junction of Martýron and Dimarchías.

⚜ Kástro
Maggiora. *Tel 22710 22819.*
☐ *daily.* ♿
The most prominent medieval feature of the town is the kástro, a Byzantine foundation improved by the Genoese after they acquired Chíos in 1346. Today the kástro lacks the southeasterly sea rampart, which fell prey to developers after the devastating earthquake in 1881. Its most impressive gate is the southwesterly Porta Maggiora; a deep dry moat runs from here around to the northwest side of the walls. Behind the walls, Ottoman-era houses line narrow lanes of what were once the Muslim and Jewish quarters of the town; after the Ottoman conquest, in 1566, Orthodox and Catholics were required to live outside the walls. Also inside, a disused mosque, ruined Turkish baths and a small

KEY

For key to map see back flap

↘ *Sámos*

SIGHTS AT A GLANCE

Avgónyma ❹
Chíos Town ❶
Mastic Villages ❷
Moní Moúndon ❻
Néa Moní ❸
Volissós ❺

0 kilometres 5

0 miles   5

For hotels and restaurants in this region see pp307–9 and pp333–4

Chíos town waterfront with the dome and minaret of the Mecidiye Mosque

the Koraïs library, situated on the ground floor, consists of a number of books and manuscripts bequeathed by the cultural revolutionary and intellectual Adamántios Koraïs (1748–1833); these include works given by Napoleon.

Environs

The fertile plain known as the **Kámpos** extends 6 km (4 miles) south of Chíos town. The land is crisscrossed by a network of unmarked lanes which stretch between high stone walls that betray nothing of what lies behind. However, through an ornately arched gateway left open, you may catch a glimpse of what were once the summer estates of the medieval Chian aristocracy.

Several of the mansions were devastated by the 1881 earthquake, but some have been restored with their blocks of multicoloured sandstone arranged so that the different shades alternate. Many of them still have their own waterwheels, which were once donkey-powered and drew water up from 30-m (98-ft) deep wells into open cisterns shaded by a pergola and stocked with fish. These freshwater pools, which are today filled by electric pumps, still irrigate the vast orange, lemon and tangerine orchards for which the region is widely known.

Ottoman cemetery can be found. The latter contains the grave and headstone of Admiral Kara Ali who commanded the massacre of 1822. He was killed aboard his flagship when it was destroyed by the Greek captain Kanáris.

Porta Maggiora, the southwesterly entrance to the kástro

🏛 Justiniani Museum

Kástro. **Tel** 22710 22819. ☐ 8am–2:30pm Tue–Sun. 🎟
This collection is devoted to religious art and includes a 5th-century AD floor mosaic rescued from a neglected Chian chapel. The saints featured on the icons and frescoes include Isídoros, who is said to have taught the islanders how to make liqueur from mastic *(see pp148–9)*, and Matrona, a martyr of Roman Ankara whose veneration here was introduced by refugees from Asia Minor after 1923.

🏛 Byzantine Museum

Plateía Vounakíou. **Tel** 22710 26866. ☐ Tue–Sat. 🎟
Though called the Byzantine Museum, this is little more than an archaeological warehouse and restoration workshop. It is housed within the only mosque to have survived intact in the East Aegean, the former Mecidiye Cami, which still retains its minaret. A number of Jewish, Turkish and Armenian gravestones stand propped up in the courtyard, attesting to the multiethnic population of the island during the medieval period.

🏛 Philip Argéntis Museum

Koraïs 2. **Tel** 22710 44246. ☐ 8am–2pm Mon–Fri (also 5–8pm Tue, 5–8:30pm Fri), 8am–12:30pm Sat. 🎟
Endowed in 1932 by a member of a leading Chian family and occupying the floor above the Koraïs library, this collection features rural wooden implements, plus examples of traditional embroidery and costumes. Also on view, alongside a number of portraits of the Argéntis family, are rare engravings of islanders and numerous copies of the *Massacre at Chíos* by Delacroix (1798–1863). This painting, as much as any journalistic dispatch, aroused the sympathy of Western Europe for the Greek revolutionary cause *(see pp42–3)*. The main core of

Detail of Delacroix's *Massacres de Chíos* (1824) in the Philip Argéntis Museum

Mastic Villages ②

Μαστιχοχώρια

The 20 settlements in southern Chíos known as the *mastichochória*, or "mastic villages", received their name from their most lucrative medieval product. Genoese overlords founded the villages well inland as an anti-pirate measure during the 14th and 15th centuries. Constructed to a design unique in Greece, they share common defensive features made all the more necessary by the island's proximity to the Turks. Though they were the only villages to be spared in the 1822 massacres (*see p151*), most have had their architecture compromised by both earthquake damage and ill-advised modernization.

Armólian pottery

MAIN MASTIC VILLAGES

Véssa
This is the one village whose regular street plan can easily be seen from above while descending from Agios Geórgios Sykoúsis or Eláta.

Pyrgí
Pyrgí is renowned for its bright houses, many patterned with xystá *("grating") decoration. Outer walls are plastered using black sand and coated with whitewash. This is then carefully scraped off in repetitive geometric patterns, revealing the black undercoat. An example of this is the church of Agioi Apóstoloi which also has medieval frescoes.*

Armólia
One of the smallest and simplest of the mastichochória, Armólia is renowned for its pottery industry.

Fortification towers guarded each corner of the village.

Houses reached three storeys, with vaulted ceilings except on the top floor.

Narrow passages were overarched by flying buttresses, to limit earthquake damage.

Streets followed an intricate grid plan designed to confuse strangers.

Flat roofs of adjacent buildings were ideally of the same height to facilitate escape.

Olýmpoi

Olýmpoi is almost square in layout. Its central tower has survived to nearly its original height, and today two cafés occupy its ground floor. Here local men and women can be seen winnowing mastic.

VISITORS' CHECKLIST

28 km (17 miles) SW of Chíos town. ﹏ Pyrgí: 1,200; Mestá: 400; Olýmpoi: 350. ﹏ Mestá. ﹏ Mávra Vólia & Kómi 5 km (3 miles) SE of Pyrgí.

A square tower in the centre of the village was the last refuge in troubled times.

Vávyloi

The 13th-century Byzantine church of Panagía tis Krínis, on the edge of the village, is famed for its frescoes and its alternating courses of stone- and brickwork.

MASTIC PRODUCTION

The mastic bush of southern Chíos secretes a resin or gum that, before the advent of petroleum-based products, formed the basis of paints, cosmetics and medicines. Today it is made into chewing gum, liqueur and even toothpaste. About 300 tonnes of gum are harvested each summer through incisions in the bark, which weep resin "tears"; once solidified a day later, the resin is scraped off and spread to air-cure on large trays.

Mastic bush bark and crystals

Crystals separated from the bark

The outer circuit of houses doubled as a perimeter wall.

MESTA

Viewed here from the southwest, Mestá is considered the best preserved of the mastic villages. It has the most even roof heights and still retains its perimeter corner towers.

Taxiárchis Church

Mestá's 19th-century church, the largest on Chíos, dominates the central square. The atmospheric interior has a fine carved altar screen.

Néa Moní ❸

Νέα Μονή

St Anne mosaic, inner narthex

Hidden in a wooded valley 11 km (7 miles) west of Chíos town, the monastery of Néa Moní and its mosaics – some of Greece's finest – both date from the 11th century. It was established by Byzantine Emperor Constantine IX Monomáchos in 1042 on the site where three hermits found an icon of the Virgin. It reached the height of its power after the fall of the Byzantine Empire, and remained influential until the Ottoman reprisals of 1822. Néa Moní has now been a convent for decades, but when the last nun dies it is to be taken over again by monks.

Néa Moní, viewed from the west

Narthex
Seen here with the main church dome in the background, the narthex contains the most complex mosaics. Twenty-eight saints are depicted, including St Anne, the only woman. The Virgin with Child *adorns the central dome.*

The belfry is a modern structure, added after the 1881 earthquake.

St Joachim mosaic

Ornate marble inlays were highly prized in the Byzantine Empire.

STAR FEATURES

★ Anástasis

★ Christ Washing the Disciples' Feet

★ **Christ Washing the Disciples' Feet**
Here Christ washes the feet of Peter, who indicates he wishes his head and hands also to be bathed.

★ Anástasis
er the Resurrection,
Christ rescues Adam
d Eve from Hell be-
re entering Heaven.

St Mark the
Evangelist
mosaic

The dome was repaired after the 1881 earthquake, though its magnificent Pantokrátor was lost.

The main apse has a mosaic of the Virgin. It is positioned above the walls and represents earthly subjects, while the dome depicts Christ.

Descent from the Cross mosaic

Altar screen

Byzantine Clock
Standing beneath the Crucifixion mosaic, this Armenian-made clock came from Smyrna after its destruction in 1922.

THE MASSACRE AT CHÍOS

After 250 years of Ottoman rule, the Chians joined the Independence uprising in March 1822, incited by Samian agitators. Enraged, the Sultan sent an expedition that massacred 30,000 Chians, enslaved almost twice that number and brutally sacked most of the monasteries and houses. Many Chians fled to Néa Moní for safety, but they and most of the 600 monks were also killed. Just inside the main gate of the monastery stands a chapel containing the bones of those who died here. The savagery of the Turks is amply illustrated by the axe-wounds visible on many skulls, including those of children.

The floor is covered with marble segments which echo the disciplined architecture of the nave.

Betrayal in the Garden
A detail of this mosaic shows Peter lopping off the ear of Malchus, the High Priest's servant, following the betrayal of Jesus in Gethsemane. Unfortunately, the Kiss of Judas has been damaged.

Cabinet containing the skulls of the Chian martyrs of 1822

Around Chíos Island

Ceiling detail at Moní Moúndon

With its verdant, semi-mountainous terrain, edged by rocky cliffs in the south and sandy beaches to the northwest, Chíos is one of the Aegean's most beautiful isles. Roads and public transport radiate in all directions from Chíos town and the best bus service is to be found on the densely populated southeast coast; to explore anywhere else you need to hire a taxi, car or powerful motorbike.

Avgónyma ❹
Αυγώνυμα

20 km (12 miles) W of Chíos town.
🏘 15. 🚌 Elínta 7 km (4 miles) W.

This is the closest settlement to Néa Moní *(see pp150–51)* and the most beautiful of the central Chian villages, built in a distinct style: less labyrinthine and claustrophobic than the mastic villages, and more elegant than the houses of northern Chíos. The town's name means "clutch of eggs", perhaps after its clustered appearance when viewed from the ridge above. Virtually every house has been tastefully restored in recent years by Greek-Americans with roots here. The medieval *pýrgos* (tower) on the main square, with its interior arcades, is home to the excellent central taverna.

Environs
Few Chian villages are as striking glimpsed from a distance as **Anávatos**, 4 km (2 miles) north of Avgónyma. Unlike Avgónyma, Anávatos has scarcely changed in recent

decades; shells of houses blend into the palisade on which they perch, overlooking occasionally tended pistachio orchards. The village was the scene of a particularly traumatic incident during the atrocities of 1822 *(see p151)*. Some 400 Greeks threw themselves into a ravine from the 300-m (985-ft) bluff above the village, choosing suicide rather than death at the hands of the Turks.

Volissós ❺
Βολισσός

40 km (25 miles) NW of Chíos town.
🏘 500. 🚌 🚢 Mánagros 2 km (1 mile) SW.

Volissós was once the primary market town for the 20 smaller villages of northwestern Chíos, but today the only vestige of its former commercial standing is a single saddlery on the western edge of town. The strategic importance of medieval Volissós is borne out by the crumbled hilltop castle, erected by the Byzantines in the 11th century and repaired

One of the many restored stone houses of Avgónyma

by the Genoese in the 14th. The town's stone houses stretch along the south and east flanks of the fortified hill; many have been bought and restored by Volissós's growing expatriate population.

Environs
Close to the village of **Agio Gála**, 26 km (16 miles) northwest of Volissós, two 15th-century chapels can be found lodged in a deep cavern near the top of a cliff. The smaller, hindmost chapel is the more interesting of the two; it is built entirely within the grotto and features a sophisticated and mysterious fresco of the *Virgin and Child*. The larger chapel, at the entrance to the cave, boasts an intricate carved *témblon* or altar screen. Agio Gála can be reached by bus from Volissós and admission to the churches should be made via the resident warden who holds the keys.

The largely deserted town of Anávatos with the few inhabited dwellings in the foreground

For hotels and restaurants in this region see pp307–9 and pp333–4

Moní Moúndon ❻
Μονή Μούνδων

35 km (22 miles) NW of Chíos town.
***Tel** 22740 21230.* 🚌 *to Volissós.*
🔲 *daily (ask for key at first house in Diefha village).*

Founded late in the 16th century, this picturesque monastery was once second in importance to Néa Moní *(see pp150–51).* The *katholikón* (or central church) has a number of interesting late-medieval murals, the most famous being the *Salvation of Souls on the Ladder to Heaven.* Although the church is only open to the public during the monastery's festival (29 August), the romantic setting makes the stop worthwhile.

Moúndon's *Salvation of Souls on the Ladder to Heaven* **mural**

Outlying Islands
Domestic architecture on the peaceful islet of **Oinoússes**, a few miles east of Chíos town, is deceptively humble, for it is the wealthiest territory in Greece. Good beaches can be found to either side of the port, and in the northwest of the island is the Evangelismoú convent, endowed by the Pateras family.

Much of **Psará**, 71 km (44 miles) to the west, was ruined in the Greek War of Independence *(see pp42–3);* as a result, the single town, built in a pastiche of island architectural styles, is a product of the last 100 years. The landscape is still desolate and infertile, though there are good beaches to visit east of the harbour, and Moní Koímisis tis Theotókou in the far north.

The remains of a Hellenistic tower near Fanári, Ikaría

Ikaría
Ικαρία

🏠 7,500. ✈ ⛴ 🚌 *Agios Kýrikos.*
ℹ *22750 22202.* ⛴ *Fanári 16 km (10 miles) NE of Agios Kírykos.*

Lying 245 km (150 miles) south of Chíos, Ikaría is named after the Ikaros of legend who flew too near the sun on artificial wings and plunged to his death in the sea when his wax bindings melted.

Agios Kírykos, the capital and main port, is a pleasant town flanked by two spas, one of them dating to Roman times and still popular with an older Greek clientele. A number of hot baths can be visited at **Thérma**, a short walk to the northeast, while at **Thérma Lefkádas**, to the southwest, the springs still well up among the boulders in the shallows of the sea.

About 2 km (1.5 miles) west of Evdilos, a village port on the north coast, lies the village of **Kámpos**. It boasts a broad, sandy beach and, beside the ruins of a 12th-century church, the remains of a Byzantine manor house can be seen. The building recalls a time when the island was considered a humane place of exile for disgraced noblemen; there was a large settlement of such officials in Kámpos. A small museum contains artifacts from the town of Oinoe, Kámpos's ancient predecessor.

Standing above Kosoíki village, 5 km (3 miles) inland, the Byzantine castle of **Nikariás** was built during the 10th century to guard a pass on the road to Oinoe. The only other well-preserved fortification is a 3rd-century BC **Hellenistic tower (Drakánou)**, once an ancient lighthouse, near Fanári.

Tiny **Armenistís**, with its surrounding forests and fine beaches, such as Livádi and Messaktí to the east, is Ikaría's main resort. The foundations of a temple to the goddess Artemis Tavropólos (Artemis incarnated as the patroness of bulls) lie 4 km (2 miles) west.

Home to the most active fishing fleet in the East Aegean, the island of **Foúrnoi**, due east of Ikaría, is far more populous and lively than its small size suggests. The main street of the port town, lined with mulberry trees, links the quay with a square well inland, where an ancient sarcophagus sits between the two cafés. Within walking distance lie Kampí and Psilí Ammos beaches.

Coastal town of Agios Kírykos, the capital of Ikaría

Sámos
Σάμος

Settled early, owing to its natural richness and ease of access from Asia, Sámos was a major maritime power by the 7th century BC and enjoyed a golden age under the rule of Polykrates (538–522 BC). After the collapse of the Byzantine Empire, most of the islanders fled from pirates and Sámos lay deserted until 1562, when Ottoman Admiral Kiliç Ali repopulated it with returned Samians and other Orthodox settlers. The 19th century saw an upsurge in fortunes made in tobacco trading and shipping. Union with Greece occurred in 1912.

VISITORS' CHECKLIST

34,000. ✈ 4 km (2 miles) W of Pythagóreio. ⛴ Vathý, Karlóvasi, Pythagóreio.
🚌 ℹ Vathý (22730 28582).
🍷 Wine Festival: August; Fishermen's Festival, Pythagóreio: June or July.

Assyrian bronaze horse figurine, Vathý Archaeological Museum

Fishermen at Vathý harbour

SIGHTS AT A GLANCE

Efpalíneio Orygma ❷
Heraion ❺
Karlóvasi ❼
Kokkári ❻
Moní Megális Panagías ❹
Mount Kerketéfs ❽
Pythagóreio ❸
Vathý ❶

Vathý ❶
Βαθύ

5,700. ⛴ ℹ Ioánnou Lekáti.
ℹ 25 Martíou 4 (22730 28582).
🛒 daily. 🏖 Psilí Ammos 8 km (5 miles) SE; Mykáli 6 km (4 miles) S.

Though the old village of Ano Vathý existed in the 1600s, today's town is recent; the harbour quarter grew up only after 1832, when the town became the capital of the island. Just large enough to provide all amenities in its bazaar, lower Vathý caters to tourists while cobble-laned Ano Vathý carries on oblivious to the commerce in the streets below.

The Sámos **Archaeological Museum** contains ar tifacts from the excavations at the Heraion sanctuary *(see p156)*. Because of the far-flung origins of the pilgrims who visited the shrine, the collection of small votive offerings is one of the richest in Greece – among them are a bronze statuette of an Urartian god, Assyrian figurines and an ivory miniature of Perseus and Medusa. The largest free-standing sculpture to have survived from ancient Greece is the star exhibit: a 5-m (16-ft) tall marble *koúros* dating from 580 BC and dedicated to the god Apollo.

🏛 Archaeological Museum
Kapetán Gymnasiárchou Katevéni.
Tel 22730 27469. ◯ Tue–Sun.
⬤ main public hols. 🎫 📷

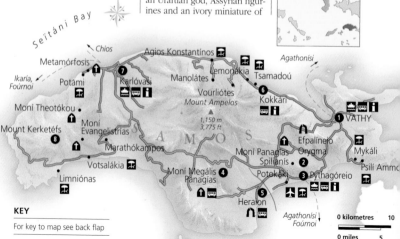

Seïtáni Bay
Chios
Metamórfosis
Ikaría, Foúrnoi
Potámi
Karlóvasi
Manolátes
Lemonákia
Agios Konstantínos
Agathonísi
Tsamadoú
Vourliótes
Mount Ampelos
Kokkári
Moní Theotókou
Moní Evangelístrias
Mount Kerketéfs
S Á M O S
1,150 m
3,775 ft
Marathókampos
Efpalíneio Orygma
VATHY
Mykáli
Votsalákia
Moní Panagías Spilianís
Psilí Ammos
Limniónas
Moní Megális Panagías
Potokáki
Pythagóreio
Heraion
Agathonísi Foúrnoi

KEY

For key to map see back flap

0 kilometres 10
0 miles 5

For hotels and restaurants in this region see pp307–9 and pp333–4

Around Sámos Island

Sámos has a paved road around the island, but buses are frequent only between Pythagóreio and Karlóvasi, via Vathý. Vehicle-hire is easy, though many points can be reached only by jeep or foot. In the south and west there are many rough dirt roads where caution is necessary.

Efpalíneio Orygma ❷
Ευπαλίνειο Όρυγμα

15 km (9 miles) SW of Vathý.
Tel 22730 61400. ◻ Tue–Sun.
◉ main public hols. ▨

Efpalíneio Orygma (Eupalinos's tunnel) is a 1,040-m (3,410-ft) aqueduct, ranking as one of the premier engineering feats of the ancient world. Designed by the engineer Eupalinos and built by hundreds of slaves between 529 and 524 BC, the tunnel guaranteed ancient Sámos a water supply in times of siege, and remained in use until this century. Eupalinos's surveying was so accurate that, when the work crews met, having begun from opposite sides of the mountain, their vertical error was nil.

Visitors may walk along the ledge used to remove rubble from the channel far below. Half the total length is open to the public, with grilles to protect you from the worst drops.

Pythagóreio ❸
Πυθαγόρειο

13 km (8 miles) SW of Vathý.
⛨ 1,300. ⛴ ⛐ ⛵ ❶ Lykoúrgou Logothéti (22730 61389).
⛴ Potokáki 3 km (2 miles) W.

Cobble-paved Pythagóreio, named after the philosopher Pythagoras who was born here in 580 BC, has long been the lodestone of Samian tourism. The extensive foundations and walls of ancient Sámos act as a brake on tower-block construction; the only genuine tower is the 19th-century manor of **Lykoúrgos Logothétis**, the local chieftain who organized a decisive naval victory over the Turks on 6 August 1824, the date of the Feast of the Transfiguration. Next to this stronghold is the church of the **Metamórfosis**, built to celebrate the victory. At the far western edge of town are the extensive remains of

Pythagoras statue (1989) by Nikoláos Ikaris, Pythagóreio

Roman Baths, still with a few doorways intact. Further west, the Doryssa Bay luxury complex stands above the silted-in area of the Archaic harbour; all that remains is Glyfáda lake, crossed by a causeway.

⋔ Roman Baths
W of Pythagóreio. **Tel** 22730 61400.
◻ variable. ♿

Environs
Polykrates protected Pythagóreio by constructing a circuit of walls enclosing Kastrí hill, to a circumference of more than 6 km (4 miles), with 12 gates. The walls were damaged by an Athenian siege of 439 BC, and today are most intact just above Glyfáda, where a fortification tower still stands. Enclosed by the walls, just above the ancient theatre, sits **Moní Panagías Spilianís** with its 100-m (330-ft) cave containing a shrine to the Virgin.

Moní Megális Panagías ❹
Μονή Μεγάλης Παναγίας

27 km (17 miles) W of Vathý. ⛐
◻ May–Oct: daily.

Founded in 1586 by Nílos and Dionýsios, two hermits from Asia Minor, the monastery of Megális Panagías is the second oldest on Sámos and contains the island's best surviving frescoes from that period. The central church is orientated diagonally within the square compound of cells, now restored, probably built directly above a temple of Artemis which it replaced. Sadly, the area was ravaged by fire in 1990, shortly after the last monk died. Visiting hours depend on the whim of the caretaker.

Fresco of Jesus washing the apostles' feet, Moní Megális Panagías

The single remaining column of Polykrates' temple, Heraion

Heraion ❺

Ηραίον

21 km (13 miles) SW of Vathý.
Tel 22730 95277. 🚌 Iraío.
🕐 Tue–Sun. 🌑 main public hols. 🎫

A fertility goddess was worshipped here from Neolithic times, though the cult only became identified with Hera after the arrival of Mycenaean colonists *(see pp28–9)*, who brought their worship of the Olympian deities with them. The sanctuary's site on flood-prone ground honoured the legend that Hera was born under a sacred osier (willow tree) on the banks of the Imvrasos and celebrated her nuptials with Zeus among the osiers here, in the dangerous pre-Olympian days when Kronos still ruled.

A 30-m (98-ft) long temple built in the 8th century BC was replaced in the 6th century BC by a stone one of the Ionic order, planned by Rhoikos, a local architect. Owing to earthquakes, or a design fault, this collapsed during the reign of Polykrates, who ordered a grand replacement designed by Rhoikos's son, Theodoros. He began the new temple in 525 BC, 40 m (130 ft) west of his father's, recycling building materials from its predecessor. Building continued off and on for many centuries, but the vast structure was never completed. The interior, full of votive offerings, was described by visitors in its heyday as a veritable art gallery.

Most of the finds on display at the Archaeological Museum in Vathý *(see p154)* date from the 8th to the 6th centuries BC, when the sanctuary was at the height of its prestige. The precinct was walled and contained several temples to other deities, though only Hera herself had a sacrificial altar. Pilgrims could approach from the ancient capital along a 4,800-m (15,750-ft) Sacred Way.

Despite diligent 20th-century German excavations, much of the sanctuary is confusing. Byzantine and medieval masons removed ready-cut stone for reuse in their buildings, leaving only one column untouched. Early in the 5th century, Christian masons built a basilica dedicated to a new mother figure: the Virgin Mary. Its foundations lie east of the Great Temple.

Plinth from Polykrates' temple, Heraion

Kokkári ❻

Κοκκάρι

10 km (6 miles) W of Vathý. 🏠 1,000.
🚌 🛈 Agiou Nikoláou (22730 92333). 🏖 Tsamadoú & Lemonákia 2 km (1 mile) W.

Built on and behind twin headlands, this charming little port takes its name from the shallot-like onions once cultivated just inland. Today it is the island's third resort after Pythagóreio and Vathý, with its windblown location turned to advantage by a multitude of windsurfers. The town's two beaches are stony and often

---

THE CULT OF HERA

Hera was worshipped as the main cult of a number of Greek cities, including Argos on the mainland, and always as out-of-town sanctuaries. Before the 1st millennium BC, she was venerated in the form of a simple wooden board which was later augmented with a copper statue. One annual rite, the Tonaia, commemorated a foiled kidnapping of the wooden statue by Argive and Etruscan pirates. During the Tonaia, the idol would be paraded to the river mouth, bound on a litter of osiers (sacred to Hera), bathed in the sea and draped with gifts. The other annual festival, the Heraia, when the copper statue was dressed in wedding finery, celebrated Hera's union with Zeus, and was accompanied by concerts and athletic contests. Housed in a special shrine after the 8th century, the statue of Hera was flanked by a number of live peacocks and sprigs from an osier tree. Both are shown on Samian coins of the Roman era stamped with the image of the richly dressed goddess.

Hera, led by peacocks, and depicted on Samian coins

The beach and harbour of **Kokkári**, flanked by its twin headlands

surf-battered, but the paved quay and its waterside tavernas are the busy focus of nightlife.

Environs
Though many of Sámos's hill-villages are becoming deserted, **Vourliótes** is an exception, thriving thanks to its orchards and vineyards. The picturesque central square is one of the most beautiful on the island, with outdoor seating at its four tavernas. Vourliótes is situated at a major junction in the area's network of hiking trails; paths come up from Kokkári, descend to Agios Konstantínos, and climb to Manolátes, which is the trailhead for the ascent of Mount Ampelos, a five-hour round trip.

Karlóvasi ⑦
Καρλόβασι

33 km (20 miles) NW of Vathý.
🏘 5,500. 🚌 🚤 🚉 *Potámi 2 km (1 mile) W.*

Sprawling, domestic Karlóvasi, gateway to western Sámos and the island's second town, divides into four separate districts. Néo Karlóvasi served as a major leather production centre between the world wars, and abandoned tanneries and ornate mansions built on shoe-wealth can still be seen down by the sea. Meséo Karlóvasi, on a hill across the river, is more attractive, but most visitors stay at the harbour of Limín, with its tavernas and lively boatyard. Above the port, Ano, or Palaió Karlóvasi is

tucked into a wooded ravine, overlooked by the landmark hilltop church of **Agía Triáda**, the only structure in Ano visible from the sea.

Environs
An hour's walk from Ano Karlóvasi, inland from Potámi beach, is the site of a medieval settlement. Its most substantial traces include the 11th-century church of **Metamórfosis**, the oldest on the island, and a Byzantine castle immediately above.

Mount Kerketéfs ⑧
Όρος Κερκετευς

50 km (31 miles) W of Vathý. 🚌 *to Marathókampos.* 🚤 *Votsalákia, 2 km (1 mile) S of Marathókampos; Limniónas, 5 km (3 miles) SW of Marathókampos.*

Dominating the western tip of Sámos, 1,437-m (4,715-ft) Mount Kerketéfs is the second highest peak in the Aegean after Sáos on Samothráki. On an island otherwise composed of smooth sedimentary rock, the partly volcanic mountain is an anomaly, with jagged rocks and bottomless chasms.

Kerketéfs was first recorded in Byzantine times, when religious hermits occupied some of its caves. Nocturnal glowings at the cave-mouths were interpreted by sailors as the spirits of departed saints, or the aura of some holy icon awaiting discovery. Today, two monasteries remain on Kerketéfs: the 16th-century **Moní Evangelistrías**, perched on the south slope, and **Moní Theotókou**, built in 1887, tucked into a valley on the northeast side.

Despite recent forest fires, and the paving of a road to remote villages west of the summit, Mount Kerketéfs still boasts magnificent scenery, with ample opportunities for hiking. At Seïtáni Bay on the north coast, a marine reserve protects the Mediterranean monk seal *(see p115).*

Mount Kerketéfs, seen from the island of Ikaría

THE DODECANESE

PATMOS · LIPSI · LEROS · KALYMNOS · KOS · ASTYPALAIA · NISYROS
TILOS · SYMI · RHODES · CHALKI · KASTELLORIZO · KARPATHOS

Scattered along the coast of Turkey, the Dodecanese are the most southerly group of Greek islands, their hot climate and fine beaches attracting many visitors. They are the most cosmopolitan archipelago, with an eastern influence present in their architecture. They were the last territories to be incorporated into modern Greece.

Due to their distance from Athens and mainland Greece, these islands have been subject to a number of invasions, with traces of occupation left behind on every island. The Classical temples built by the Dorians can be seen on Rhodes. The Knights of St John were the most famous invaders, arriving in 1309 and staying until they were defeated by Suleiman the Magnificent in 1522.

Ottoman architecture is most prominent on larger, wealthier islands, such as Kos and Rhodes. After centuries of Turkish rule, the Italians arrived in 1912 and began a regime of persecution. Mussolini built many imposing public buildings, notably in the town of Lakkí on Léros. After years of occupation, the islands were finally united with the Greek state in 1948.

A statue at Mandráki harbour in Rhodes

Geographically, the Dodecanese vary dramatically in character: some are dry, stark and barren, such as Chálki and Kásos, while Tílos and volcanic Nísyros are fertile and green. Astypálaia and Pátmos, with their whitewashed houses, closely resemble Cycladic islands; the pale houses of Chóra, on Pátmos, are spectacularly overshadowed by the dark monastery of St John. Rhodes is the capital of the island group, and is one of the most popular holiday destinations due to its endless sandy beaches and many sights.

The climate of these islands stays hot well into the autumn, providing a long season in which to enjoy the beaches. These vary from black pebbles to silver sands, and deserted bays to shingle strips packed with sunbathers.

One monk's method of travelling around on the holy island of Pátmos

◁ A façade on the waterfront of Sými town's harbour

Exploring the Dodecanese

The Dodecanese offer an unparalleled range of landscapes and activities. There are beautiful beaches with all kinds of watersports, safe yachting harbours, lush valleys and barren mountains, caves and fjords, and even the semi-active volcano on Nísyros. Historical sights in the group are just as diverse, including the 11th-century Monastery of St John on Pátmos, the Hellenistic Asklepieion of Kos, the medieval walled city of the Knights of Rhodes and the unique traditional village of Olympos on Kárpathos. This island group divides neatly into north and south. Kos in the north and Rhodes, the group's capital, in the south make good bases for air and ferry travel.

Agathonísi

↑ Sámos

∩ Pátmos — *Arkí*

Lámpi

Skála — *Lipsí*

Gríkos — Lipsí Town 🔟

Piraeus, Mýkonos

Alínda 🏛 *Léros*

Lakkí

🎿∩ **Kálymnos**

Arginónta

Myrtiés — Rína *Psérimos*

Póthia

Kos Town

Piraeus → Tigkáki — As...

Mastichári

Piraeus, Sýros, Naxos → Antimácheia — Kardámain

Kos — Kamári

🎿∩⚓

Mandráki — Páloi

○Vathý — Nikiá

○Maltezána — *Nísyros* 🎿

Livádi○ — Astypálaia Town

Astypálaia — *Piraeus* ↗ — *Kandeliousa* — Me...

⛰🎿 — *Tílos* 🔟

Piraeus, Amorgós — *Sýrna*

ISLANDS AT A GLANCE

Sariá

Olympos○ — ○Di

Kárpathos ∩🏛

○ Apélla

Lefkós○

○Apéri

Arkása○ — Kâ... To...

Mene...

✈

Armáthia

○ Frý

Crete ↙

Kasos

Crete ↙

The domed entrance to the New Market in Rhodes town

GETTING AROUND

Kos, Rhodes and Kárpathos have international airports; those at Léros, Astypálaia and Kásos are domestic. Travelling by sea, it is wise to plan where you want to go, as some islands do not share direct connections even when quite close. Also journeys can be long – it takes nine hours from Rhodes to Pátmos. If possible allow time for changes in the weather. The cooling *meltémi* wind is welcome in the high summer but, if strong, can mean ferries will not operate and even leave you stranded. Bus services are good, especially on the larger islands, and there are always cars and bikes for hire or taxis available, though the standard of roads can vary.

LOCATOR MAP

An aerial view of Sými town with its Neo-Classical houses

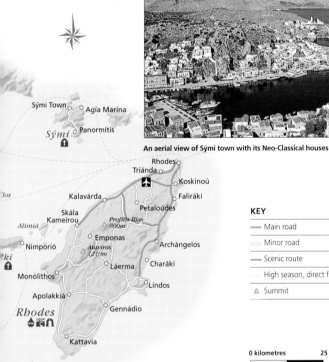

Sými Town — Agía Marína
Sými — Panormítis

los

Rhodes
Triánda — Koskinoú
Kalavárda — Faliráki
Skála Petaloúdes
Kameírou — *Profítis Ilías 800m*
Alimiá — Emponas
Nimporió *Atáviros 1210m* — Archángelos
ki — Láerma — Charáki
Monólithos — Líndos
Apolakkiá
Rhodes — Gennádio
Kattavía

KEY

— Main road

— Minor road

— Scenic route

--- High season, direct ferry route

△ Summit

0 kilometres 25

0 miles 15

Ro — Kastellórizo
Kastellórizo
Strongylí
Rhodes

SEE ALSO

Pátmos
Πάτμος

Known as the Jerusalem of the Aegean, Pátmos's religous significance dates from St John's arrival in AD 95 and the founding of the Monastery of St John *(see pp164–5)* in 1088. Monastic control declined as the islanders grew rich through ship-building and trade, and in 1720 the laymen and monks divided the land. Today Pátmos tries to maintain itself as a centre for both pilgrims and tourists.

Skála

Ferries, yachts and cruise ships dock at Skála, the island's port and main town, which stretches around a wide sheltered bay. As there are many exclusive gift shops and boutiques, Skála has a smart, up-market feel. There are several travel and shipping agencies along the harbourfront.

Skála's social life centres on the café-bar *Aríon*, a Neo-Classical building that doubles as a meeting place and waiting point for ferries. From the harbourfront caïques and small cruise boats leave daily for the island's main beaches.

Environs

The sandy town beach can get very crowded. To the north, around the bay, lies the shingly, shaded beach at **Melói**. There is an excellent campsite and taverna, and taxi boats also run back to Skála. Above Skála lie the ruins of the ancient acropolis at **Kastélli**.

KEY

For key to map see back flap

0 kilometres   2

0 miles   1

The remains include a Hellenistic wall. The little chapel of **Agios Konstantínos** is perched on the summit where the wonderful views at sunset make the hike up from Mérichas Bay well worthwhile.

Chóra

From Skála an old cobbled pathway leads up to the Monastery of St John *(see pp164–5)*. The panoramic views to Sámos and Ikaría are ample reward for the long trek. A maze of white narrow lanes with over 40 monasteries and chapels, Chóra is a gem of Byzantine architecture. Many of the buildings have distinctive window mouldings, or *mantómata*, decorated with a Byzantine cross. Along the twisting alleys, some doorways lead into vast sea captains' mansions, or *archontiká*, that were built to keep marauding

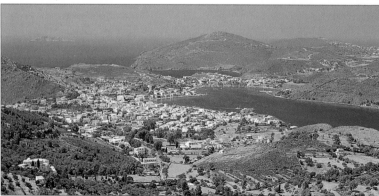

View of Skála from the Monastery of St John

Souvenirs on sale on the pathway to the Monastery of St John

pirates at bay. Down the path to Skála is the church of **Agía Anna**. Steps decked with flowers lead down from the path to the church (1090) which is dedicated to the mother of the Virgin Mary. Inside the church is the **Holy Cave of the Apocalypse**, where St John saw the vision of fire and brimstone and dictated the book of *Revelation* to his disciple, Próchoros. On view is the rock where the book of *Revelation* was written, and the indentation where the saint is said to have rested his head. There are 12th-century wall paintings and icons from 1596 of St John and the Blessed Christodoulos (*see p164*) by the Cretan painter Thomás Vathás. St John is said to have heard the voice of God coming from the cleft in the rock, still visible today. The rock is divided into three, symbolizing the Trinity.

Near Plateía Xánthou is an *archontikó*, **Simantíris House**, preserved as a Folk Museum. Built in 1625 by Aglaïnós Mousodákis, a wealthy merchant, it still has the original furnishings and contains objects from Mousodákis's travels, such as Russian samovars.

Nearby, the tranquil convent of **Zoödóchou Pigís**, built in 1607, has some fine frescoes and icons and is set in peaceful gardens.

🔒 Holy Cave of the Apocalypse
Between Skála and Chóra.
Tel 22470 31234. ☐ daily.

🏛 Simantíris House
Chóra. ☐ daily. 📷

Votive offerings from
pilgrims to Pátmos

Around the Island
Pátmos has some unspoiled beaches and a rugged interior with fertile valleys. Excursion boats run to most beaches and buses from Skála serve Kámpos, Gríkos and Chóra.

The island's main resort is **Gríkos**, set in a magnificent bay east of Chóra. It has a shingly beach with fishing boats, watersports facilities and a handful of tavernas. From here the bay curves past the uninhabited Tragonísi islet south to the bizarre Kallikatsoús rock, perched on a sand spit, which looks like the cormorant it is named after. The rock has

been hollowed out to make rooms, possibly by 4th-century monks, or it could have been the 11th-century hermitage mentioned in the writings of Christodoulos.

On the southwestern coast is the island's best beach, **Psilí Ammos**, with its stretch of fine sand and sweeping dunes. It is the unofficial nudist beach and is also popular with campers. Across the bay, the Rock of Genoúpas is marked by a red buoy. This is where, according to legend, the evil magician Genoúpas challenged St John to a duel of miracles. Genoúpas plunged into the sea to bring back effigies of the dead, but God then turned him to stone. Cape Genoúpas has a grotto that is said to be where the wizard lived.

Situated in the more fertile farming region in the north of the island, **Kámpos** beach, reached via the little hill-village of Kámpos, is another popular beach with watersports and a few tavernas. From Kámpos a track leads eastwards to the good pebble beaches at **Vagiá**, **Geranoú** and **Livádi**.

Windy **Lámpi** on the north coast is famous for its coloured and multipatterned pebbles. There are two garden tavernas and a little chapel set back from the reed-beds. You can walk here from the hamlet of Christós above Kámpos.

Holy Cave of the Apocalypse where St John lived and worked

Pátmos: Monastery of St John
Μονή του Αγίου Ιωάννου του Θεολόγου

The 11th-century Monastery of St John is one of the most important places of worship among Orthodox and Western Christian faithful alike. It was founded in 1088 by a monk, the Blessed Christodoulos, in honour of St John the Divine, author of the book of *Revelation*. One of the richest and most influential monasteries in Greece, its towers and buttresses make it look like a fairy-tale castle, but were built to protect its religious treasures, which are now the star attraction for the thousands of pilgrims and tourists.

Monastery of St John above Chóra

Chapel of John the Baptist

Kitchens

Inner courtyard

The Hospitality of Abraham
This is one of the most important of the 12th-century frescoes that were found in the chapel of the Panagía. They had been painted over but were revealed after an earthquake in 1956.

The monks' refectory has two tables made of marble taken from the Temple of Artemis, which originally occupied the site.

The Chapel of Christodoulos contains the tomb and silver reliquary of the Blessed Christodoulos.

★ **Icon of St John**
This 12th-century icon is the most revered in the monastery and is housed in the katholikón, *the monastery's main church.*

STAR FEATURES
★ Main Courtyard
★ Icon of St John

For hotels and restaurants in this region see pp309–12 and pp334–6

Chapel of the Holy Cross
This is one of the monastery's ten chapels built because church law forbade Mass being heard more than once a day in the same chapel.

VISITORS' CHECKLIST

Chóra, 4 km (2.5 miles) S of Skála. *Tel* 22470 31398.
Monastery & Treasury
⏱ 8am–1:30pm daily (4–6pm Tue, Thu & Sun). 🔲 treasury only.
ℹ www.monipatmou.gr

Chrysobull
This scroll of 1088 in the treasury is the monastery's foundation deed, sealed in gold by the Byzantine Emperor Alexios I Comnenos.

The treasury houses over 200 icons, 300 pieces of silverware and a dazzling collection of jewels.

★ Main Courtyard
Frescoes of St John from the 18th century adorn the outer narthex of the katholikón, *whose arcades form an integral part of the courtyard.*

The Chapel of the Holy Apostles lies just outside the gate of the monastery.

NIPTIR CEREMONY
The Orthodox Easter celebrations on Pátmos are some of the most important in Greece. Hundreds of people pack Chóra to watch the *Niptír* (washing) ceremony on Maundy Thursday. The abbot of the Monastery of St John publicly washes the feet of 12 monks, re-enacting Christ's washing of His disciples' feet before the Last Supper. The rite was once performed by the Byzantine emperors as an act of humility.

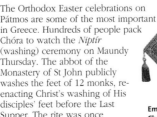

Embroidery of Christ washing the disciples' feet

The main entrance has slits for pouring boiling oil over marauders. This 17th-century gateway leads up to the cobbled main courtyard.

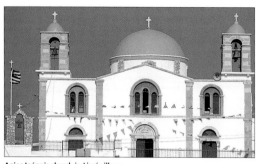

Agios Ioánnis church in Lipsí village

Lipsí
Λειψοί

🏠 700. 🚢 Lipsi town. ℹ️ Town hall, Lipsi (22470 41185). 🚌 Platýs Gialós 4 km (2.5 miles) N of Lipsí town.

Little Lipsí is a magical island characterized by green hills dotted with blue and white chapels, and village houses painted in a riot of colours. It is one of many islands claiming to be the enchanted place where Calypso beguiled Odysseus. Officially owned by the monastery at Pátmos since Byzantine times, Lipsí has excellent beaches, and is popular for day excursions from Pátmos and Kálymnos.

The island is only 10 sq km (4 sq miles) and remains a haven for traditional Greek island life, producing some good local wines and cheeses.

The main settlement, **Lipsí town** is based around the harbour. The blue-domed church of **Agios Ioánnis** holds a famous icon of the Panagía. Ancient lilies within the frame miraculously spring into bloom on 23 August, the feast of the Yielding of the Annunciation. In the town hall the **Nikofóreion Ecclesiastical Museum** features an odd collection of finds, from neatly labelled bottles of holy water to traditional costumes.

These sights are all signposted from the harbour, and there are informal taxi services to the more distant bays and beaches at **Platýs Gialós, Monodéntri** and the string of sandy coves at **Katsadiás**.

🏛️ **Nikofóreion Ecclesiastical Museum**
⭕ May–Sep: am only.

Léros
Λέρος

🏠 8,000. ✈️ Panthéni. 🚢 Lakkí, Agía Marína (hydrofoils). 🚌 Plateía Plátanos, Plátanos. ℹ️ Harbourfront, Lakkí (22470 22109).

Once famous as the island of Artemis, Léros's more recent history, as the home of Greece's prison camps and later mental hospitals, has kept tourism low-key. The hospitals still provide the main source of employment for the locals. However, life here is traditional, and the people are very welcoming and friendly.

Neo-Classical façade of Maliamate villa, Agía Marína

The island was occupied by the Knights of St John in 1309, by the Turks from 1522 to 1831, and by the Italians in 1912 when they built naval bases in Lakkí bay. Under German rule from 1943 until the Allied liberation, Léros was eventually united with Greece in 1948. When the military Junta took power in 1967 they exiled political dissidents to Léros's prison camps.

Today, Léros is keen to emphasise its strong cultural and educational heritage. Famous for its musicians and poets, the island has preserved traditional folk dance and music through Artemis, the youth cultural society.

Lakkí

Lakkí, the main port and former capital, has one of the best natural harbours in the Aegean, and served as an anchorage point in turn for the Italian, German and then the British fleets. Today it resembles a disused film set full of derelict Art Deco buildings, the remains of Mussolini's vision of a Fascist dream town. Lakkí is a ghost town during the day, but the seafront cafés come to life in the evening. Around the bay at Lépida, the former Italian naval base now houses the State Therapeutical Hospital and within the complex is a mansion once used as Mussolini's summer

---

THE ART DECO ARCHITECTURE OF LAKKI

Mussolini's vision of a new Roman Empire took shape here in 1923 when Italian architects and town planners turned their energies to building the new town. A quite remarkable example of Art Deco architecture, Lakkí was built around wide boulevards by the engineers Sardeli and Caesar Lois, an Austrian. The model town was all curves and featured a saucer-shaped market building with a clocktower, completed in 1936; a cylindrical Town Hall and Fascist centre, dating to 1933–34; and the vast Albergo Romana, later the Léros Palace Hotel. The Albergo, with the cinema and theatre complex, was completed in 1937 for visiting Italian performers. These days the majority of the buildings are crumbling and neglected.

Lakki's Art Deco cinema building

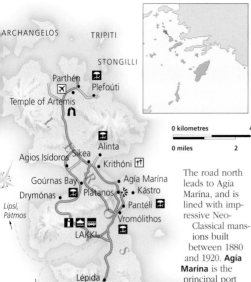

ARCHANGELOS   TRIPITI

STONGILLI

Parthéni   Plefoúti

Temple of Artemis

Alinta

Agios Isídoros   Síkea

Krithóni

Goúrnas Bay   Agía Marína

Drymónas   Plátanos   Kástro

Lipsí,   Pantéli

Pátmos   Vromólithos

LAKKÍ

Lépida

Kálymnos,   Palaiókastro   Xirókampos

Psérimos   PIGANOUSSA

0 kilometres   4

0 miles   2

KEY

For key to map see back flap

residence. Also in Lépida is the 11th-century church of **Agios Ioánnis Theológos** (St John the Divine), built over the remains of a Byzantine church by the monk Christodoulos *(see p164)*.

Around the Island

Léros is a pretty, green island with an indented coastline sweeping into vast gulfs, the "four seas" of Léros. With craggy hills and fertile valleys, it is good walking country.

To defy the Italians, the Lerians abandoned Lakkí and made the village of **Plátanos** the capital. Straddling a hilltop, its houses spill down to the little port of Pantéli and to the fishing village of Agía Marína.

Perched above Plátanos, the Byzantine kástro offers fine views. Renovated by the Venetians and the Knights of St John, it houses the church of **Megalóchari or Kyrá tou Kástrou** (the Madonna of the Castle) famous for its miraculous icon. Nearby Pantéli is a fishing village with a tree-fringed beach and harbour.

The road north leads to Agía Marína, and is lined with impressive Neo-Classical mansions built between 1880 and 1920. **Agía Marína** is the principal port for hydrofoils. Following the coastal road north to Krithóni, the **British War Cemetery** is a site of pilgrimage for those who lost relatives in the 1943 Battle of Léros.

Beaches line the road leading further north to **Álinda**, the island's main resort, which has a long beach with watersports and seafront cafés. Alinta's **Historic and Folk Museum** is housed in the twin-towered Belénis Castle, built by an expatriate benefactor, Paríssis Belénis. Little remains of the once-powerful Temple of

Artemis, now overlooking the airport at Parthéni in the north. There are a few carved blocks of stone and fragments of pillars. The goddess still has some influence in Léros, however, as property passes down the female family line.

Early Christian basilicas have been found in the area, and south of the airport the 11th-century church of **Agios Geórgios**, built by the monk Christodoulos *(see p164)* using temple columns, has a fresco of the saint.

Agios Isídoros, on the west coast above sandy Goúrnas Bay, has a white chapel on an islet that can be reached by means of a narrow causeway.

At Drymónas, with its coves and oleander gorge, is the church of the **Panagía Gourlomáta**, which translates as the "goggle-eyed Virgin". Reconstructed in 1327 from an 11th-century chapel, the church takes its name from the wide-eyed expression of the Madonna seen in one of its frescoes.

The resort of **Xirókampos**, lying in a bay to the south of the island, is overlooked by ancient Palaiókastro, the former site of the 3rd-century castle of Lépida. The huge Cyclopean walls remain, and within them is the church of Panagía, that is home to some fine mosaics.

🏛 Historic and Folk Museum
Belénis Castle, Alínda
◻ *May–Sep: daily.*

Plátanos village with the kástro in the background

Kálymnos
Κάλυμνος

Famous today as the sponge-fishing island, Kálymnos's history can be traced back to a Neolithic settlement in Vothýnoi, near Póthia; it was colonized after the 1450 BC devastation of Crete. The people have been known for their resilience since the 11th-century massacre by the Seljuk Turks, which a few survived in fortified Kastélli.

KEY

For key to map see back flap

| 0 kilometres | 5 |
| 0 miles | 3 |

Póthia

The capital and main port of the island is a busy working harbour. Wedged between two mountains, the town's brightly painted houses curve around the bay.

Póthia is home to Greece's last sponge fleet and there is a sponge-diving school on the eastern side of the harbour. The waterfront is lined with cafés and the main landmarks are the pink, domed Italianate buildings, including the old **Governor's Palace**, which now houses the market, and the silver-domed cathedral of **Agios Christós** (Holy Christ).

lavishly reconstructed and there is a collection of Neolithic and Bronze-Age finds from the island plus local memorabilia. The **Sponge Factory**, just off Plateía Eleftherías, has a complete history of sponges.

🏛 Archaeological Museum
Near Plateía Kýprou. **Tel** 22430 23113.
◯ Tue–Sun. ● main public hols.

🏛 Sponge Factory
Off Plateía Eleftherías. **Tel** 22430 28501. ◯ daily. 📷

Around the Island

Kálymnos is easy to get around with a good bus service to the villages and numerous taxis. This rocky island has three mountain ranges, the peaks offset by deep fjord-like inlets.

Northwest of Póthia the suburb of Mýloi, with its three derelict windmills, blends into **Chorió**, the pretty white town and former capital. On the way, standing to the left, is the ruined **Castle of the Knights**, and above, via steps from Chorió, is the citadel of **Péra Kástro**. Following a Turkish attack, this fortified village was inhabited from the 11th to the 18th century. It has good views and nine white chapels stand on the crags.

The **Cave of Seven Virgins** (Eptá Parthénon) shows traces of nymph worship. Legend has it that the seven virgins hid here from pirates, but disappeared in the bottomless channel below.

The main resorts on the island are strung out along the

This 19th-century cathedral has a reredos (screen) behind the altar by Giannoúlis Chalepás (*see p44*). The *Mermaid* at the harbour is one of 43 works that were donated to the island by local sculptors Irene and Michális Kókkinos.

The **Archaeological Museum**, housed in a Neo-Classical mansion, has been

The Mermaid at Póthia harbour

View of Póthia and harbour

The deep Vathý inlet with the settlement of Rína at its head

VISITORS' CHECKLIST

16,000. *Póthia.*
behind marketplace, *Póthia.*
*Plateía Taxi, Póthia (22430
59141). Póthia: Mon–Sat.
Easter celebrations around
island: Easter Sat; Sponge week
at Póthia: week following
Greek Easter.*

Caïques from Rína visit the **Daskalió Cave** in the side of the sheer inlet, and Armiés, Drasónia and Palaiónissos beaches on the east coast.

Outlying Islands
Excursion boats leave Póthia daily for **Psérimos** and the islet of **Nerá** with its Moní Stavroú. Psérimos has an often busy, sandy beach and a popular festival of the Assumption on 15 August.

Télendos, reached from Myrtiés, is perfect for a hide-away holiday, with a few rooms to rent and a handful of tavernas, plus shingly beaches. There are Roman ruins, a derelict fort and the ruined Moní Agíou Vasileíou, dating from the Middle Ages. The Byzantine castle of Agios Konstantínos also stands here.

west coast. The sunset over the islet of Télendos from **Myrtiés** is one of Kálymnos's most famous sights. Although Myrtiés and neighbouring Masoúri have now grown into noisy tourist centres, the Armeós end of Masoúri is less frenetic. To the north is the fortified **Kastélli**, the refuge of survivors from the 11th-century Turkish massacre. The coast road from here is spectacular, passing fish farms, inlets and the fjord-like beach at **Arginónta**. A visit to the northernmost fishing hamlet, **Emporeiós** makes a good day out and is in craggy walking country. You can walk to **Kolonóstilo** (the Cyclops Cave), which is named after its massive stalactites.

In the southeast is the most beautiful area of Kálymnos: the lush Vathý valley which has three small villages at the head of a stunning blue inlet. Backed by citrus groves, **Rína**, named after St Irene, is a pretty hamlet with a working boatyard. **Plátanos**, the next village, has a huge plane tree and the remains of Cyclopean walls. There is a 3-hour trail from here via **Metóchi**, the third Vathý village, across the island to Arginónta.

SPONGE FISHING AROUND KALYMNOS

Sea sponge

Kálymnos has been a sponge-fishing centre from ancient times, although fishing restrictions and sponge blight have hit the trade in recent years. Once in great demand, sponges were used for the Sultan's harem, for padding in armour and later for cosmetic and industrial purposes. Divers were weighed down with rocks or used crude air apparatus, and many men were drowned or died of the bends. The week before Kálymnos's fleet sets out to fish is the *Ipogros* or Sponge Week Festival. Divers are given a celebratory send off with food, drink and dancing in traditional costume.

A stone was used to weigh divers to keep them near to the seabed.

Diving equipment *varied greatly over the years. Early diving suits were made from rubber and canvas with huge helmets. You can see some on display in the sponge factory at Póthia and on stalls where divers sell their wares.*

This black-figure *Greek vase depicts an early sponge-diving scene. The diver, pictured standing at the front of the boat, is preparing to enter the sea to search for sponges. The vase dates back to around 500 BC.*

Kos
Κως

The second largest of the Dodecanese, Kos has a pleasant climate and fertile land, famous for producing the kos lettuce. Kos has attracted settlers since 3000 BC, and Hippocrates' teachings *(see p172)* increased the island's renown. By the 4th century BC Kos was a strong trading power, though it declined after the Romans arrived in 130 BC. The Knights of St John ruled from 1315, and the Turks governed from 1522–1912. Italian and German occupation followed until unification with Greece in 1948.

Mastichári
Limnióna
Kéfalos
Agios Ioánnis
Theológos
Agios Stéfanos
Astypálaia
Kamári
Aspri Pétra
Paradise Beach
Antimácheia 7
Antima
Moni Agiou Ioánni
Kardáma
Chelóna

0 kilometres   5
0 miles   3

Yachts moored in the harbour at Kos town

Kos Town 1
Κως

15,000. Aktí Koudourio-tou. Vasiléos Georgíou 1 (22420 28724). daily. Kos town.

Dominated by its Castle of the Knights, old Kos town was destroyed in the 1933 earthquake. This revealed many ancient ruins which the Italians excavated and restored.

The harbour bristles with yachts and excursion boats, and pavement cafés line the street. At night in high season you can almost get swept along by the crowds. There are palm trees, pines and gardens full of jasmine. Ancient and modern sit oddly side by side: Nafklírou, the "street of bars", runs beside the ancient agora,

at night lit up by strobes and lasers. Hippocrates' ancient plane tree, in Plateía Platánou, is said to have been planted by him 2,400 years ago. Despite its 14-m (46-ft) diameter the present tree is only about 560 years old and is

The water fountain near Hippocrates' plane tree

For hotels and restaurants in this region see pp309–12 and pp334–6

KEY

For key to map see back flap

probably a descendent of the original. The nearby fountain was built in 1792 by the Turkish governor Hadji Hassan, to serve the Mosque of the Loggia. The water gushed into an ancient marble sarcophagus.

♣ Castle of Knights
Platánou. **Tel** 22420 27927. Jun–Sep: 8am–7:30pm; Oct–May: 8am–2:30pm.
The 16th-century castle gateway is carved with gargoyles and an earlier coat of arms of Fernández de Heredia, the Grand Master from 1376 to 1396. The outer keep and battlements were built between 1450 and 1478 from stone and marble, including blocks from the Asklepieíon *(see p168)*. The fortress was an important defence for the Knights of Rhodes against Ottoman attack

♦ Ancient Agora
South of Plateía Platánou.
This site is made up of a series of ruins; from the original Hellenistic city to Byzantine buildings. Built over by the Knights, the ancient remains were revealed in the 1933 earthquake. Highlights include

the 3rd-century BC stoa Kamára tou Fórou (Arcade of the Forum), the 3rd-century 3C Temple of Herakles, mosaic floors depicting Orpheus and Herakles, and ruins of the Temple of Pándemos Aphrodite. A 5th-century Christian basilica was also discovered, along with the Roman Agora.

Archaeological Museum
Plateía Eleftherías. **Tel** 22420 28326. Tue–Sun.
The museum has an excellent collection of the island's Hellenistic and Roman finds, including a 4th-century BC marble statue of Hippocrates. The main hall displays a 3rd-century AD mosaic of Asklepios surrounded by 2nd-century statues of Dionysos with Pan and a satyr. The east wing exhibits Roman statues and the north Hellenistic finds, while the west room has later gigantic statuary.

Roman Remains
Grigoríou E. Tue–Sun.
The most impressive of these ruins is the Casa Romana (closed until 2007), built in the Pompeiian style. It had 26 rooms and three pools surrounded by shady court-yards lined with Ionian and Corinthian columns. There are mosaics of dolphins, lions and leopards. The dining room has decorated marble walls and several rooms are painted. In the grounds are the excavated thermal baths and part of the main Roman road, covered with ancient capitals and Hellenistic fragments. Set back off the road down an avenue of cypresses

VISITORS' CHECKLIST

31,000. 27 km (16 miles) W of Kos town. Aktí Koudouriótou, Kos town. Kos town. Kos town (22420 28724, kosinfo@ kos.forthnet.gr). Hippocrates Cultural Festival: Jul–Sep; Panagía at Kardámaina: 8 Sep; Agios Geórgios Festival at Palaió Pylí: 23 Apr.

Kos lettuce on a market stall in Plateía Eleftherías

is the ancient odeion or theatre. It has rows of marble benches (first class seats) and limestone blocks for the plebeians.

The western excavations opposite reveal a mix of historical periods. There are Mycenaean remains, a tomb dating from the Geometric period and Roman houses with some fine mosaics. One of the most impressive sights is the gym or *xystó* with its 17 restored Doric pillars.

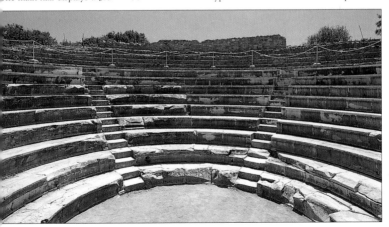

Rows of marble benches for the Roman audiences that came to the ancient odeion

Around Kos Island

Carving at the Asklepieíon

Mainly flat and fertile, Kos is known as the "Floating Garden". It has a wealth of archaeological sites and antiquities, Hellenistic and Roman ruins, and Byzantine and Venetian castles. Most visitors, however, come for Kos's sandy beaches. Those on the southwest shore are some of the finest in the Dodecanese, while the northwest bays are ideal for watersports. Much of the coast has been developed, but inland you can still see remnants of Kos's traditional lifestyle.

The seven restored columns of the Temple of Apollo at the Asklepieíon

Asklepieíon ❷

Ασκληπιείο

4 km (2.5 miles) NW of Kos town.
Tel *22420 28763.* *Jul–Oct 8am–7pm Tue–Sun; Nov–Jun: 8:30am–3pm Tue–Sun.*

With its white marble terraces cut into a pine-clad hill, the Asklepieíon site was chosen in the 4th century BC for rest and recuperation and still exudes an air of tranquillity. The views from the sanctuary are breathtaking and it is one of Greece's most important Classical sites.

Temple, school and medical centre combined, it was built after the death of Hippocrates and was the most famous of ancient Greece's 300 asklepieia dedicated to Asklepios, god of healing. The doctors, priests of Asklepiados, became practitioners of Hippocrates' methods. The cult's symbol was the snake, once used to seek healing herbs, and is the emblem of modern western medicine. There are three levels: the lowest has a 3rd-century BC porch and 1st-century AD Roman baths; the second has a 4th-century BC Altar of Apollo and a 2nd to 3rd-century AD Temple of Apollo; on the third level is the Doric Temple of Asklepios from the 2nd century BC.

Asfendíou Villages ❸

Χωριά Ασφενδίου

14 km (9 miles) W of Kos town.

The Asfendíou villages of Zía, Asómatos, Lagoúdi, Evangelístria and Agios Dimítrios are a cluster of picturesque hamlets on the wooded slopes of Mount Dikaíos. These mountain villages have managed to retai their traditional character, with whitewashed houses and attractive Byzantine churches. The highest village, **Zía**, has become the epitome of a traditional Greek village, at least to the organizers of the many coach tours that regularly descend upon it. The more adventurous traveller can take the very rough track from the Asklepieíon via tiny Asómato to Zía. The lowest village, **Lagoúdi**, is less commercialized and a road leads from here to Palaió Pylí.

Tigkáki ❹

Τιγκάκι

12 km (7 miles) W of Kos town.
Tigkáki.

The popular resorts of Tigkák and neighbouring Marmári have long white sand beache ideal for windsurfing and othe watersports. Boat trips are available from Tigkáki to the island of Psérimos opposite. The nearby **Alykés Saltpans** are a perfect place for birdwatching. The many wetland species here include small waders like the avocet, and the black-winged stilt with its long pink legs.

HIPPOCRATES

The first holistic healer and "father of modern medicine", Hippocrates was born on Kos in 460 BC and died in Thessaly in about 375 BC. He supposedly came from a line of healing demigods and he learned medicine from his father and grandfather: his father was a direct descendant of Asklepios, the god of healing, his mother of Herakles. He was the first physician to classify diseases and introduced new methods of diagnosis and treatment. He taught on Kos before the Asklepieíon was established, and wrote the Hippocratic Oath, to cure rather than harm, still sworn by medical practitioners worldwide.

Palaió Pylí castle perched precariously on a cliff's edge

Palaió Pylí 🟤
Παλαιό Πυλί

5 km (9 miles) W of Kos town.
 to Pylí.

The deserted Byzantine town of Palaió Pylí is perched on a crag 4 km (2 miles) above the farming village of Pylí, with the remains of its castle walls built into the rock. Here the Blessed Christodoulos built the 11th-century church of the Ypapandís (Presentation of Jesus), before he went to Pátmos (see p160). In Pylí lies the Classical *thólos* tomb of the mythical hero-king Chármylos It has 12 underground crypts, which are now surmounted by the church of Stavrós.

Kardámaina 🟤
Καρδάμαινα

26 km (16 miles) SW of Kos town.
 Kardámaina.

Once a quiet fishing village noted for its ceramics, Kardámaina is the island's biggest resort – brash, loud and packed with young British and Scandinavian tourists. It has miles of crowded golden sands and a swinging nightlife. It is quieter further south with some exclusive developments. Sights include a Byzantine church and the remains of a Hellenistic theatre.

Antimácheia 🟤
Αντιμάχεια

25 km (16 miles) W of Kos town.

The village of Antimácheia is dominated by its Venetian castle and windmills. The castle, located near the airport, was built by the Knights of Rhodes (see pp184–5) as a prison in the 14th century, and was constantly bombarded by pirates. Its massive crenellated battlements and squat tower now overlook an army base, and there are good views towards Kardámaina. The inner gateway still bears the coat of arms of the Grand Master Pierre d'Aubusson (1476–1503) and there are two small chapels within the walls.

Antimácheia castle battlements

Environs
The road north from Antimácheia leads to the charming port of **Mastichári**. There are good fish tavernas here and a long sandy beach that sweeps into dunes at the western end. On the way to the dunes, the ruins of an early Christian basilica, with good mosaics, can be seen.

Kamári 🟤
Καμάρι

15 km (9 miles) SW of Kos town.
 Paradise 7 km (4 miles) E.

Kamári is a good base for exploring the southwest coast, where the island's best beaches can be found. Mostly reached via steep tracks from the main road, the most famous is Paradise beach with fine white sands. Kamári beach leads to the 5th-century AD Christian basilica of Agios Stéfanos which has mosaics and Ionic columns.

Environs
Kéfalos, on the mountainous peninsula inland from Kamári, is known for its thyme, honey and cheeses. Sights include the ruined Castle of the Knights, said to be the lair of a dragon. According to legend, Hippocrates' daughter was transformed into a dragon by Artemis, and awaits the kiss of a knight to resume human form. Above Kéfalos is the windmill of Papavasílis, and nearby at Palátia are the remains of **Astypálaia**, the birthplace of Hippocrates. Neighbouring Aspri Pétra cave has yielded remains. The journey to **Moní Agíou Ioánni**, 6 km (4 miles) south of Kéfalos, passes through dramatic scenery, and a track leads to the beach of Agios Ioánnis Theológos.

Music bars and clubs in the resort of Kardámaina

Chóra overlooking Astypálaia's main harbour, Skála

Astypálaia
Αστυπάλαια

🏠 1,200. ✈ 11 km (7 miles) E of Astypálaia town. 🚢 🚌 Astypálaia town. 🛈 near Kástro, Astypálaia town (22430 61778).

With its dazzling white fortified town of Chóra and its scenic coastline, the island of Astypálaia retains an exquisite charm. A backwater in Classical times, Astypálaia flourished in the Middle Ages when the Venetian Quirini family ruled from 1207 to 1522.

The most westerly of the Dodecanese, it is a remote island with high cliffs and a hilly interior. There are many coves and sandy bays along the coast, which was once the lair of Maltese pirates.

Astypálaia town incorporates the island's original capital, Chóra, which forms its maze-like upper town. The splendid Venetian kástro of the Quirini family is on the site of the ancient acropolis. Houses were built into the kástro's walls for protection, and the Quirini coat of arms can still be seen on the gateway. Within its walls are two churches: the silver-domed, 14th-century Panagía Portaïtissa (Madonna of the Castle Gates), and the 14th-century Agios Geórgios (St George), built on the site of an ancient temple.

A two-hour hike westwards from the derelict windmills above Chóra leads to **Agios Ioánnis** and its gushing waterfall. **Livádi**, the main resort, lies south of Chóra in a fertile

valley with citrus groves and cornfields. It has a long beach. The nudist haunt of **Tzanáki** lies a short distance to the south. From Livádi a dirt track leads north to **Agios Andréas**, a remote and attractive cove, an hour and a half's trek away.

North of Chóra, on the narrow land bridge between the two sides of the island, lies **Maltezána** (also known as Análipsi), the fastest-growing resort on the island. Named after the marauding pirates who once frequented it, Maltezána was where the French Captain Bigot set fire to his ship in 1827 to prevent it being captured.

On the northeastern peninsula is the "lost lagoon", a deep inlet at the hamlet of **Vathý**. From here you can visit the caves of Drákou and Negrí by boat, or the Italian Kastellano fortress, built in 1912, 3 km (2 miles) to the south.

A typical housefront in Mandráki on Nísyros

Nísyros
Νίσυρος

🏠 1,000. ✈ 🚢 🚌 Mandráki harbour. 🛈 22420 31203. ☎ Gialiskári 2 km (1 mile) E of Mandráki; Páloi 4 km (2 miles) E of Mandráki. **www**.nisyros.gr

Almost circular, Nísyros is on a volcanic line which passes through Aígina, Póros, Mílos and Santoríni. In 1422 there was a violent eruption and its 1,400-m (4,593-ft) high peak exploded, leaving a huge caldera *(see p176)*. Everything flourishes in the volcanic soil and there is some unique flora and fauna.

According to mythology, Nísyros was formed when the enraged Poseidon threw a chunk of Kos on the warring giant, Polyvotis, who was submerged beneath it, fiery and fuming. In ancient times, it was famous for its millstones, often known as the "stones of Nísyros". Now the island prospers from pumice mining on the islet of Gyalí.

Mandráki
Boats dock at Mandráki, the capital, with quayside tavernas, ticket agencies and buses shuttling visitors to the volcano. Mandráki's narrow two-storey houses have brightly painted wooden balconies, often hung with strings of drying tomatoes and onions. A maze of lanes congregates at Plateía Iróön, with its war memorial. Other roads weave south, away from the sea, past the *kípos* (public

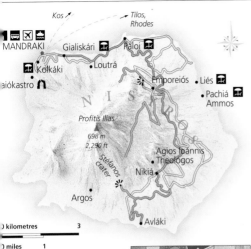

KEY

For key to map see back flap

orchard) to the main square,
Plateía Ilikioménon. At night,
the area is bustling: shops that
resemble houses are open,
with traditional painted signs
depicting their wares. The
lanes become narrow and
more winding as you
approach the medieval
Chóra district. In the nearby
Langádi area, the balconies
on the houses almost touch
across the street.

The major attractions in
Mandráki are the 14th-century
kástro and the monastery. The
former is the castle of the
Knights of St John (*see
pp188–9*), built in 1325 high
up the cliff face. The monastery, **Moní Panagías Spilianís**,
lies within the kástro and
dates from around 1600.
Inside, a finely carved iconostasis holds a Russian-style
icon, decked in gold and silver
offerings, of the Virgin and
Child. The fame of the church
grew after Saracens failed to
find its treasure of silver, hidden by being worked into the
Byzantine icons. The library
holds rare editions and a number of ecclesiastical treasures.

**The main square in Nikiá with
its *choklákia* mosaic**

The **Historical and Folk
Museum**, on the way up to the
kástro, has a reconstructed
traditional island kitchen,
embroideries and a small collection of local photographs.

Excursion boats offer trips
from Mandráki to **Gyalí** and
the tiny **Agios Antónios** islet
beyond. Both destinations
have white sandy beaches.

**⌂ Historical and Folk
Museum**
Kástro. ☐ *May–Sep: daily.*

Around the Island

Nísyros is lush and green with
terraces of olives, figs and
almond trees contrasting with
the strange grey and yellow
moonscape of the craters.
No visit would be complete
without an excursion to the
volcano and by day the island
is swamped with visitors
from Kos. However, it is
quiet when the excursion
boats have left.

Above Mandráki lies the
Palaiókastro, the acropolis
of ancient Nísyros, dating
back 2,600 years. Remains
include Cyclopean walls
made from massive blocks
carved from the volcanic
rock, and Doric columns.

Nísyros is pleasant for
walking. Visits to the volcano
must include the pretty village
of **Nikiá** (*see p176*), with its
choklákia mosaic in the round
"square", and abandoned
Emporeiós which clings to
the rim of the crater.

To the east of Mandráki,
Páloi is a pretty fishing village
with good tavernas and a string
of dark volcanic sand beaches.
Two kilometres (1 mile) west
of the village, at **Loutrá**, an
abandoned spa can be found.

The *meltémi* wind blows
fiercely on Nísyros in high
season, and the beaches
east of Páloi can often be
littered with debris.

View of Mandráki, the capital of Nísyros

The Geology of Nísyros

Crystals in a steam vent

Fuming and smelling of rotten eggs, the centre of Nísyros is a semi-active caldera – a crater formed by an imploded mountain. Its eruption, around 24,000 years ago, was accompanied by an outpouring of pumice, forming a blanket 100 m (328 ft) thick on the upper slopes of the island. When formed, the caldera was 3 km (2 miles) in diameter. It is now occupied by two craters and five solidified lava domes, forced upwards in the last few thousand years, including Profítis Ilías, the largest in Europe. Further eruptions in 1873 built cones of ash 100 m (328 ft) high.

Steep paths *descend to the crater floor, where the surface is hot enough to melt rubber-soled shoes. Gas vents let off steam, at 98° C (208° F), which bubbles away beneath the earth's crust.*

Paths lead visitors around the caldera.

Profítis Ilías dome is almost 700 m (2,300 ft) high.

Ash cones have been produced in the recent life of the caldera.

Original caldera wall

Lava dome

The Stéfanos crater, which is 300 m (985 ft) wide and 25 m (82 ft) deep, was created by an explosion of pressurized water and superheated steam.

NISYROS CALDERA

This huge caldera contains several water-filled mini craters. The largest is the still-active Stéfanos crater, which has a number of hot springs, boiling mud pots and gas vents. There is a stench of sulphur and numerous pure sulphur crystals are eagerly snapped up by would-be geologists.

Nikiá *is the more appealing of Nísyros's two rim villages with its brightly painted houses and* choklákia *pebble mosaics. There are good views from Nikiá of the crater, and a path down to the caldera.*

The oldest volcanic *minerals found on Nísyros date back 200,000 years. There are vast amounts of pumice around the caldera and rich deposits of sulphur and kaoline.*

Sulphur

Kaoline

Pumice

Tílos
Γήλος

🏛 500. 🚢 🚌 Livádia. 🛈 Megálo Choríó (22460 44212). 🚗 Eristós 10 km (6 miles) NW of Livádia.

Remote Tílos is a tranquil island, with good walking and, as a resting stop on migration paths, it offers rich rewards for birdwatchers. Away from the barren beaches, Tílos has a lush heartland, with small farms growing everything from tobacco to almonds. Its hills are scattered with chapels and ruins of Crusader castles, outposts of the Knights of St John, who ruled from 1309 until 1522.

There is a strong tradition of music and poetry on the island – the poet Erinna, famous for the *Distaff*, was born here in the 4th century BC. In the 18th and 19th centuries Tílos was known for weaving cloth for women's costumes, still worn by some islanders today.

Livádia

Livádia, the main settlement, has a tree-fringed pebble beach sweeping round its bay. The blue and white church of **Agios Nikólaos** dominates the waterfront, and has an iconostasis carved in 1953 by Katasáris from Rhodes. On the beach road, the tiny, early Christian basilica of **Agios Panteleïmon kai Polýkarpos** has an attractive mosaic floor.

The pebble beach at Livádia

Around the Island

Buses run from Livádia to Megálo Choríó and Erystos, and mopeds can be hired; otherwise you are on foot.

Built on the site of the ancient city of Telos, **Megálo Choríó** is 8 km (5 miles) uphill from Livádia. The kástro was built by the Venetians who incorporated a Classical gateway and stone from the ancient acropolis. The **Palaeontological Museum** has midget fossilized mastodon (elephant) bones from the Misariá region, and a gold treasure trove, found in a Hellenistic tomb in the Kená region of the island.

Detail of the War Memorial at Livádia

The church of Archángelos Michaíl (1827) was built against the kástro walls. It has silver icons from the original Taxiárchis church, a gilded 19th-century iconostasis and the remains of 16th-century frescoes.

South of Megálo Choríó lies **Erystos**, a long sandy beach. **Agios Antónis** beach to the west of Megálo Choríó has the petrified remains of human skeletons. These "beach rocks" are thought to be of sailors caught in the lava when Nísyros erupted in 600 BC.

Perched on a cliff on the west coast, the Byzantine **Moní Agíou Panteleïmonos** is the island's main sight. In a cluster of trees, this fortified monastery with red pantiled roofs is famous for its sunset views. Built in 1470 it has circular chapels, a mosaic courtyard and medieval monks' cells. The dome of the church has a vision of *Christ Pantokrátor* (1776) by Gregory of Sými. Other important artifacts include 15th-century paintings of Paradise and the apostles, and a carved iconostasis that dates from 1714.

The fossilized bones of mini mastodons from 7000 BC were discovered in the **Charkadió Grotto**, a ravine in the Misariá area. The ruined fortress of Misariá marks the spot.

Mikró Choríó, below Misariá, has about 220 roof-less, abandoned houses. Those residents who had stone roofs took them with them to Livádia when the population abandoned the village in the 1950s. Quiet during the day, at night the ruins are illuminated, and one house has been restored as a bar. There is also the mid-17th-century church of **Timía Zóní**, which has 18th-century frescoes, and the chapels of **Sotíros, Eleoúsas** and **Prodrómou**, with 15th-century paintings.

🏛 **Palaeontological Museum**
Megálo Choríó. ⬤ daily; request key at town hall.

One of many almond orchards on Tílos

Sými
Σύμη

Ever since classical times, rocky, barren Sými
has thrived on the success of its
sponge-diving fleet and boat-building
industry, which once launched 500
ships a year. By the 17th century it
was the third-richest island in the
Dodecanese. The Italian
occupation in 1912 and
the arrival of artificial
sponges and steam

*A prayer in a
bottle at Moní
Taxiárchi*

power ended Sými's
good fortunes. Its popu-
lation had fallen from
23,000 to 6,000 by World War
II, and the mansions built
in its heyday crumbled.

Sými Town

The harbour area, Gialós, is
one of the most beautiful in
Greece, surrounded by Neo-
Classical houses and elaborate
churches built on the hillside.
Gialós is often busy with day
trippers, particularly late
morning and early afternoon.

  A clock tower (1884) stands
on the western side of the harbour where the ferries dock;
beyond is the shingle bay of
Nos beach. Next door to the
town hall, the **Maritime
Museum** has an interesting
record of Sými's seafaring past.

  Gialós is linked to the upper
town, Chorió, by a road and
also by 375 marble steps.
Chorió comprises a maze of
lanes and distinctive houses,
often with traditional interiors.
The late 19th-century church
of **Agios Geórgios** has an
unusual pebble mosaic of
fierce mermaids who, in

Greek folklore, are responsible for storms that sink ships.
The **Sými Museum**, high up
in Chorió, has a small but interesting collection of costumes
and traditional items. Beyond
the museum is the ruined Byzantine **kástro** and medieval
walls. Megáli Panagía church,
the jewel of the kástro, has an
important post-Byzantine icon

of the *Last Judgment*, from
the late 16th century, by the
painter Geórgios Klontzás.

Maritime Museum
Plateía Ogdóis Maḯou. **Tel** 22460
72363. ☐ Apr–Oct: daily.
☐ Nov–Apr.

Sými Museum
Chorió. **Tel** 22460 71114.
☐ Tue–Sun. ☐ main public hols.

Map labels

Tilos
NIMOS
Rhodes
Agía Marí
Empoërós
Nos
Noúlia
Moní Agíou
Michaḯl
Roukouniótí
SÝMI
TOWN
250 m
820 ft
Agi
Nik
Pédi
Agios Aimilianós
Agio
Geó
Diss
Cape
Kefála
Agios Vasílios
Nanoú
PIDIMA
GIALESINO
MEGALONISI
Marathoú
Panormítis
Moní Taxiárchi
Michaḯl
Panormíti
TEFTLOYSA

KEY
For key to map see back flap

0 kilometres 4
0 miles 2

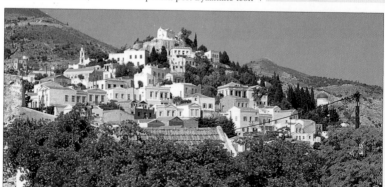
The pastel-coloured houses of Chorió on the ancient acropolis overlooking Sými's harbour

For hotels and restaurants in this region see pp309–12 and pp334–6

The traditional craft of boat building in Sými town

Environs

The road from Gialós to Chorió passes the hill of **Noúlia**, also known as Pontikókastro. On the hill are the remains of 20 windmills and an ancient tomb monument believed to have been erected by the Spartans in 412–411 BC.

Around the Island

Sými's road network is limited but there are plenty of tracks over its rocky terrain. East of Sými town, an avenue of eucalyptus trees leads down through farmland to **Pédi** bay, a beach popular with local families. From here taxi boats run to **Agios Nikólaos** beach and there are paths to Agios Nikólaos and **Agía Marína**.

The 18th-century church of **Moní Agíou Michaíl Roukou-nióti**, 3 km (2 miles) west of Sými town, is built like a desert fortress in Gothic and folk architecture. It houses 14th-century frescoes and a rare 15th-century, semicircular icon of the *Hospitality of Abraham* by Cretan artist Stylianós Génis.

Sými's most popular sight is **Moní Taxiárchi Michaíl Panormíti** in Panormítis bay, a place of pilgrimage for Greek sailors worldwide. Its white buildings, spanning the 18th to 20th centuries, line the water's edge. The pleasant horseshoe-shaped harbour is dominated by the elaborate mock-Baroque bell tower, a 1905 copy of the famous bell tower of Agía Foteiní in Izmir.

The monastery is famous for its icon of the Archangel Michael, Sými's patron saint and guardian of seafarers. Despite being removed to Gialós, it mysteriously kept

returning to Panormítis so the monastery was founded here. The single-nave *katholikón* was built in 1783 on the remains of an early Byzantine chapel also dedicated to the saint.

According to tradition, if you ask a favour of St Michael, you must vow to give something in return. As a result, the interior is a dazzling array of

The mock-Baroque belltower of Moní Taxiárchi Michaíl Panormíti

votive offerings, or *támata*, from pilgrims, including small model ships in silver and gold.

The intricate Baroque iconostasis by Mastrodiákis Taliadoúros is a remarkable piece of woodcarving. The walls and ceiling are covered in smoke-blackened 18th-century frescoes by the two Sýmiot brothers Nikítas and Michaíl Karakostís.

The sacristy museum is full of treasures, including a post-Byzantine painting of the ten saints, Agioi Déka, by the Cretan Theódoros Poulákis. There are prayers in bottles, which have floated miraculously into Panormítis, containing money for the monastery from faithful sailors. The cloister has a *choklákia* courtyard of zigzag pebble mosaics *(see p198)* and an arcaded balcony.

West of the monastery, past the taverna, is a memorial to the former abbot, two monks and two teachers executed by the Germans in 1944 for running a spy radio for British commandos. Small Panormítis beach is here and there are woodland walks to **Marathoúnta**.

Moní Taxiárchi Michaíl Panormíti
Panormítis bay. Tue–Sun.

THE TREATY OF THE DODECANESE

A plaque outside Les Katerinettes Restaurant, on the quayside in Gialós, marks the end of Nazi occupation on 8 May 1945, when the islands were handed over to the Allies at the end of World War II. The islands officially became part of Greece on 7 March 1948, having been under Italian rule since 1912. Further along the quayside a bas-relief of an ancient trireme commemorates the liberation of the islands. It is a copy of an original at the base of the Acropolis at Líndos, on Rhodes island *(see pp196–7)*.

The bas-relief of a trireme on the quayside at Sými town

Rhodes

Ρόδος

Rhodes, the capital of the Dodecanese, was an
important centre in the 5th to 3rd centuries BC.
It was part of both the Roman and Byzantine empires,
before being conquered by the Knights of St John.
They occupied Rhodes from 1306 to 1522, and their
medieval walled city still dominates Rhodes town.
Ottoman and Italian rulers followed. Fringed by sandy
beaches, and with good hiking and lively nightlife,
Rhodes attracts thousands of tourists each year.

KEY

For key to map see back flap

Ancient Kámeiros
*The stunning ruins of this
once-thriving Doric city
include a 6th-century
BC Temple of Athena
Polias* **5**

Skála Kameírou
*A pleasant place to relax, Skála Kameírou
is an attractive harbour that once served
the ancient city of Kámeiros* **6**

Kritinía castle, built by the
Knights of Rhodes, was one of
their larger strongholds (see p193).

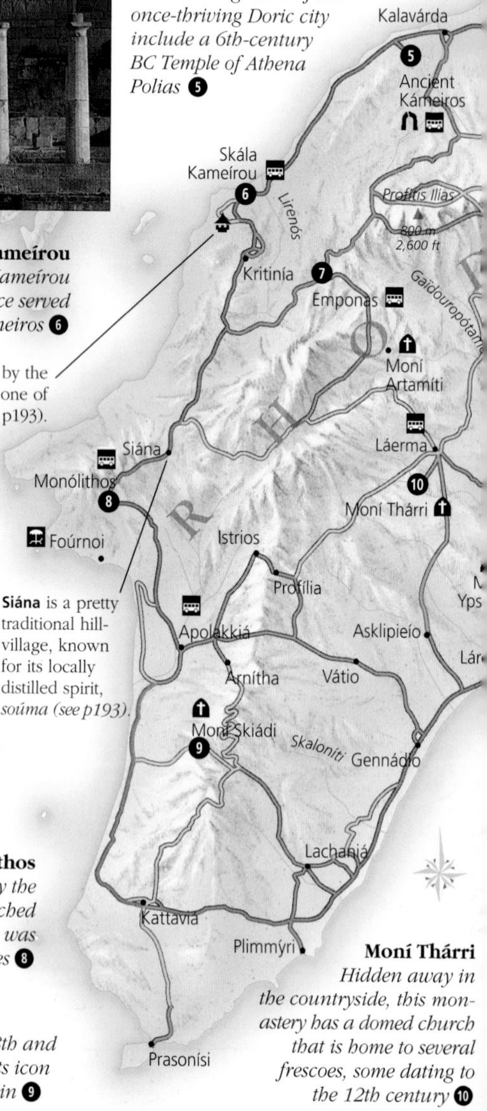

Siána is a pretty
traditional hill-
village, known
for its locally
distilled spirit,
soúma (see p193).

Emponas
*The slopes around this traditional
town have been cultivated with
vines by the Emery winery
since the 1920s* **7**

Monólithos
*The village is dominated by the
15th-century castle, perched
high on a massive rock. It was
built by the Knights of Rhodes* **8**

Moní Skiádi
*This monastery was built in the 18th and
19th centuries and is famous for its icon
of the Panagía, or the Blessed Virgin* **9**

Moní Thárri
*Hidden away in
the countryside, this monastery has a domed church
that is home to several
frescoes, some dating to
the 12th century* **10**

Kalavárda

Ancient
Kámeiros **5**

Profítis Ilías
800 m
2,600 ft

Skála
Kameírou **6**

Líndenós

Kritinía **7**

Emponas

Gaidouropótama

Moní
Artamíti

Láerma

Moní Thárri **10**

Siána

Monólithos **8**

Foúrnoi

Istríos

Profília

Apolakkiá

Asklipieío

Arnítha

Vátio

Láre

Moní Skiádi **9**

Skaloníti

Gennádio

Lachaniá

Kattavía

Plimmýri

Prasonísi

Petaloúdes
Called butterfly valley, this tranquil place is, in fact, home to thousands of moths during the summer ④

Moní Filerímou
The monastery is set on the beautiful hillsides of Mount Filérimos. The main church dates back to the 14th century ③

Chálki, Piraeus, Astypálaia

Sými, Kos

Kastellórizo

RHODES TOWN ①

Triánda

Paradísi

Moní Filerímou

Ancient Ialyssós ②

Réni Koskinoú

Koskinoú ⑮

Thérmes Kalithéas

Kalithéa

Kalythiés

Faliráki ⑭

Petaloúdes

Psínthos

Ladikó Bay

Loútari

Afántou

Kolympia

Eptá Pigés ⑬

Moní Tsampíkas

Tsampíka

Archángelos ⑫

Stégna

Charáki

Líndos ⑪

Péfkoi

Ancient Ialyssós
Set on a plateau with commanding views, this ancient site dates back to 2500 BC. The ruins include remains of a 3rd-century BC acropolis ②

Faliráki
This fun-packed resort offers all sorts of nightlife and watersports, and is particularly popular with the young ⑭

★ **Rhodes Town**
Mandráki harbour is at the centre of Rhodes town, which is one of Greece's most popular tourist destinations ①

Koskinoú
This small village offers visitors the opportunity to see traditional Rhodian houses and choklákia pebble mosaics (see p198) ⑮

Faraklós was once used by the Knights of Rhodes as a prison. Today it overlooks Charáki village (see pp194–5).

Eptá Pigés
This is an enchanting beauty spot that takes its name from the "seven springs" that are the source for the area's central reservoir ⑬

Archángelos
A popular place to visit, Archángelos is set in attractive countryside, and maintains a tradition of handicraft production ⑫

| kilometres | 10 |
| miles | 6 |

★ **Líndos**
One of the island's most visited sites, the acropolis at Líndos towers over the town from its clifftop position ⑪

Street-by-Street: Rhodes Old Town ❶
Παλιά Πόλη Ρόδου

The town of Rhodes has been inhabited for more than 2,400 years. A city was first built here in 408 BC, and when the Knights of St John arrived in 1309 they built their citadel over these ancient remains. The Knights' medieval citadel, dominated by the towers of the Palace of the Grand Masters, forms the centre of the Old Town. The new town *(see pp190–91)* lies beyond the original walls. Of the walls' 11 gates, Koskinoú (St John's) gate, which leads into the Bourg quarter *(see p185)*, has the best view of the city's defences.

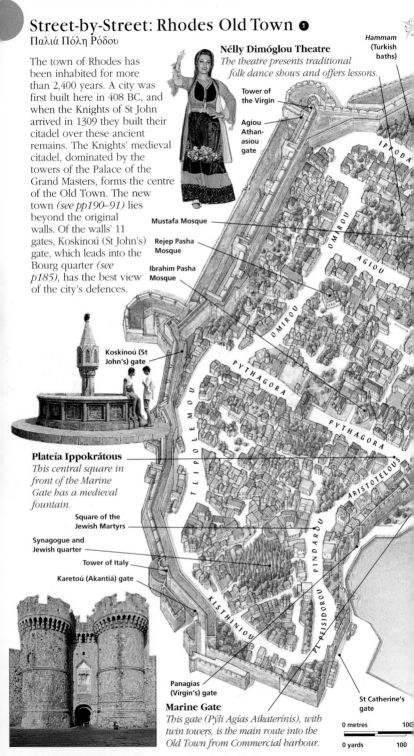

Hammam (Turkish baths)

Nélly Dimóglou Theatre
The theatre presents traditional folk dance shows and offers lessons.

Tower of the Virgin

Agiou Athanasiou gate

Mustafa Mosque

Rejep Pasha Mosque

Ibrahim Pasha Mosque

IPPODA

OMIROU

AGIOU

OMIROU

PYTHAGORA

PYTHAGORA

TLIPOLEMOU

ARISTOTELOUS

Koskinoú (St John's) gate

Plateía Ippokrátous
This central square in front of the Marine Gate has a medieval fountain.

Square of the Jewish Martyrs

Synagogue and Jewish quarter

Tower of Italy

Karetoú (Akantiá) gate

PINDAROU

KISTHINIOU

PL. PEISIDOROU

Panagías (Virgin's) gate

Marine Gate
This gate (Pýli Agías Aikaterínis), with twin towers, is the main route into the Old Town from Commercial harbour.

St Catherine's gate

| 0 metres | 100 |
| 0 yards | 100 |

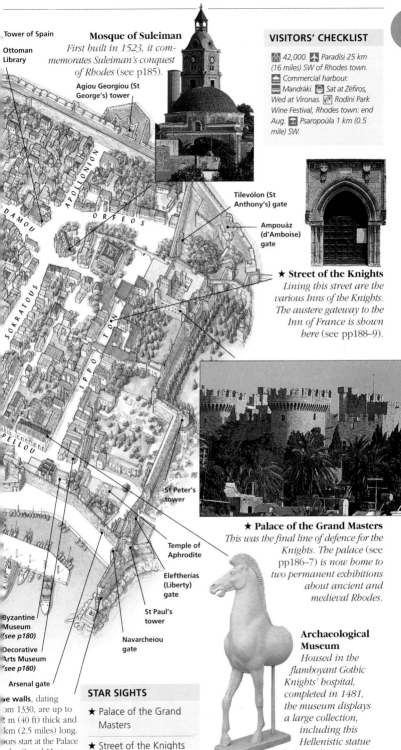

Tower of Spain

Ottoman Library

Mosque of Suleiman
*First built in 1523, it com-
memorates Suleiman's conquest
of Rhodes (see p185).*

Agiou Georgíou (St George's) tower

APOLLONION

ORFEOS

DAMOU

SOKRATOUS

IPPOTON

PELLOU

Byzantine
Museum
(see p180)

Decorative
Arts Museum
(see p180)

Arsenal gate

e walls, dating
om 1330, are up to
m (40 ft) thick and
km (2.5 miles) long.
urs start at the Palace
the Grand Masters.

St Peter's tower

Temple of Aphrodite

Eleftherías (Liberty) gate

St Paul's tower

Navarcheíou gate

Tilevólon (St Anthony's) gate

Ampouáz (d'Amboise) gate

VISITORS' CHECKLIST

👥 42,000. ✈ Paradísi 25 km
(16 miles) SW of Rhodes town.
⚓ Commercial harbour.
🚌 Mandráki. 🍴 Sat at Zéfiros,
Wed at Víronas. 🎭 Rodíni Park
Wine Festival, Rhodes town: end
Aug. 🚉 Psaropoúla 1 km (0.5
mile) SW.

★ Street of the Knights
*Lining this street are the
various Inns of the Knights. The
austere gateway to the
Inn of France is shown
here (see pp188–9).*

★ Palace of the Grand Masters
*This was the final line of defence for the
Knights. The palace (see
pp186–7) is now home to
two permanent exhibitions
about ancient and
medieval Rhodes.*

Archaeological Museum
*Housed in the
flamboyant Gothic
Knights' hospital,
completed in 1481,
the museum displays
a large collection,
including this
Hellenistic statue
of a horse (see p184).*

STAR SIGHTS

★ Palace of the Grand Masters

★ Street of the Knights

Exploring Rhodes Old Town

Dominated by the Palace of the Grand Masters, this medieval citadel is surrounded by moats and 4 km (2.5 miles) of walls. Eleven gates give access to the Old Town, which is divided into the Collachium and the Bourg. The Collachium was the Knights' quarter, and dates from 1309. The Bourg housed the rest of the population, which included Jews and Turks as well as Greeks. As one of the finest walled cities in existence, the Old Town is now a World Heritage Site.

An arched street in the Old Town

The imposing 16th-century d'Amboise gate

The Collachium

This area includes the Street of the Knights *(see pp188–9)* and the Palace of the Grand Masters *(see pp186–7)*. The main gates of entry from the new town are d'Amboise gate and the Eleftherías (Liberty) gate. The former was built in 1512 by Grand Master d'Amboise, leading from Dimokratías to the palace. The Eleftherías gate was built by the Italians and leads from Eleftherías to Plateía Sýmis. An archway leads from here into Apelloú.

⚱ Archaeological Museum

Plateía Mouseíou. **Tel** 22410 25500. ◯ 8:30am–7pm Tue–Sun. ● main public hols. ▨ ◉

The museum is housed in the Gothic Hospital of the Knights, built in 1440–81. Most famous of the exhibits is the 1st-century BC marble *Aphrodite of Rhodes*. Other gems include a 2nd-century BC head of Helios the Sun God, discovered at the Temple of Helios on the nearby hill of Monte Smith. The grave *stelae* from the necropolis of Kámeiros give a good insight into 5th-century BC life. Exhibits also include coins, jewellery and ceramics from the Mycenaean graves at nearby Ialyssós.

Aphrodite of Rhodes, Archaeological Museum

⚱ Decorative Arts Museum

Plateía Argyrokástrou. **Tel** 22410 25500. ◯ Tue–Sun. ● main public hols. ▨ ♿

This is an excellent folk museum featuring Lindian plates and tiles, a wide range of island costumes and a reconstructed traditional Rhodian house.

⚱ Medieval Rhodes and Ancient Rhodes Exhibitions

Palace of the Grand Masters. **Tel** 22410 23359. ◯ Tue–Sun. ● main public hols. ▨ ♿

Both of these permanent exhibitions can be seen as part of a tour of the Palace of the Grand Masters *(see pp186–7)*. The Medieval Rhodes exhibition is titled: Rhodes from the 4th century AD to the Turkish Conquest (1522). It gives an insight into trade and everyday life in Byzantine and medieval times, with Byzantine icons, Italian and Spanish ceramics, armour and militaria. The Ancient Rhodes exhibition, entitled Ancient Rhodes: 2,400 years, is situated off the inner court. It details 45 years of archaeological investigations on the island with a marvellous collection of finds.

⚱ Byzantine Museum

Apéllou. **Tel** 22410 27657. ◯ Tue–Sun. ● main public hols. ▨

Dating from the 11th century, this Byzantine church became the Knights' cathedral, but was converted under Turkish rule into the Mosque of Enderum, known locally as the Red Mosque. Now a museum, it houses a fine collection of icons and frescoes. Among the exhibits are striking examples of 12th-century paintings in the dynamic Comnenian style from Moní Thárri *(see p194)* and late 14th-century frescoes from the abandoned church of Agios Zacharías on Chálki.

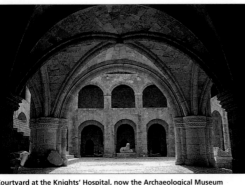

Courtyard at the Knights' Hospital, now the Archaeological Museum

⌂ Medieval City Walls

Tours from the Palace of the Grand Masters. ⬚ *Tue & Fri: 2.45pm.* 🖼

A masterpiece of medieval military architecture, the huge walls run for 4 km (2.5 miles) and display 151 escutcheons of Grand Masters and Knights.

The Bourg

The Bourg's clocktower

Close to d'Amboise gate is the restored clock tower, which has excellent views. It was built in 1852 on the site of a Byzantine tower and marks the end of the Collachium. The Bourg's labyrinth of streets begins at Sokrátous, the Golden Mile of bazaar-style shops, off which lie shady squares with pavement cafés and tavernas. The architecture is a mix of medieval, Neo-Classical and Levantine. Between the houses, with rickety wooden balconies, Ottoman mosques can be found.

Other than the major sights listed below, the Hospice of the Tongue of Italy (1392) on Kisthiníou is worth a visit, as is the Panagía tis Níkis (Our Lady of Victory). It stands near St Catherine's gate, and was built by the Knights in 1480 after the Virgin had appeared to them, inspiring victory over the Turks.

🕌 Mosque of Suleiman the Magnificent

Orféos Sokrátous. ◖ *under renovation.*

The pink mosque was constructed in 1522 to commemorate the Sultan's victory over the Knights. Rebuilt in 1808, using material from the original mosque, it remains one of the town's major landmarks. Its superb, but unsafe, minaret had to be removed in 1989, and the once-mighty mosque is now crumbling. It is sadly closed to the public.

🏛 Library of Ahmet Havuz

44 Orféos. **Tel** *22410 74090.* ⬚ *Mar–Oct: 9:30am–4pm Mon–Sat.* ◖ *Nov–Feb; main public hols.*

The Library of Ahmet Havuz (1793) houses the chronicle of the siege of Rhodes in 1522. This is a collection of very rare Arabic and Persian manuscripts, including beautifully illuminated 15th- and 16th-century Korans, which were restored to the library in the early 1990s, having been stolen then rediscovered in London.

🎭 Nélly Dimóglou Theatre

7 Andrónikou. **Tel** *22410 20157.* ⬚ *mid-May–mid-Oct: Mon, Wed & Fri.* 🖼 ♿

The Nélly Dimóglou Theatre offers lessons in authentic Greek folk dancing. Its gardens are open all day for refreshments, and performances begin at 9:20pm every evening from Monday to Friday.

🏛 Hammam

Plateía Aríonos. **Tel** *22410 27739.* ⬚ *10am–5pm Mon–Fri, 8am–5pm Sat.* 🖼

The *hammam* were built by Mustapha Pasha in 1765. For decades a famous place of rest and relaxation for Eastern nobility, it is now used by Greeks, tourists and the Turkish minority. Your own soap and towels are essential, and sexes are segregated.

🕌 Mosque of Ibrahim Pasha

Plátanos. **Tel** *22410 73410.* ⬚ *daily.* 🖼 *donation.*

Situated off Sofokléous, the Mosque of Ibrahim Pasha was built in 1531 and refurbished in 1928. The mosque has an exquisite interior.

🕌 Mosque of Rejep Pasha

Ekátonos. ◖ *under renovation.*

Built in 1588, Rejep Pasha is one of the most striking of the 14 or so mosques to be found in the Old Town. The mosque, which has a fountain made from Byzantine and medieval church columns, contains the sarcophagus of the Pasha. The tiny Byzantine church of Agios Fanoúrios is situated close by.

The Jewish Quarter

East from Hippocrates Square, the Bourg embraces Ovriakí. This was the Jewish Quarter from the 1st century AD until German occupation in 1944, when the Jewish population was transported to Auschwitz.

East along Aristotélous is **Plateía Evraíon Mart'yron** (Square of the Jewish Martyrs), named in memory of all those who perished in the concentration camps. There is a bronze sea horse fountain in the centre, and to the north is Admiralty House, an imposing medieval building. The Synagogue is on Simíou.

The dome of the Mosque of Suleiman the Magnificent

Rhodes: Palace of the Grand Masters
Παλάτι του Μεγάλου Μαγίστρου

Gilded angel
candleholder

A fortress within a fortress, this was the seat of 19 Grand Masters, the nerve centre of the Collachium, or Knights' Quarter, and last refuge for the population in times of danger. Built in the 14th century, it survived earthquake and siege, but was blown up by an accidental explosion in 1856. It was restored by the Italians in the 1930s for Mussolini and King Victor Emmanuel III. The palace has some priceless mosaics from sites in Kos, after which some of the rooms are named. It also houses two exhibitions: Medieval, and Ancient Rhodes *(see p184)*.

Chamber with Colonnades
Two elegant colonnades support the roof an there is a 5th-century AD early Christian mosaic.

Thyrsus Chamber

Chamber of
Sea Horse a
Nymph

The Second Cross-Vaulted Chamber, once used as the governor's office, is paved with an intricately decorated, early Christian mosaic of the 5th century AD from Kos.

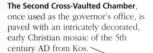

First Cross-
Vaulted
Chamber

★ **Medusa Chamber**
The mythical Gorgon Medusa, with hair of writhing serpents, forms the centrepiece of this important late Hellenistic mosaic. The chamber also features Chinese and Islamic vases.

Laocoön Chamber
A copy of the sculpture of the death of the Trojan, Laocoön, and his sons dominates the hall. The 1st-century BC original by Rhodian masters Athenodoros, Agesandros and Polydoros is in the Vatican.

The battlements and heavy fortifications of the palace were to be the last line of defence in the event of the city walls being reached.

★ **Central Courtyard**
The palace is built around a courtyard paved with geometric marble tiles. The north side is lined with Hellenistic statues taken from the Odeion in Kos (see p171).

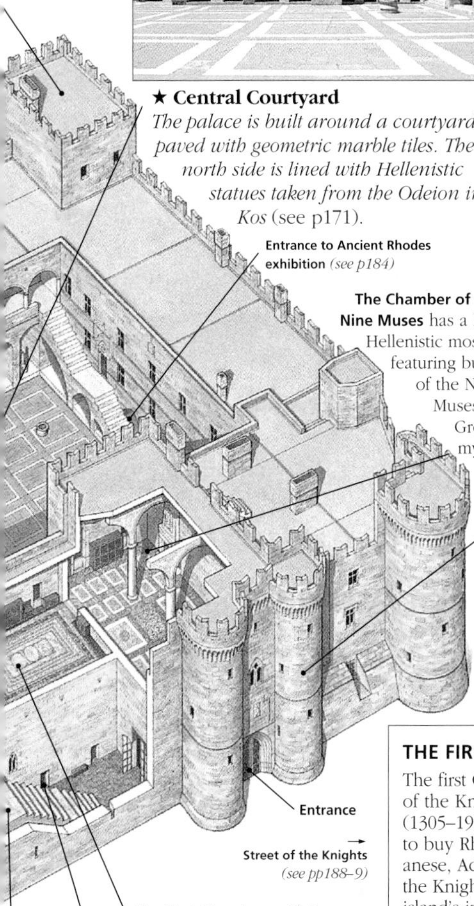

Entrance to Ancient Rhodes exhibition *(see p184)*

The Chamber of the Nine Muses has a late Hellenistic mosaic featuring busts of the Nine Muses of Greek myth.

Entrance

→
Street of the Knights *(see pp188–9)*

The First Chamber, with its 16th-century choir stalls, features a late Hellenistic mosaic.

Grand staircase

Entrance to Medieval Rhodes exhibition *(see p184)*

The Second Chamber has a late Hellenistic mosaic and carved choir stalls.

VISITORS' CHECKLIST

Ippotón. **Tel** 22410 23359.
⬚ Aug–Sep: 12:30–4pm Mon, 8am–7pm Tue–Sun; Oct–Jul: 8:30am–3pm Tue–Sun; 12–3pm Good Fri. ⬤ 1 Jan, 25 Mar, Easter Sun, 1 May, 25, 26 Dec. ⬚
⬚ ⬚ limited. ⬚

★ **Main Gate**
This imposing entrance, built by the Knights, has twin horseshoe-shaped towers with swallowtail turrets. The coat of arms is that of Grand Master del Villeneuve, who ruled from 1319 to 1346.

THE FIRST GRAND MASTER

The first Grand Master, or Magnus Magister, of the Knights was Foulkes de Villaret (1305–19), a French knight. He negotiated to buy Rhodes from the Lord of the Dodecanese, Admiral Vignolo de Vignoli. This left the Knights with the task of conquering the island's inhabitants. The Knights of Rhodes *(see pp188–9)*, as they became, remained here until their expulsion in 1522. The Villaret name lives on in Villaré, one of the island's white wines.

Foulkes de Villaret

MAGNUS | MAGISTER
FRATER FULCUS | DE VILLARET
1305 | 1319

STAR FEATURES

★ Central Courtyard

★ Medusa Chamber

★ Main Gate

Rhodes: Street of the Knights

One of the old town's most famous sights, the medieval Street of the Knights (Odos Ippotón) is situated between the harbour and the Palace of the Grand Masters *(see pp188–9)*. It is lined by the Inns of the Tongues, or nationalities, of the Order of St John. Begun in the 14th century in Gothic style, the Inns were used as meeting places for the Knights. The site of the German Inn is unknown, but the others were largely restored by the Italians in the early 20th century.

This residence was built for the head of the Tongue of Aragon, Diomede de Vilaragut.

SOUTH SIDE

←
To Inn of England

The Archaeological Museum *(see p180)*, was originally the New Hospital of the Knights.

Access to the Turkish garden

The Inn of Provence *has coats of arms set in the wall. They represent the Order of the Knights of St John, the Royal House of France, Grand Master del Carretto and the Knight de Flota.*

Agía Triáda, or French Chapel

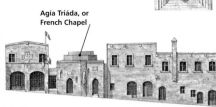

NORTH SIDE

Palace of the Grand Masters
←

Arched bridge connecting Inn of Spain and Inn of Provence

The Knights of Rhodes

Coat of arms of Foulkes de Villaret, first Grand Master

Founded in the 11th century by merchants from Amalfi, the Order of Hospitallers of the Knights of St John guarded the Holy Sepulchre and tended Christian pilgrims in Jerusalem. They became a military order after the First Crusade (1096–9), but had to take refuge in Cyprus when Jerusalem fell in 1291. They then bought Rhodes from the Genoese pirate Admiral Vignoli in 1306, and eventually conquered the Rhodians in 1309. A Grand Master was elected for life to govern the Order, which was divided into seven Tongues, or nationalities: France, Italy, England, Germany, Provence, Spain and Auvergne. Each Tongue protected an area of city wall known as a Curtain. The Knights fortified the Dodecanese with around 30 castles and their defences are some of the finest examples of medieval military architecture.

The Knights *were drawn from noble Roman Catholic families. Those who entered the Order of the Knights of St John swore vows of chastity, obedience and poverty. Although knights held all the major offices, there were also lay brothers.*

Odos Ippotón, *the Street of the Knights, lies along a section of ancient road that led all the way down to the harbour. It was here that the Knights would muster in times of attack.*

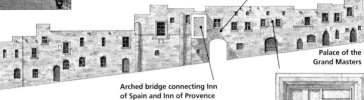

Archway to Ippárchou

Palace of the → Grand Masters

Arched bridge connecting Inn of Spain and Inn of Provence

Archway to Láchitos

The Inn of Spain *is one of the largest inns. Its assembly hall was over 150 sq m (1,600 sq ft). On the exterior there is a small and simple coat of arms of the Spanish Tongue.*

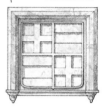

The Inn of France's *armorial bearings are the French royal fleur-de-lys, and those of Grand Master Petrus d'Amboise.*

The Inn of Italy *has a marble escutcheon bearing the arms of the Grand Master Fabricius del Carretto.*

Palace of Grand Master Villiers de l'Isle Adam (1521–34)

Inn of Auvergne →

The Great Siege of Rhodes *in 1522 resulted in the Knights being defeated by the Turks. From a garrison of 650 Knights, only 180 survived. They negotiated a safe departure, although the Rhodians who fought with them were slaughtered. Seven years later, the Knights found sanctuary on the island of Malta. Their final defeat came in 1798 when Malta was annexed by Napoleon.*

Pierre d'Aubusson, *Grand Master from 1476 to 1503, is featured in this market scene. He oversaw a highly productive time in terms of building in Rhodes, including completion of the Hospital (now the Archaeological Museum).*

Exploring Rhodes New Town

The new town grew steadily over the last century, and became firmly established during the Italian Fascist occupation of the 1920s with the construction of the grandiose public buildings by the harbour. The New Town is made up of a number of areas including Néa Agora and Mandráki harbour in the eastern half of town. The Italian influence remains in these areas with everything from pizzerias to Gucci shops. The town's west coast is a busy tourist centre, with lively streets and a crammed beach.

Government House, previously the Italian Governor's Palace

Mandráki harbour with the two statues of deer at its entrance

Mandráki Harbour

The Harbour is the hub of life, the link between the Old and New towns where locals go for their evening stroll, or *vólta*. It is lined with yachts and excursion boats for which you can book a variety of trips in advance.

A bronze doe and stag guard the harbour entrance, where the Colossus was believed to have stood. The harbour sweeps round to the ruined 15th-century fortress of **Agios** Nikólaos, now a lighthouse, on the promontory past the three medieval windmills.

Elegant public buildings, built by the Italians in the 1920s, line Mandráki harbour: the post office, law courts, town hall, police station and the National Theatre all stand in a row. The **National Theatre** often shows Rhodian character plays based on folk customs.

Nearby, on Plateía Eleftherías, is the splendid church of the **Evangelismós** (Annunciation), a 1925 replica of the Knight's Church of St John, which has a lavishly decorated interior. The Archbishop's Palace is next door beside a giant fountain, which is a copy of the Fontana Grande in Viterbo, Italy. Further along, the mock Venetian Gothic **Government House** (Nomarchía) is ornately decorated and surrounded by fine vaulted arcades. Unfortunately there is no access for tourists or the general public.

At the north end of Plateía Eleftherías is the attractive **Mosque of Murad Reis**, with its graceful minaret. It was named after a Turkish admiral serving under Suleiman who was killed during the 1522 siege of Rhodes. Situated within the grounds is the Villa Kleoboulos, which was the home of the British writer Lawrence Durrell between 1945 and 1947. Also in the grounds is a cemetery reserved for Ottoman notables.

Heading north from the area around Mandráki harbour, a pleasant stroll along the waterfront via the crowded Elli beach leads to the northern tip of the New Town. The Hydrobiological Institute is situated on the coastal tip, housing the **Aquarium**. Set in a subterranean grotto, this is the only major aquarium in Greece, displaying nearly 40 tanks of fish. Opposite, on the north point of the island is Aquarium Beach, which is particularly good for windsurfing and paragliding.

The minaret of the Mosque of Murad Reis

🗽 Aquarium
Hydrobiological Institute, Kássou.
Tel 22410 27308. ◯ daily. ⬤ main public hols. 🈺 ♿

THE COLOSSUS OF RHODES

Painting of the Colossus by Fischer von Erlach, 1700

One of the Seven Wonders of the Ancient World, the Colossus was a huge statue of Helios, the sun god, standing at 32–40 m (105–130 ft). Built in 305 BC to celebrate Rhodian victory over Demetrius, the Macedonian besieger, it was sculpted by Chares of Líndos. It took 12 years to build, using bronze from the battle weapons, and cost 9 tons (10 imperial tons) of silver. Traditionally pictured straddling Mandráki harbour, it probably stood at the Temple of Apollo, now the site of the Palace of the Grand Masters in the Old Town (*see pp182–3*). An earthquake in 227 BC caused it to topple over.

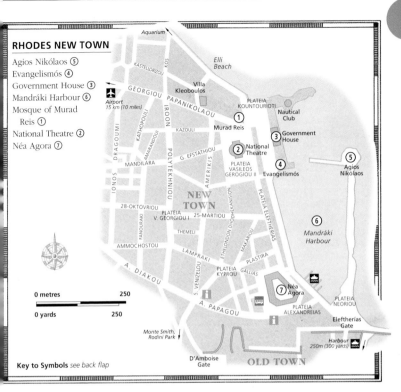

RHODES NEW TOWN

Agios Nikólaos ⑤
Evangelismós ④
Government House ③
Mandráki Harbour ⑥
Mosque of Murad
  Reis ①
National Theatre ②
Néa Agora ⑦

Key to Symbols see back flap

0 metres   250
0 yards   250

Néa Agora

Mandráki is backed by the New Market or Néa Agora with its Moorish domes and lively cafés. Inside the market are food stalls, gift shops, small *souvláki* bars and cafés. It is popular as a meeting place for people coming from outlying

A view of the domed centrepiece of the New Market from Mandráki Harbour

villages and islands. Behind the Néa Agora, in the grounds of the Palace of the Grand Masters, a sound and light show is held. This takes place daily in one of four languages and tells the story of the overthrow of the Knights by Suleiman the Magnificent in 1522.

Monte Smith

Monte Smith, a hill to the west of town, offers panoramic views over Rhodes town and the coast. It is named Monte Smith after the English Admiral Sir Sidney Smith who kept watch from there for Napoleon's fleet in 1802. It is also known as Agios Stéphanos.

The hill is the site of a 3rd-century BC Hellenistic city which was excavated by the Italians. They restored the 3rd-century BC stadium, the 2nd-century BC

acropolis and a small theatre or odeion. This was built in an unusual square shape and is used for performances of ancient drama in the summer. Only three columns remain of the once-mighty Temple of Pythian Apollo, and there are other ruins of the temples of Athena Polias and Zeus. Nearby, on Voreíou Ipeírou, are the remains of the Asklepieíon, a temple dedicated to the god of healing, Asklepios.

Rodíni Park

The beautiful Rodíni Park, 3 km (2 miles) to the south of Rhodes town, is now home to the Rhodian deer sanctuary, and perfect for a break away from the crowded centre. It is the site where the orator Aeschines built the School of Rhetoric in 330 BC, attended by both Julius Caesar and Cassius, although there are no remains to visit. Sights include a 3rd-century BC necropolis with Doric rock tombs and several Ptolemaic, rock-cut tombs. In medieval times the Knights grew their herbs at Rodíni.

Exploring Western Rhodes

The windswept west coast is a busy strip of hotels, bars and restaurants, along shingly beaches from Rhodes town to the airport at Paradísi. But head south and the landscape becomes green and fertile, with vineyards and wooded mountain slopes, dotted with traditional farming villages. The attractions include Moní Filerímou, Ancient Kámeiros, the wine-making village of Emponas, and the enchanting valley of Petaloúdes, the place that gives Rhodes its name as the "Island of Butterflies". Further south is a dramatic mix of scenery with castle-topped crags and sea views to the islands of Chálki and Alimiá.

An icon at Our Lady of Filérimos

Ancient Ialyssós ❷
Αρχαία Ιαλυσός

15 km (9 miles) SW of Rhodes town. 🚌 to Triánda. ⏱ 8am–7:10pm Tue–Sat, 8:30am–2:40pm Sun. 🌙 main public hols.

Ialyssós fused with two other Doric city-states, Líndos and Kámeiros, to create one capital, Rhodes, in 408 BC. As this new centre grew, Ialyssós, Líndos and Kámeiros lost their former importance. However, Ialyssós proved a much fought-over site: the Byzantines were besieged by the Genoese there in 1248; the Knights *(see pp188–9)* used it as a base before taking Rhodes in 1309; and it was Suleiman's head-quarters before his assault on the Knights in 1522. The Italians used it again for gun positions during World War II.

The only remnant of the acropolis is the 3rd-century BC **Temple of Athena Polias and Zeus Poliefs** by the church of Agios Geórgios. The restored lion-head fountain, to the south, is 4th century BC.

Moní Filerímou ❸
Μονή Φιλερήμου

15 km (9 miles) SW of Rhodes town. **Tel** 22410 92202. 🚌 to Triánda. ⏱ 8am–7:10pm Tue–Sat, 8:30am–2:40pm Sun. 🌙 📷

One of Rhodes' beauty spots, the hillsides of Filérimos are home to cypresses and pines. Among the trees sits Moní Filerímou, its domed chapels decorated with the cross of the Knights and the coat of arms of Grand Master Pierre d'Aubusson. A place of worship for 2,000 years, layers of history and traditions can be seen, from Phoenician to Byzantine, Orthodox and Catholic.

The main attraction is Our Lady of Filérimos, the Italian reconstruction of the Knights' 14th-century church of the Virgin Mary. It is a complex of four chapels: the main one, built in 1306, leads to three others. The innermost chapel has a Byzantine floor decorated with a red mosaic fish.

The Italians erected a Calvary from the entrance of the monastery, in the form of an avenue with the Stations of the Cross illustrated on plaques. On the headland stands a giant 18-m (59-ft) cross.

Petaloúdes ❹
Πεταλούδες

26 km (16 miles) SW of Rhodes town. 🚌

Petaloúdes, or Butterfly Valley is a narrow leafy valley with a stream crisscrossed by wooden bridges. It teems, not with butterflies, but with Jersey tiger moths from June to September. Thousands are attracted by the golden resin of the storax trees, which exude vanilla-scented gum used for incense. Cool and pleasant, Petaloúdes attracts walkers as well as lepidopterists, and is at its most peaceful in the early morning before all the tour buses arrive.

There is a walk along the valley to the **Moní Panagías Kalópetras**. This rural church, built in 1782, is a tranquil resting place, and the fine views are well worth the climb.

Jersey tiger moth

Ancient Kámeiros ❺
Αρχαία Κάμειρος

36 km (22 miles) SW of Rhodes town. **Tel** 22410 40037. 🚌 ⏱ Tue–Sun. 🌙 main public hols. 📷 📷 ♿ to lower sections only.

Discovered in 1859, this Doric city was a thriving community during the 5th century BC. Founded by Althaemenes of Crete, the city was probably destroyed in a large earthquake in 142 BC. In spite of this, it remains one of the best preserved Classical Greek cities.

There are remains of a 3rd-century BC Doric temple, an altar to Helios, public baths and a 6th-century BC cistern, which supplied 400 families. The 6th-century BC Temple of Athena Polias is on the top terrace, below which are remains of the Doric stoa, 206 m (675 ft) long.

Moní Filerímou in its woodland setting

For hotels and restaurants in this region see pp309–12 and pp334–6

Monólithos castle in its precarious position overlooking the sea

Skála Kameírou ⑥
Σκάλα Καμείρου

50 km (30 miles) SW of Rhodes town. 🏠 100. 🚌

The fishing harbour of Skála Kameírou makes a good place for lunch. It was the Doric city of an ancient port, and the outline of a Lycian tomb remains on the cliff side. Nearby, **Kritinía castle** is one of the Knights' more impressive ruins. Its three levels are attributed to different Grand Masters. Clinging to the hillside, a cluster of white houses form the picturesque village of **Kritinía**.

Emponas ⑦
Έμπωνας

55 km (34 miles) SW of Rhodes town. 🏠 1,500. 🚌

Situated in the wild foothills of Mount Attávyros, the atmospheric village of Emponas has been home to the Cair

winery since the 1920s and is also famous for its folk dancing and festivals. Although the village is popular for organized Greek nights, Emponas has maintained its traditional ways.

Monólithos ⑧
Μονόλιθος

80 km (50 miles) SW of Rhodes town. 🏠 250. 🚌 🚕 Foúrni 5 km (3 miles) SW.

Named after its Monolith, a crag with a dramatic 235-m (770-ft) drop to the sea, Monólithos is the most important village in the southwest.
Situated at the foot of Mount Akramýtis, the village is 2 km (1 mile) from **Monólithos castle**. This impregnable 15th-century fortress, built by Grand Master d'Aubusson, is perched spectacularly on the vast grey rock. Its massive walls enclose two small 15th-century chapels, Agios Panteleḯmon

and Agios Geórgios, both decorated with frescoes. Views from the top are impressive.
Down a rough road south from the castle is the sheltered sandy beach of **Foúrni**, which has a seasonal taverna.

Environs
Between Emponas and Monólithos, the pretty hill village of **Siána** is famous for its honey and fiery *soúma* – a kind of grape spirit, like the Cretan raki. The villagers were granted a licence by the Italians to make the spirit, and you can sample both the firewater and honey at the roadside cafés. The village houses have traditional clay roofs, and the domed church of **Agios Panteleḯmon** has restored 18th-century frescoes.

Moní Skiádi ⑨
Μονή Σκιάδι

8 km (5 miles) S of Apolakkiá. **Tel** 22440 46006. 🚌 to Apolakkiá. 🕘 9am–6pm daily. 📷 ♿

Moní Skiádi is famous for its miraculous icon of the Panagía or the Blessed Virgin. When a 15th-century heretic stabbed the Virgin's cheek it was supposed to have bled, and the brown stains are still visible. The present monastery was built during the 18th and 19th centuries around the 13th-century church of Agios Stavrós, or the Holy Cross. At Easter the holy icon is carried from village to village until finally coming to rest for a month on the island of Chálki.

Sunset over the village of Emponas and Mount Attávyros

Exploring Eastern Rhodes

The sheltered east coast has miles of beaches and rocky coves, the crowded holiday playgrounds of Faliráki and Líndos contrasting with the deserted sands in the southeast. For sightseeing purposes the way east divides into two sections: from the southern tip of the island at Prasonísi up to Péfkoi, and then from Líndos up to Rhodes town. The landscape is a rich patchwork, from the oasis of Eptá Pigés and the orange groves near Archángelos, to the stretches of rugged coastline and sandy bays.

Fountain in Lárdos village

Moní Thárri ⑩
Μονή Θάρρι

40 km (25 miles) S of Rhodes town.
🚌 to Laérma. ⬜ daily.

From the inland resort of Lárdos follow signs to Láerma, which is just north of Moní Thárri, famous for its 12th-century frescoes. Reached through a forest, the domed church was hidden from view in order to escape the attention of marauding pirates.

According to legend, it was built in the 9th century by a mortally ill Byzantine princess, who miraculously recovered when it was completed.

The 12th-century north and south walls remain, and there are vestiges of the 9th-century building in the grounds. The nave, apse and dome are covered with frescoes. Some walls have four layers of paintings, the earliest dating as far back as 1100, while there are three layers in the apse dating from the 12th–16th centuries. These

Asklipieío village

are more distinct, and depict a group of prophets and a horse's head. The monastery has been extended and has basic accommodation for visitors.

About 8 km (5 miles) south along a rough track is the pleasant village of **Asklipieío**, with the frescoed church of Kímisis tis Theotókou.

Líndos ⑪
See pp196–7.

Archángelos ⑫
Αρχάγγελος

33 km (20 miles) S of Rhodes town.
🏠 3,000. 🚌 🚕 Stégna 3 km (2 miles) E.

The island's largest village, Archángelos lies in the Valley of Aíthona, which is renowned for its oranges. The town itself is famous for pottery, hand-woven rugs and leather boots. Traditionally worn as protection from snakes while in the fields working, they are made of sturdy cowhide for the feet, with soft goatskin leggings. The townspeople have their own dialect and are fiercely patriotic – some graves are even painted blue and white.

In the centre, the church of **Archángeloi Michaïl and Gavriíl**, the village's patron saints, is distinguished by a tiered bell tower and pebble-mosaic courtyard.

Above the town are ruins of the **Crusader castle**, built by Grand Master Orsini in 1467 as part of the Knights' defences against the Turks. Inside, the chapel of Agios Geórgios has a modern fresco of the saint in action against the dragon. To the east of the town lies the bay of **Stégna**, a quiet and sheltered stretch of sand.

Environs
South past Malónas is the castle of **Faraklós**. It was a pirate stronghold before the Knights saw them off and turned it into a prison. The fortress overlooks Charáki, a

Charáki village with the castle of Faraklós in the background

The sandy beach at Tsampíka

pleasant fishing hamlet, now growing into a holiday resort, with a pebble beach that is lined with fish tavernas.

Eptá Pigés ⑬
Επτά Πηγές

26 km (16 miles) S of Rhodes town.
🚌 to Kolýmpia. 🚉 Tsampíka 5 km (3 miles) SE.

Eptá Pigés, or Seven Springs, is one of the island's leading woodland beauty spots. Peacocks strut beside streams and waterfalls, where the seven springs feed a central reservoir. The springs were harnessed to irrigate the orange groves of Kolýmpia to the east. The lake can be reached either by a woodland trail, or you can shuffle ankle-deep in water through a 185-m (605-ft) tunnel.

Peacock at Eptá Pigés

Environs
Further east along the coast, the Byzantine **Moní Tsampíkas** sits on a mountain top at 300 m (985 ft). Legend has it that the 11th-century icon in the chapel was found by an infertile couple, who later conceived a child. The chapel hence became a place of pilgrimage for childless women come to pray to the icon of the Virgin. They also pledge to name their child Tsampíka or Tsampíkos, names unique to the Dodecanese.

Below the monastery lies **Tsampíka** beach, a superb stretch of sand that becomes very crowded in the tourist season. Various watersports are also available here.

Faliráki ⑭
Φαληράκι

15 km (9 miles) S of Rhodes town.
🏘 400. 🚌

Faliráki, one of the island's most popular resorts, consists of long sandy beaches surrounded by whitewashed hotels, holiday apartments and restaurants. Also a good base for families who like a lively holiday with plenty of activities, it is a brash and loud resort that caters mostly for a younger crowd. As well as a huge waterside complex, **Faliráki Water Park**, there are all types of watersports to enjoy. There are bars and discos, and numerous places to eat, from fish and chips to Chinese. Other diversions include bungee-jumping.

🎢 Faliráki Water Park
Faliráki. **Tel** 22410 84403.
◯ May–Oct: daily. 🎫 🍴

Environs
Slightly inland, the village of **Kalythiés** offers a more traditional break. Its attractive Byzantine church, **Agía Eleoúsa**, contains some interesting frescoes. Further southeast,

rocky **Ladikó Bay** is worth a visit. It was used as a location for filming *The Guns of Navarone*.

Golfers can visit the 18-hole course at **Afántou** village, with its pebbly coves and beaches, popular for boat trips from Rhodes town. Set in apricot orchards, Afántou means the "hidden village", and it is noted for its hand-woven carpets.

Koskinoú ⑮
Κοσκινού

10 km (6 miles) S of Rhodes town.
🏘 1,200. 🚌 🚉 Réni Koskinoú 2 km (1 mile) NE.

The old village of Koskinoú is characterized by its traditional Rhodian houses featuring the *choklákia* pebble mosaic floors and courtyards. There is an attractive church of **Eisódia tis Theotókou**, which has a multi-tiered bell tower. Nearby, **Réni Koskinoú** has good hotels, restaurants and beaches.

Environs
South of Koskinoú lies **Thérmes Kalithéas**, Kalithea Spa, once frequented for its healing waters. Now abandoned, the domed pavilions, pink-marbled pillars and Moorish archways look quite bizarre. Often used in films, the spa is set in lovely gardens, reached through pinewoods. There is now a busy lido here, and the rocky coves are popular for scuba-diving and snorkelling.

A church with a tiered bell tower in Koskinoú village

Líndos ⑪

Λίνδος

Líndos was first inhabited around 3000 BC. Its twin harbours gave it a head start over Rhodes' other ancient cities of Kámeiros and Ialyssós as a naval power. In the 6th century BC, under the benevolent tyrant Kleoboulos, Líndos thrived and grew rich from its many foreign colonies. With its dazzling white houses, Crusader castle and acropolis dramatically overlooking the sea, Líndos is a magnet for tourists. Second only to Rhodes town as a holiday resort, it is now a National Historic Landmark, with development strictly controlled.

Carved stones of stoa

A traditional Líndian doorway

Exploring Líndos Village

Líndos is the most popular excursion from Rhodes town, and the best way to arrive is by boat. The narrow cobbled streets can be shoulder to shoulder with tourists in high summer, so spring or autumn are more relaxed times to visit. Líndos is a sun trap, and is known for consistently recording the highest temperatures on the island. In winter, the town is almost completely deserted.

Traffic is banned so the village retains much of its charm and donkeys carry people up to the acropolis (be warned that they proceed rather quickly downhill). It is

very busy, with a bazaar of gift shops and fast-food outlets. Happily there are also several good tavernas and, at the other end of the scale, there are a number of stylish restaurants offering international cuisine. Some quiet, romantic little places can be found, with views of the bay and the sea.

The village's winding lanes are fronted by imposing doorways which lead into the flower-filled courtyards of the unique Líndian houses. Mainly built by rich sea-captains between the 15th and 18th centuries, these traditional houses are called *archontiká*. They have distinctive carvings on the stonework, like ship's cables or chains (the number of chains supposedly corresponds to the number of ships owned), and are built round *choklákia* pebble mosaic courtyards *(see p198)*. A few of them are open to the public for viewing. The older houses mix Byzantine and Arabic styles and a few have small captain's rooms built over the doorway.

Some of the *archontiká* have been converted into apartments and restaurants.

In the centre of the village lies the Byzantine church of the **Panagía**, complete with its graceful bell tower and pantiled domes. Originally a 10th-century basilica, it was rebuilt between 1489 and 1490. The frescoes inside were painted by Gregory of Sými in 1779.

On the path leading to the acropolis, are a number of women selling the lace for which Líndos is renowned. Lindian stitchwork is sought after by museums throughout

Líndos Stoa
This colonnade or stoa was built in the Hellenistic period around 200 BC.

The battlements were built in the 13th century by the Knights of Rhodes.

A trireme warship is carved into the rock.

Líndos lace seller on the steps to the acropolis

THE ACROPOLIS AT LINDOS

Perched on a sheer precipice 125 m (410 ft) above the village, the acropolis is crowned by the 4th-century BC Temple of Lindian Athena, its remaining columns etched against the skyline. The temple was among the most sacred sites in the ancient world, visited by Alexander the Great and supposedly by Helen of Troy and Herakles. In the 13th century, the Knights Hospitallers of St John fortified the city with battlements much higher than the original walls.

The acropolis overlooking Líndos town and bay

the world; it is said that even Alexander the Great wore a cloak stitched by Lindian women. The main beach at Líndos, **Megálos Gialós**, is where the Líndian fleet once anchored, and it sweeps north of the village round Líndos bay. It is a popular beach and it tends to get very crowded in summer, but a wide selection of watersports are available. It is also safe for children, and several tavernas can be found along the beachfront.

VISITORS' CHECKLIST

1 km (0.5 mile) E of Líndos village. **Tel** 22440 31900. Jul–Sep: 8am–7pm Tue–Sun, noon–7pm Mon; Oct–Jun: 8:30am–3:20pm Tue–Sun. main public hols.

Environs

Tiny, trendy **Pallás** beach is linked to Líndos's main beach by a walkway. Nudists make for the headland, around which is the more exclusive **St Paul's Bay**, where the Apostle landed in AD 43, bringing Christianity to Rhodes. An idyllic, almost enclosed cove, it has azure waters and a white chapel dedicated to St Paul, with a festival on 28 June.

Although called the **Tomb of Kleoboulos**, the stone monument on the promontory north of the main beach at Líndos bay had nothing to do with the great Rhodian tyrant. The circular mausoleum was constructed around the 1st century BC, several centuries after his death. In early Christian times the tomb was converted into the church of Agios Aimilianós, though who was originally buried here still remains a mystery.

Péfkos, 3 km (2 miles) south of Líndos, has small sandy beaches fringed by pine trees, and is fast developing as a popular resort.

Lárdos is a quiet inland village, 7 km (4 miles) west of Líndos. **Lárdos Bay**, 1 km (0.5 mile) south of the village, has sand dunes bordered by reeds, and is being developed with upmarket village-style hotels.

RECONSTRUCTION OF THE ACROPOLIS (C.AD 300)

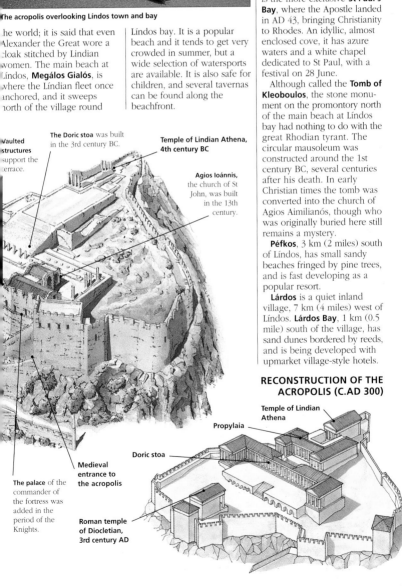

The Doric stoa was built in the 3rd century BC.

Vaulted structures support the terrace.

Temple of Lindian Athena, 4th century BC

Agios Ioánnis, the church of St John, was built in the 13th century.

Medieval entrance to the acropolis

The palace of the commander of the fortress was added in the period of the Knights.

Roman temple of Diocletian, 3rd century AD

Temple of Lindian Athena

Propylaia

Doric stoa

Nimporió with Agios Nikólaos church towering above the surrounding buildings

Chálki

Χάλκη

🏛 280. 🚢 Nimporió. ℹ Piátsa, Nimporió (22460 45333). 🏺 Chorió: Panagía 15 Aug. 🚢 Nimporió. **www**.chalki.gr

Chálki was once a thriving sponge-fishing island, but was virtually abandoned when its sponge divers emigrated to Florida in search of work in the early 1900s. Tourism has grown steadily as the island has been smartened up. Once fertile, Chálki's water table was infiltrated by sea water and the island is now barren with fresh water shipped in by tanker. Sheep and goats roam the rocky hillside, there is little cultivation and produce is imported from Rhodes.

Nimporió

Chálki's harbour and only settlement, Nimporió is a quiet and picturesque village with a Neo-Classical flavour.

A goat farmer in Chálki on his journey home

The main sight in Nimporió is the church of **Agios Nikólaos** with its elegant bell tower, the highest in the Dodecanese, tiered like a wedding cake. The church is also known for its magnificent black and white *choklákia* pebble mosaic courtyard depicting birds and the tree of life. The watchful eye painted over the main door is to ward off evil spirits.

A row of ruined windmills stands above the harbour, which also boasts an Italianate town hall and post office plus a fine stone clock tower. Nearby is sandy **Póntamos** beach, which is quiet and shallow and suitable for children.

Around the Island

The island is almost traffic-free so it is ideal for walkers. An hour's walk uphill from Nimporió is the abandoned former capital of **Chorió**. Its Crusader castle perches high on a crag, worth a visit for the coat of arms and Byzantine

---

CHOKLAKIA MOSAICS

A distinctive characteristic of the Dodecanese, these decorative mosaics were used for floors from Byzantine times onwards. An exquisite art form as well as a functional piece of architecture, they were made from small sea pebbles, usually black and white but occasionally reddish, wedged together to form a kaleidoscope of raised patterns. Kept wet, the mosaics also helped to keep houses cool in the heat.

Early examples featured abstract, formal and mainly geometric designs such as circles. Later on the decorations became more flamboyant with floral patterns and symbols depicting the lives of the householders with ships, fish and trees. Aside from Chálki, the houses of Líndos also have fine mosaics *(see pp 196–7)*. On Sými the church of Agios Geórgios *(see p 178)* depicts a furious mermaid about to dash a ship beneath the waves.

A *choklákia* mosaic outside Moní Taxiárchi in Sými

Circular *choklákia* mosaic in Chálki

---

escoes in the ruined chapel. On a clear day you can see Crete. The Knights of St John (*see pp188–9*) built it on an ancient acropolis, using much of the earlier stone.

The Byzantine church of the **Panagía** below the castle has some interesting frescoes and is the centre for a giant festival on 15 August. Clinging to the mountainside opposite is the church of **Stavrós** (the Cross).

From Chorió you can follow the road west to the Byzantine **Moní Agíou Ioánnou Prodrómou** (St John the Baptist). The walk takes about three to five hours, or it is a one hour drive. The monastery has an attractive shaded courtyard. It is best to visit in the early morning or to stay overnight: the caretakers will offer you a cell. You can walk from Nimporió to the pebbly beaches of **Kánia** and **Dyó Gialí** or take a taxi boat.

The interior of Moní Agíou Ioánnou Prodrómou

Outlying Islands

Excursions run east from Nimporió to deserted **Alimiá** island, where Italy berthed some submarines in World War II. There are several small chapels and a ruined castle.

Kastellórizo
Καστελλόριζο

Remote Kastellórizo is the most far-flung Greek island, just 2.5 km (1.5 miles) from Turkey but 120 km (75 miles) from Rhodes. It was very isolated until the airport opened up tourism in 1987. Kastellórizo has no beaches, but clear seas full of marine life, including monk seals, and it is excellent for snorkelling. Known locally as Megísti (the biggest), it is the largest of 14 islets.

The island's population has declined from 15,000 in the 19th century to nearly 300 today. From 1920 it was severely oppressed by the Italians who occupied the Dodecanese, and in World War II it was evacuated and looted.

Despite hardships, the waterside bustles with tavernas and sometimes impromptu music and dancing. It is a strange backwater but the indomitable character of the islanders is famous throughout Greece.

Kastellórizo town is the island's only settlement, with reputedly the best natural harbour between Piraeus and Beirut. Above the town is the ruined fort or **kástro** with spectacular views over the islands and the coast of Turkey. It was named the Red Castle (Kastello Rosso) by the Knights of St John due to its red stone, and this name was adopted by the islanders. The **Castle Museum** contains costumes, frescoes and photographs. Nearby, cut into the rock, is Greece's only **Lycian Tomb**, from the ancient Lycian civilization of Asia Minor. It is noted for its Doric columns.

Most of the old Neo-Classical houses stand in ruins, blown up during World War II or destroyed by earthquakes.

A traditional housefront in Kastellórizo town

However, buildings are being restored as tourism develops. The Italian film *Mediterraneo* was set here and since then the island has attracted many Italian tourists.

Highlights worth seeing include the elegant cathedral of **Agioi Konstantínos kai Eléni**, incorporating granite columns from the Temple of Apollo in Patara, Anatolia.

From town a path leads up to four white churches and the **Palaiókastro**. This Doric fortress and acropolis has a 3rd-century BC inscription on the gate referring to Megísti.

A boat trip southeast from Kastellórizo town to the spectacular **Parastá Cave** should not be missed; it is famed for its stalactites and the strange light effects on the vivid blue waters.

🏛 **Castle Museum**
Kastellórizo town. **Tel** 22460 49283.
🕐 8:30am–2:30pm Tue–Sun.

Kastellórizo town with Turkey in the background

Kárpathos
Κάρπαθος

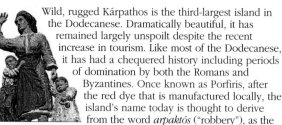

Wild, rugged Kárpathos is the third-largest island in the Dodecanese. Dramatically beautiful, it has remained largely unspoilt despite the recent increase in tourism. Like most of the Dodecanese, it has had a chequered history including periods of domination by both the Romans and Byzantines. Once known as Porfiris, after the red dye that is manufactured locally, the island's name today is thought to derive from the word *arpaktós* ("robbery"), as the island was a pop-

Folk reliefs on a taverna in Diafáni

ular pirate lair in medieval times.

KEY

For key to map see back flap

Kárpathos Town

Kárpathos town, also known as Pigádia, is the island's main port and capital, sheltered in the southeast of Vróntis bay. Once an ordinary working town, it now has hotels strung out all around the bay. The waterfront is bustling with cafés and restaurants that serve international fare. Opposite the Italianate town hall, **Kárpathos park** has an open-air display of ancient objects. Exhibits include an early Christian marble font and objects discovered in 5th-century BC Mycenaean tombs on the island.

Environs

South of Kárpathos town there is a pretty walk through olive groves to the main resort of **Amoopí**, 7 km (4 miles) away, with its string of sandy beaches. Above Amoopí, the village of **Menetés**, nestling at 350 m (1,150 ft) on the slopes of Mount Profítis Ilías, has quaint traditional pastel-coloured houses with attractive court-yards and gardens. Inside the village church is a carved wooden iconostasis.

SARIA

Chálki, Rhodes

• Vroukoúnda

Avlóna •

Olympos •   • Diafár

Apélla

Lefkós   *Kali Limni*

685 m / 2,250 ft   • Kyrá Panagiá

Apéri •

Othos •

KÁRPATHOS TOWN

Menetés

Arkása •   Profitis Ilías   Amoopí

510 m / 1,670 ft

Crete (Siteía)

ARMATHIA

Frý

Agía Marína •

KASOS

Crete (Siteía, Irákleio)

Chélathros Bay

0 kilometres 5

0 miles 3

Around the Island

A mountainous spine divides the wild north from the softer, fertile south. On the west coast, 8 km (5 miles) from Menetés, the village of **Arkása** has been transformed into a resort. In 1923, the 4th-century church of Agía Anastasía was discovered. It contained some fine early Byzantine mosaics, the best of which depicts two deer gazing into a water jug, now in the Rhodes's Archaeological Museum *(see p184)*.

Apéri, 8 km (5 miles) north of Kárpathos town, was the island's capital until 1892, and is said to be one of the richest villages in Greece. It sits 300 m (985 ft) up Mount Kalí Límni and has fountains and fine houses with exquisite gardens dating from the 1800s.

Othos, just to the west of Apéri, is the highest village on the island, at 450 m (1,500 ft) above sea level. It is also one

The white mansions of Apéri, clustered on the hillside

◁ **The historical village of Olympos, sitting high in the hills of northern Kárpathos**

Windmills in the traditional village of Olympos

VISITORS' CHECKLIST

5,000. 17 km (11 miles) S
of Kárpathos town. Kárpathos
town, Diáfani. corner of
28th Oktovríou & Dimokratías,
Kárpathos town. Kárpathos
town (22450 22222).
Panagía at Olympos: 15 Aug.

of the oldest, with traditional Karpathian houses. One of the houses is a **Folk Museum** with textiles and pottery on show. There is also a family loom and tools for traditional crafts.

The west coast resort of **Lefkós** is considered to be the jewel of the island by the Karpathians, with its three horseshoe bays of white sand. On the east coast, **Kyrá Panagiá**, with its pink-domed church, is another beautiful cove of fine white sand. **Apélla**, the next beach along, is a stunning crescent of sand with azure water.

Diafáni, a small, colourful village on the northeast coast, has a handful of tavernas and hotels and both sand and shingle beaches. A 20-minute bus-ride away is the village of **Olympos**, which spills down from a bleak ridge 600 m (1,950 ft) up. Founded in 1420, and virtually cut off from the rest of the island for centuries by its remote location, this village is now a strange mix of medieval and modern. The painted houses huddle together in a maze of steps and alleys just wide enough for mules. One traditional house, with just a single room containing many embroideries and bric-a-brac, is open to visitors. Customs and village life are carefully preserved and traditional dress is daily wear for the older women who still bake their bread in outdoor ovens.

From Olympos a rough track leads north to **Avlóna**, inhabited only in the harvest season by local farmers. From here, **Vroukoúnda**, the site of a 6th-century BC city, is a short walk away. Remains of the protective city walls can be seen, as can burial chambers cut into the cliffs.

Folk Museum
Othos village. **Tel** 22460 49283.
Apr–Oct: Tue–Sun. Nov–Mar.

Outlying Islands
North of Avlóna is the island of **Sariá**, site of ancient Nísyros, where the ruins of the ancient city can be seen. Excursion boats go there from Diáfani.

Barely touched by tourism, **Kásos**, off the south coast of Kárpathos, was the site of a massacre by the Turks in 1824, commemorated annually on 7 June in the capital, Frý. Near the village of Agía Marína are two fine caves, Ellinokámara and Selái, both with stalactites and stalagmites. Chélathros Bay is ideal for sun lovers, as are the quiet beaches of the tiny offshore islet of **Armáthia**.

Matriarch at the
Olympos windmills

THE TRADITIONS OF OLYMPOS

The costume of the women of Olympos consists of white pantaloons with an embroidered tunic or a dark skirt with a long patterned apron. Fabrics are heavily embroidered in lime green, silver and bright pinks. Daughters wear a collar of gold coins and chains to indicate their status and attract suitors. The society was once strictly matriarchal. Today the mother passes on her property to the first-born daughter and the father to his son, ensuring that the personal fortunes of each parent are preserved through the generations.

Traditional houses in Olympos often have decorative balconies and the initials of the owners sculpted above the entrance. Consisting of one room built around a central pillar with fold-away bedding, they are full of photographs and souvenirs. People flock to Olympos from all over the world for the Festival of the Assumption of the Virgin Mary, from 15 August, one of the most important festivals in the Orthodox church. The village celebrations of music and

Interior of an Olympos house

dance last three days. Traditional instruments are played, including the *lýra*, which stems from the ancient lyre, the bagpipe-like goat-skin *tsampourás*, and the *laoúto*, which is similar to a mandolin.

THE CYCLADES

ANDROS · TINOS · MYKONOS · DELOS · SYROS · KEA
KYTHNOS · SERIFOS · SIFNOS · PAROS · NAXOS · AMORGOS
IOS · SIKINOS · FOLEGANDROS · MILOS · SANTORINI

*D*eriving their name from the word "kyklos", meaning circle, because they surround the sacred island of Delos, the Cyclades are the most visited island group. They are everyone's Greek island ideal, with their dazzling white houses, twisting cobbled alleyways, blue-domed churches, hilltop windmills and stunning beaches.

The islands were the cradle of the Cycladic civilization (3000–1000 BC). The early Cycladic culture developed in the Bronze Age and has inspired artists ever since with its white marble figurines. The Minoans from Crete colonized the islands during the middle Cycladic era, making Akrotíri on Santoríni a major trading centre. During the late Cycladic period the Mycenaeans dominated, and Delos became their religious capital. The Dorians invaded the islands in the 11th century BC, a calamity that marked the start of the Dark Ages.

Venetian rule (1204–1453) had a strong influence, evident today in the medieval kástra seen on many islands and the Catholic communities on Tínos, Náxos and Sýros.

Traditional mule transport

There are 56 islands in the group, 24 inhabited, some tiny and undisturbed, others famous holiday playgrounds. They are the ultimate islands for sun, sea and sand holidays, with good nightlife on Mýkonos and Ios. Sýros, the regional and commercial capital, is one of the few islands in the group where tourism is not the mainstay. Cycladic life is generally centred on the village, which is typically divided between the harbour and the upper village, or Chóra, often topped with a kástro.

Most of the Cyclades are rocky and arid, with the exceptions of wooded and lush-valleyed Andros, Kéa and Náxos. This variety ensures the islands are popular with artists, walkers and those seeking quiet relaxation.

The sandy cove of Kolympíthres beach, Páros

◁ The tiered, whitewashed houses of Triandáros village, Tínos

Exploring the Cyclades

The Cyclades are best known for their beaches and whitewashed clifftop villages with stunning views; most famously, Firá on Santoríni. Mýkonos and Ios are well-established beach destinations, while more remote islands such as Mílos and Amorgós also have beautiful stretches of sand. Packed in July and August, these usually arid islands are beautiful in spring when they are carpeted with wild flowers. Varying in character, some of the islands, such as Síkinos, are quiet and traditional whereas others, such as Ios, are more nightlife-orientated. The Cyclades also offer a rich ancient history, evident in the ruins of ancient Delos.

GETTING AROUND

Páros and Sýros are the travel hub of the Cyclades. Ferries serve most of the islands from here and link to Crete and the Dodecanese. The islands are buffeted by the strong *meltémi* wind from July to September. It provides relief from the heat but can play havoc with ferry timetables.

Mýkonos and Santoríni have international airports, and islands with domestic airports include Sýros, Mílos, Páros and Náxos.

SEE ALSO

- *Where to Stay* pp313–17
- *Where to Eat* pp336–8
- *Travel Information* pp366–9

Boat garages in Mandrákia, Mílos

KEY

═══	Minor road
━━━	Scenic route
---	High-season, direct ferry route

ISLANDS AT A GLANCE

LOCATOR MAP

The cliff top town of Firá on Santoríni

Chíos,
karía

Vólos,
Thessaloníki

Kolympíthres

Pátmos,
Rafína,
Ikaría

Exómpourgo

Agios Ioánnis

Mýkonos

Mýkonos — Ano Merá

Platýs Gialós

Ríneia   *Delos*

Ikaría

Kos

Náxos

Apóllon

Komiakí

Náxos
Glinádo
Náousa
efkes
Píso
Livádi
Alyki
Chalkí
Apeíranthos
Filóti

Donoússa
Donoússa

Koufonísi
Chóra

Kéros

Astypálaia

Shinoússa
Agios Geórgios   Mesariá

Irákleia

Tholária
Aigiális

Katápola   Amorgós

Amorgós

Ios

Gialós

ikinos
Aloprónoia

Manganári

Anidro

Oía

Thirasía
Néa Kaméni

Firá

Santoríni

Akrotíri   Períssa

Anáfi

Anáfi

Astypálaia

Crete

0 kilometres   20

0 miles   10

Andros
Άνδρος

The northernmost of the Cyclades, Andros is lush and green in the south, scorched and barren in the north. The fields are divided by distinctive dry-stone walls. The island was first colonized by the Ionians in 1000 BC. In the 5th century BC, Andros sided with Sparta during the Peloponnesian War *(see p24)*. After Venetian rule, the Turks took power in 1566 until the War of Independence. Andros has long been the holiday haunt of wealthy Athenian shipping families.

Andros Town ❶
Χώρα

🏠 1,680. 🚌 Plateía Agías Olgas. 🛈 22820 22316.

The capital, Andros town, or Chóra, is located on the east coast of the island 20 km (12 miles) from the island's main port at Gávrio.

An elegant town with magnificent Neo-Classical buildings, it is the home of some of Greece's wealthiest shipowners. The pedestrianized main street is paved with marble slabs and lined with old mansions converted into public offices among the *kafeneía* and small shops.

The Hermes of Andros, in the Archaeological Museum

exhibits include the *Matron of Hercula-neum*, which was found with the *Hermes*, and finds from the 10th-century BC city at Zagorá. There are also finds from Ancient Palaiópoli *(see p210)* near Mpatsí, architectural illustrations and a large collection of ceramics.

The **Museum of Modern Art**, which was endowed by the Goulandrís family, has an excellent collection of paintings by 20th-century artists such as Picasso and Braque and leading Greek artists such as Alékos Fasianós. The sculpture garden has works by Michális Tómpros (1889–1974).

Plateía Kaïri
This is the main square in the town's Ríva district and is home to the **Archaeological Museum**, built in 1981. The museum's most famous exhibit is the 2nd-century BC *Hermes of Andros*, a fine marble copy of the 4th-century BC bronze original. Other

🏛 **Archaeological Museum**
Plateía Kaïri. **Tel** 22820 23664.
⭕ Tue–Sun. ● Main public hols.
🈚 📷

🏛 **Museum of Modern Art**
Plateía Kaïri. **Tel** 22820 22444.
⭕ Wed–Mon. ● main public hols. 🈚 except Sun. ♿

Káto Kástro and Plateía Ríva
From Plateía Kaïri an archway leads into the maze of streets that form the medieval city, Káto Kástro, wedged between Parapórti and Nimporió bays. The narrow lanes lead to wind-swept Plateía Ríva at the end of the peninsula, jutting into the sea and dominated by the heroic statue of the *Unknown Sailor* by Michális Tómpros. Just below, a precarious stone bridge leads to the islet opposite site, with the Venetian castle, **Mésa Kástro**, built in 1207–1233. The **Maritime Museum** has model ships, photographs and a collection of nautical instruments on display, is situated inside the town hall.

On the way back to the centre of the town is the church of **Panagía Theosképasti**, built in 1555 and dedicated to the Virgin Mary. Legend has it that the priest could not afford the wood for the church roof, so the ship delivering the wood set sail again. It ran into a storm and the crew prayed to the Virgin for help, promising to return the cargo to Andros. The seas were miraculously calmed and the church became known as Theosképasti, meaning "sheltered by God".

Statue of Unknown Sailor

🏛 **Maritime Museum**
Plateía Ríva. **Tel** 22820 22275.
⭕ Tue–Sun. ● main public hols. 🈚

Environs
Steniés, 6 km (4 miles) northwest of Andros town, is very beautiful and popular with wealthy shipping families. Fifteen minutes' walk south-west of Steniés, the 17th-century Mpístis-Mouvelás tower is a fine example of an Andriot house.

Below Steniés lies **Giália** beach where there is a fish taverna and trees for shade. In **Apoíkia**, 3 km (2 miles) west, mineral water is bottled from the Sáriza spring. You can taste the waters at the spring.

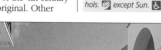

Typical white houses and a small church in Káto Kástro

Around Andros Island

Lion's head ntain in Ménites

Prosperous, neat, and dotted with many white dovecotes first built by the Venetians, Andros retains its traditional charm while playing host to international holiday-makers. There are a number of unspoiled sandy beaches, watersport facilities, wild mountains and a good network of footpaths. However, unless you are a keen trekker, car or bike hire is essential as the bus service is quite limited.

Mesariá ❷
Μεσαριά

8 km (5 miles) SW of Andros town.
🏠 850. 🚍

From Andros town the road passes through the medieval village of Mesariá with ruined tower-houses and the pantiled Byzantine church of the **Taxiárchis**, built by Emperor Emanuel Comnenus in 1158 and recently restored.

Springs gush from marble lion's head fountains in the leafy village of **Ménites**, just above Mesariá. Ménites is

known both for its nightingales and for the taverna overlooking a stream. Steps lead up to the pretty restored church of **Panagía i Koúmoulos** (the Virgin of the Plentiful) thought to be built on the site of an ancient Temple of Dionysos.

Moní Panachrántou overlooking the valley

SIGHTS AT A GLANCE

Andros Town ❶
Gávrio ❻
Mesariá ❷
Moní Panachrántou ❸
Mpatsí ❺
Palaiókastro ❹

VISITORS' CHECKLIST

🏠 10,000. 🛳 Gávrio. 🚍
ℹ 22820 22275. 🎊 Agios Panteleímon Festival at Moní Panachrántou: 27 Jul.

Moní Panachrántou ❸
Μονή Παναχράντου

12 km (7 miles) SW of Andros town.
Tel 22820 51090. ◯ daily.

This spectacular monastery is perched 230 m (755 ft) above sea level in the mountains southwest of Andros town. It can be reached either by a two-hour steep walk from Mesariá or a three-hour trek from Andros town.

It was founded in 961 by Nikifóros Fokás, who later became Byzantine Emperor as reward for his help in the liberation of Crete from Arab occupation. The fortified monastery is built in Byzantine style and today houses just three monks. The church holds many treasures, including the skull of Agios Panteleímon, believed to have healing powers. Visitors flock here to see the skull on the saint's annual festival day.

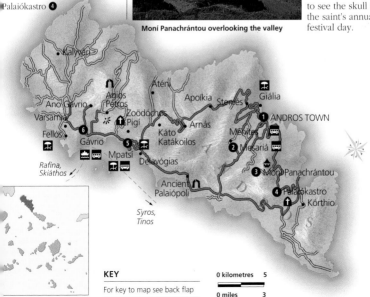

KEY

For key to map see back flap

0 kilometres 5

0 miles 3

Palaiókastro ❹
Παλαιόκαστρο

18 km (11 miles) SW of Andros town.
⬡ unrestricted access.

High on a rocky plateau inland is the ruined Venetian Palaiókastro built between 1207 and 1233. Its alternative name, the Castle of the Old Woman, is after a woman who betrayed the Venetians to the Turks in the 16th century. After tricking her way inside the castle, she opened the gates for the Ottoman Turks. Appalled by the bloody massacre that followed, she hurled herself off the cliffs near Kórthio, 5 km (3 miles) to the southeast, in remorse. The rock from which she jumped is known as Tis Griás to Pídima, or Old Lady's Leap.

Mpatsí ❺
Μπατσί

8 km (5 miles) S of Gávrio.
🏨 200. 🚌

Built around a sweeping sandy bay, Mpatsí is a pretty resort. It has a small fishing harbour and a maze of narrow lanes reached by white steps from the café-lined seafront. Despite the lively nightlife Mpatsí has retained its village atmosphere. The main beach is popular with families while **Delavógias** beach, south along the coastal track, is a favourite with naturists. Agía Marína, further along, has a friendly, family-run taverna.

Environs
South of Mpatsí the original capital of Andros, **Ancient Palaiópoli**, was inhabited until around AD 1000 when the

people moved to Mesariá *(see p209)*. It was largely destroyed in the 4th century AD by an earthquake, but part of the acropolis is still visible, as are the remains of some of the temples under the sea.

Inland lies **Káto Katákoilos** village, known for its island music and dance festivals. A rough track leads north from here to remote **Aténi**, a hamlet at the head of a lush valley. Two beautiful beaches lie further to the windy northeast, in the bay of Aténi. The garden village of **Arnás**, high on the slopes of the Kouvára mountain range, has flowing springs and is one of the island's greenest spots. The area has many dry-stone walls and is spectacular walking country.

🏛 Ancient Palaiópoli
9 km (6 miles) S of Mpatsí.
⬡ unrestricted access. ♿ limited.

Gávrio ❻
Γαύριο

🏨 450. 🚢 🚌 🚕 *Fellós 4 km (2.5 miles) NW.*

Gávrio is a rather characterless port which, at weekends, becomes packed with Athenians heading for their holiday homes. There is a beach, a good campsite and plenty of tavernas. During the high season it can be the only place with rooms available as

Agios Pétros tower near Gávrio

Mpatsí is often pre-booked by package companies.

Environs
From Gávrio, it takes an hour or so to walk up to the tower of **Agios Pétros**, the island's best-preserved ancient monument. Dating from the Hellenistic era, the tower stands 20 m (66 ft) high in an olive grove below the hamlet of Káto Agios Pétros. The upper storeys of the tower were reached by footholds and an internal ladder, and its inner hall was once crowned by a corbelled dome. The purpose of the tower remains a mystery, although it may have been built to serve as a watchtower to guard the nearby mines from attack by marauding pirates.

North of Gávrio there are good beaches beyond the village of Varsamiá, which has two sandy coves. **Fellós** beach is the best, but is fast being developed with holiday villas.

A turn-off from the coastal road, 8 km (5 miles) south of Gávrio, leads to the 14th-century convent, **Zoödóchos Pigí**, the Spring of Life. Only a handful of nuns remain where there were 1,000 monks, but they are happy to show visitors their collection of icons and Byzantine tapestries

The beach at Mpatsí Bay on Andros

Cycladic Art

With their simple geometric shapes and purity of line, Cycladic marble figurines are the legacy of the islands' Bronze-Age civilization *(see pp28–9)* and the first real expression of Greek art. They all come from graves and are thought to represent, or be offerings to, an ancient deity. The earliest figures, from before 3000 BC, are slim and violin-shaped. By the time of the Keros-Sýros culture of 2700–2300 BC, the forms are recognizably human and usually female. They range from palm-sized up to life-size, the proportions remaining consistent. Obsidian blades, marble bowls prefiguring later Greek art, abstract jewellery and pottery, including the strange "frying pans", also survive. The examples of Cycladic art shown here are from the Museum of Cycladic Art in Athens *(see p291)*. Cycladic artefacts are also in many museums throughout the Cyclades.

"Violin" figurines, such as this one, date from the early Cycladic period of 3300–2700 BC. Often no bigger than a hand, the purpose of these highly schematic representations of the human form is unknown. In some graves up to 14 of these figurines were found; other graves had none.

"Frying pan" *pottery vessels take their name from their shape but their function is unknown. They may have been used in religious rituals. Decorated with spirals or suns, they belong to the mature phase of Cycladic art.*

Collared vases, *or kandelas, carved from marble, are one of the high points of Cycladic art. Probably used for food storage, the four lugs on the sides would have allowed them to be hung from a support.*

INFLUENCE ON MODERN ART

Considered crude and ugly when first discovered in the 19th century, the simplicity of both form and decoration of Cycladic art exerted a strong influence on 20th-century artists and sculptors such as Picasso, Modigliani, Henry Moore and Constantin Brancusi.

This male figurine, *found together with a female figurine, is one of the few male figures to have been found. He is also atypical in having one arm raised and a band slung across his chest.*

This female figurine *with folded arms is typical of Cycladic sculpture. The head is slightly tipped back, with only minimal markings for arms, legs and features.*

Henry Moore's *Three Standing Figures*

The Kiss **by Brancusi**

Tínos
Τήνος

A craggy yet green island, Tínos was first settled by Ionians in Archaic times. In the 4th century BC it became known for its Sanctuary of Poseidon and Amphitrite. Under Venetian rule from medieval times, Tínos became the Ottoman Empire's last conquest in 1715. Tínos has over 800 chapels, and in the 1960s the military Junta declared it a holy island. Many Greek Orthodox pilgrims come to the church of the Panagía Evangelístria (Annunciation) in Tínos town. The island is also known for its many dovecotes (peristeriónes), scattered across the landscape.

Archaeological Museum exhibit from Exómpourgo

Christians. Tínos becomes very busy during the festivals of the Annunciation and the Assumption when the icon is paraded through the streets (see pp48–9) and the devout often crawl to Panagía Evangelístria.
The church is a treasury of offerings, such as an orange tree made of gold and silver, from pilgrims whose prayers have been answered. The icon itself is so smothered in gold and jewels it is hard to see the painting. The crypt where it was found is known as the chapel of Evresis, or Discovery. Where the icon lay is now lined with silver and the holy spring here, Zoödóchos Pigí, is said to have healing powers.
The vestry has gold-threaded ecclesiastical robes, and valuable copies of the gospels.

Tínos town and the small harbourfront

Tínos Town
A typical island capital, Tínos town has narrow streets, whitewashed houses and a bustling port lined with restaurants and hotels.

🏠 Panagía Evangelístria
Church & museums ☐ daily.
Tel 22830 22256. 🅰
Situated at the top of Megalóchari, the main street that runs up from the ferry, Panagía Evangelístria, the church of the Annunciation, dominates Tínos town. The pedestrianized Evangelistrías, which runs parallel to Megalóchari, is packed with stalls full of icons and votive offerings. Built in 1830,

the church houses the island's miraculous icon. In 1822, during the Greek War of Independence, Sister Pelagía, a nun at Moní Kechrovouníou, had visions of the Virgin Mary showing where an icon had been buried. In 1823, acting on the nun's directions, excavations revealed the icon of the Annunciation of the Archangel Gabriel, unscathed after 850 years underground. Known in Greece as the Megalóchari (the Great Joy) the icon was found to have healing powers, and the church became a pilgrimage centre for Orthodox

Pilgrim crawling to the Panagia Evangelístria

KEY

For key to map see back flap

0 kilometres 5
0 miles 3

The pretty village of Pýrgos in the north of the island

lso within the complex is museum with items by ocal sculptors and painters, ncluding works by sculptors ntónios Sóchos, Geórgios itális and Ioánnis Voúlgaris. he art gallery has works of he Ionian School, a Rubens, a embrandt and 19th-century vorks by international artists.

Archaeological Museum

Iegalóchari **Tel** 22830 22670.
Je–Sun. main public hols.

On Megalóchari, near the hurch, is the Archaeological Iuseum which has displays f sculptures of nereids (sea-ymphs) and dolphins found t the Sanctuary of Poseidon nd Amphitrite. There is also 1st-century BC sundial by ndronikos Kyrrestes, who esigned Athens' Tower of the Vinds *(see p287)*, and some uge 8th-century BC storage rs from ancient Tínos on e rock of Exómpourgo.

nvirons

ast of town, the closest each is shingly **Agios Fokás**. o the west is the popular each at **Stavrós**, with a jetty hat was built in Classical mes. To the north near Kiónia re the foundations of the 4th-entury BC **Sanctuary of oseidon and Amphitrite**, his ea-nymph bride. The excava-ons here have yielded many olumns, or *kiónia*, after which e surrounding area is named.

round the Island

ínos is easy to explore as there re plenty of taxis and a good us service around the island. orth of Tínos town is the 2th-century walled **Moní**

Kechrovouníou, one of the largest convents in Greece. You can visit the cell where Sister Pelagía had her visions and the chest where her em-balmed head is kept.

At 640 m (2,100 ft) high, the great rock of **Exómpourgo** was the site of the Archaic city of Tínos and later became home to the Venetian fortress

The interior of the 12th-century Moní Kechrovouníou

of St Elena. Built by the Ghisi family after the Doge handed over the island to them in 1207, the fortress was the toughest stronghold in the Cyclades, until it surrendered to the Turks in 1714. You can see remains of a few ancient walls on the crag, medieval houses, a fountain and three churches.

From Kómi, to the north, a valley runs down to the sea at **Kolympíthres**, with two sandy bays: one is deserted; the other has rooms and tavernas.

Overlooking the harbour of Pánormos in the northwest of the island, the pretty village of **Pýrgos** is famous for its sculpture school. The area is known for its green marble, and the stonework here is among the finest in the islands. Distinctive, carved marble fan-lights and balconies decorate the island villages. There are examples at the Giannoúlis Chalepás Museum, housed in the former home of the island's renowned sculptor (1851–1938). The old grammar school is now the School of Fine Arts, and a shop in the main square exhibits and sells works by the students.

Giannoúlis Chalepás Museum

Pýrgos. daily. Oct–Apr.

THE PERISTERIONES (DOVECOTES) OF TINOS

The villages of Tínos are studded with around 1,300 beautiful white dovecotes *(peristeriónes)*, all elaborately decorated. They have two storeys: the lower floor is for storage, the upper houses the doves and is usually topped with stylized winged finials or mock doves. The breeding of doves was introduced by the Venetians. Although also found on the islands of Andros and Sífnos, the *peristeriónes* of Tínos are considered the finest.

A dovecote in Kámpos with traditional elaborate patterns

Mýkonos
Μύκονος

Pétros the Pelican, the island mascot

Although Mýkonos is dry and barren, its sandy beaches and dynamic nightlife make this island one of the most popular in the Cyclades. Under Venetian rule from 1207, the islanders later set up the Community of Mykonians in 1615 and flourished as a self-sufficient society. Visited by intellectuals in the early days of tourism, today Mýkonos thrives on its reputation as the glitziest island in Greece.

Mýkonos harbour in the early morning

Mýkonos Town

Mýkonos town (or Chóra) is the supreme example of a Cycladic village – a tangle of dazzling white alleys and cube-shaped houses. Built in a maze of narrow lanes to defy the wind and pirate raids, the bustling port is one of the most photographed in Greece. Many visitors still get lost around the lanes today.

Taxi boats for the island of Delos *(see pp 218–19)* leave from the quayside. The island's mascot, Pétros the Pelican, may be seen near the quay, hunting for fish.

Adjacent to the harbour is Plateía Mavrogénous, overlooked by the bust of revolutionary heroine Mantó Mavrogénous (1796–1848). She was awarded the rank of General for her victorious battle against the Turks on Mýkonos during the War of Independence in 1821.

The **Archaeological Museum**, housed in a Neo-Classical building south of the ferry port, has a large collection of Roman and Hellenistic carvings, 6th- and 7th-century BC ceramics, jewellery and gravestones, as well as many finds from the ancient site on Delos.

Kástro, the oldest part of the town, sits high up above the waterside district. Built on part of the ancient castle wall is the excellent **Folk Museum**, one of the best in Greece. It is housed in an elegant sea-captain's mansion and has a fine collection of ceramics, embroidery and ancient and modern Mykonian textiles. Among the more unusual exhibits is the original Pétros the Pelican, now stuffed, who was the island's mascot for 29 years. The 16th-century Voní̱s Windmill is part of the Folk Museum and has been restored to full working order. It was one of the 30 windmills that were used by families all over the island to grind corn. There is also a small threshing floor and a dovecote in the grounds around the windmill.

Mantó Mavrogénous

The most famous church on the island, familiar from postcards, is the extraordinary **Panagía Paraportianí**, in the Kástro. Built on the site of the postern gate *(parapórti)* of the medieval fortress, it is made up of four chapels at ground level with another above. Part of it dates from 1425 while the rest was built in the 16th and 17th centuries.

7th-century amphora in Archaeological Museum

From Kástro, the lanes run down into Venetía, or **Little Venice** (officially known as Alefkándra), the artists' quarter. The tall houses have painted balconies jutting out over the sea. The main square, Plateía Aléfkandras, is home to the large Orthodox cathedral of Panagía Pigadiótissa (Our Lady of the Wells).

The **Maritime Museum of the Aegean**, at the end of Matogiánni, features a collection of model ships from pre-Minoan times to the 19th century, maritime instruments, paintings and 5th-century BC coins with nautical themes.

Next door, **Lena's House**, a 19th-century mansion, evokes the life of a Mykonian lady, Léna Skrivánou. Everything is preserved, from her needlework to her chamber pot.

Works of Greek and international artists are on show at the **Municipal Art Gallery** on Matogiánni, and include an exhibition of works by local Mykonian painters.

Working 16th-century windmill, part of the Folk Museum

For hotels and restaurants in this region see pp313–7 and pp336–8

The famous Paraportianí church

Archaeological Museum
Harbourfront. **Tel** 22890 22325.
Tue–Sun. main public hols.

Folk Museum
Harbourfront. **Tel** 22890 22591.
Apr–Oct: 5–8pm Mon–Sat.

Maritime Museum of the Aegean
Enóplon Dynámeon. **Tel** 22890 22700. Apr–Oct: daily. main public hols.

Lena's House
Enóplon Dynámeon. Apr–Oct: daily. limited.

Municipal Art Gallery
Matogiánni. **Tel** 22890 22615.
Jun–Oct: daily.

Around the Island

Mýkonos is popular primarily for its beaches; it has no lush countryside. The best ones are along the south coast. At stylish **Platýs Gialós**, 3 km (2 miles) south of the town, regular taxi boats are available to ferry sun-worshippers from bay to bay. Backed by hotels and restaurants, this is the main family beach on the island, with watersports and a long sweep of sand. Serious sun-lovers head southeast to the famous nudist beaches. First is **Parágka**, or Agía Anna, a quiet spot with a good taverna. Next is **Paradise**, with its neighbouring camping site, disco

music and water-sports. The lovely cove of **Super Paradise** is gay and nudist. **Eliá**, at the end of the boat line is also nudist and busy in high season.

In contrast to Mýkonos town, the inland village of **Ano Merá**, 7.5 km (4.5 miles) east, is traditional and largely unspoilt by tourism. The main attraction is the 16th-century **Panagía i Tourlianí**, dedicated to the island's protectress. Founded by two monks from Páros, the red-domed monastery was restored in 1767. The ornate marble tower was sculpted by Tíniot craftsmen. The monastery houses some fine 16th-century icons, vestments and embroideries. Northwest

of the village is Palaiókastro hill, once crowned by a Venetian castle. It is thought to be the site of one of the ancient cities of Mýkonos. Today it is home to the 17th-century working **Moní Palaiokástrou**. To the northwest, in the pretty village of **Maráthi**, is Moní Agíou Panteleïmona, founded in 1665. From here, the road leads to **Pánormos Bay** and **Fteliá**, a windsurfers' paradise.

Platýs Gialós beach, one of the best on Mýkonos

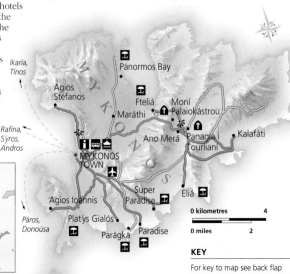

Ikaría, Tínos
Pánormos Bay
Agios Stéfanos
Fteliá
Moní Palaiokástrou
Maráthi
Rafína, Sýros, Andros
Ano Merá
Panagía i Tourlianí
Kalafáti
MÝKONOS TOWN
Agios Ioánnis
Super Paradise
Eliá
Páros, Donoúsa
Platýs Gialós
Parágka
Paradise

0 kilometres 4
0 miles 2

KEY

For key to map see back flap

The old houses of Little Venice, Mýkonos town ▷

Delos

Δήλος

Tiny, uninhabited Delos is one of the most important archaeological sites in Greece. According to legend, Leto gave birth to Artemis and Apollo here. The Ionians arrived in about 1000 BC, bringing the worship of Apollo and founding the annual Delia Festival, during which games and music were played in his honour. By 700 BC, Delos was a major religious centre. First a place of pilgrimage, it later became a thriving commercial port particularly in the 3rd and 2nd centuries BC. It is now an open-air archaeological museum with mosaics and marble ruins covered in wild flowers in spring.

Artemis of Delos

Archaeological Mus
This displays most of
finds from the
island, inclu
ing storage p
used for offeri
and koúroi d*
from the 7th
century BC.

The Sanct
of Apollo has t
temples: one da
from the 6th century
and two dating from
5th century

Stadium and Gymnasium

The Sanctuary of Dionysos
has remains of huge phallic monuments dating back to 300 BC.

The Sacred Lake,
now dried up, was so called because it had witnessed Apollo's birth. A wall marks the lake's Hellenistic boundaries.

★ Lion Terrace
The famous lions (now replaced by replicas) were set up to over-look and protect the Sacred Lake. They were carved from Naxian marble at the end of the 7th century BC. Originally there were nine, but now only five remain.

TIMELINE

3000 BC	1000 BC	750	500	250	AD 1
	1000 BC Ionians arrive on Delos and introduce Apollo worship	**422 BC** Athens exiles Delians to Asia Minor; Delians return the following year		**88 BC** Delos sacked by Mithridates	
		426 BC Second purification			
		478 BC Athenians make Delos the centre of the first Athenian League	**166 BC** Romans return Delos to Athens. Trade flourishes		
	700 BC Naxians in control of Sanctuary of Apollo		**314 BC** Delos declares independence from Athens	**250 BC** Romans settle in Delos	
2000 BC Earliest settlement on Mount Kýnthos		**550 BC** Polykrates, the tyrant of Sámos, conquers the Cyclades, but respects the sanctity of Delos	**543 BC** First purification (removal of tombs) of Delos by Athenians	**69 BC** Romans fortify Delos after sack by pirates	

House of the Dolphins
This house of the 2nd century BC contains a mosaic of two dolphins with an elaborate Greek key design and waved borders.

VISITORS' CHECKLIST

2.5 km (1 mile) SW of Mýkonos town. **Tel** 22890 22259.
8–10am daily from Mýkonos town returning 12–2pm.
8:30am–3pm Tue–Sun.
1 Jan, 25 Mar, Good Fri am, Easter Sun, Mon, 1 May, 25, 26 Dec.

Mount Kýnthos

House of the Masks
Probably a hostelry for actors, this house contains a 2nd-century BC mosaic of Dionysos, god of theatre, riding a panther.

★ Theatre
Built in 300 BC to hold 5,500 spectators, the theatre was sited in a natural amphitheatre. On its west side, a huge, vaulted cistern collected rainwater draining from the theatre and supplied part of the town.

★ Theatre Quarter
In Hellenistic and Roman times the wealthy built houses near the theatre, many with opulent, colonnaded courtyards.

House of Dioscourides and Cleopatra
Two statues represent the couple Cleopatra and Dioscourides, who lived here in the 2nd century BC.

KEY

☐ Theatre quarter

House of Dionysos
Inside the house is a mosaic depicting Dionysos riding a leopard. Twenty-nine tesserae *are used just to make up the animal's eye.*

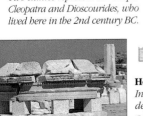

STAR SIGHTS

★ Theatre

★ Lion Terrace

★ Theatre Quarter

Sýros
Σύρος

Rocky Sýros, or Sýra, is the commercial, administrative and cultural centre of the Cyclades. Archaeological digs have revealed finds of the Cycladic civilization dating from 2800 to 2300 BC. The inhabitants converted to Catholicism under the French Capuchins in the Middle Ages. The 19th century saw Sýros become a wealthy and powerful port in the eastern Mediterranean. Though Sýros does not live off tourism, more visitors arrive each year attracted by its traditional charm.

Town hall, designed by Ernst Ziller

Andréas Miaoúlis. The square is dominated by the vast Neo Classical **town hall** (1876), designed by the German architect Ernst Ziller.

The **Archaeological Museum**, up the steps to the left of the town hall, houses bronze and marble utensils from the 3000 BC Cycladic settlement of Chalandrianí. Also on display are Cycladic statuettes and Roman finds. Left of the town hall is the **Historical Archives Office**.

Nearby, on Plateía Vardáka, is the **Apollo Theatre**, designed in 1864 by French architect Chabeau as a copy of La Scala, Milan. The first opera house in Greece, it is noted for its

The twin peaks of Ermoúpoli: Ano Sýros and Vrondádo

Ermoúpoli ❶
Ερμούπολη

🏠 *13,000.* 🚢 🚌 *Akti Ethnikis Antistassis.* ℹ️ *Thymáton Sperchion 11 (22810 86725).*

Elegant Ermoúpoli, named after Hermes, the god of commerce, is the largest city in the Cyclades. In the 19th century it was Greece's leading port and a major coaling station with a huge natural harbour and thriving shipyard. Crowned by the twin peaks of Catholic Ano Sýros to the north, and the Orthodox Vrontádo to the south, the city is built like an amphitheatre around the harbour.

The Lower Town
The architectural glories of central **Plateía Miaoúli** have led to the town becoming a National Historical Landmark. Paved with marble and lined with palm-shaded cafés and pizzerias, the grand square is the city's hub and meeting place, especially for the evening stroll, or *vólta*. There is also a marble bandstand and a statue dedicated to the revolutionary hero Admiral

SIGHTS AT A GLANCE
Ermoúpoli ❶
Galissás ❸
Kíni ❷
Poseidonía ❹
Vári ❺

KEY

For key to map see back flap

0 kilometres   4
0 miles   2

Statue of Andr Miaoúlis

Statue of Andr Miaoúlis

MARKOS VAMVAKARIS

One of the greatest exponents of *rempétika*, the Greek blues, Márkos Vamvakáris (1905–72) was born in Ano Sýros. Synonymous with hash dens and the low-life, *rempétika* was the music of the urban underclass. With strong Byzantine and Islamic influences, it is often played on the *baglama* or the bouzouki. Vamvakáris was a master of the bouzouki as well as a noted composer. Over 20 recordings have been made of his music, the earliest of which dates back to the 1930s. A bust of Vamvakáris looks out to sea from the small square named after him in Ano Sýros.

fine wall paintings of Mozart and Verdi and is still used for plays and concerts.

Across the street the 1871 **Velissarópoulos Mansion**, now housing the Labour Union, has an elaborate marble façade and splendid painted ceilings and murals. Beyond here is the church of **Agios Nikólaos** (1848) with a marble iconostasis by the 19th-century sculptor Vitális. Also by Vitális is the world's first monument of the unknown soldier, in front of the church.

The Upper Town

The twin bell towers and distinctive blue and gold dome of Agios Nikólaos mark the start of the **Vapória** district. Here Sýros's shipowners built their Neo-Classical mansions, with some of the finest plasterwork, frescoes and marble carvings in

Marble iconostasis by Vitális, in the church of Agios Nikólaos

Greece. The houses cling to the coastline above the town's quays and moorings at Tálira, Evangelístria and Agios Nikólaos.

The charming district of **Vrontádo**, on the eastern peak, has a number of excellent tavernas spread out on its slopes at night. The Byzantine church of the **Anástasis** on top of the hill has views to Tínos and Mýkonos.

A half-hour's climb along Omiroú, or a brief bus ride, is the fortified medieval quarter of **Ano Sýros**, on the western peak. It is also known as Apáno Chóra or Kástro. On the way is the Orthodox cemetery of **Agios Geórgios** with its elaborate marble mausoleums. Ano Sýros is a maze of whitewashed passages, arches and steps forming a huddle of interlinking houses. The architecture is unique, making the most of minimal space with *stegádia* (slate or straw roofs) and tight corners. The main entrance to Ano Sýros is Kamára, an ancient passageway leading into the main road, or Piatsa. The **Vamvakáris Museum**, dedicated to the life and work of Márkos Vamvakáris, is situated just off this road. At the top of Ano

A ceiling in one of Ermoúpoli's mansions

Sýros, the Baroque **Aï-Giórgis**, known as the cathedral of St George, was built on the site of a 13th-century church. The basilica contains fine cloister. The Jesuit cloister was founded in 1744 around the church of Our Lady of Karmilou (1581), and houses 6,000 books and manuscripts in its library. Below it, the Capuchin convent of **Agios Ioánnis** was a meeting place and a refuge from pirates. Its church was founded by Louis XIII of France as a poorhouse.

🏛 **Archaeological Museum**
Plateía Miaoúli. **Tel** 22810 88487. ◯ Tue–Sun. ⬤ main public hols.

🏛 **Historical Archives Office**
Plateía Miaoúli. **Tel** 22810 86891. ◯ 8:30am–2:30pm Mon–Fri. ⬤ main public hols. 📷

🏛 **Vamvakáris Museum**
Plateía Vamvakári, Ano Sýros. **Tel** 22813 60900. ◯ Jun–Sep: daily. ⬤ main public hols. 📷

A typical street in the Ano Sýros quarter

Around Sýros Island

Sýros has numerous attractive coves as well as popular resorts like Galissás and Kíni. The landscape is varied with palm trees and terraced fields. In the northern region of Apáno Meriá the traditional farms built to house both families and animals are in total contrast to the Italianate mansions and holiday homes of the south. Sýros has good roads, especially in the south, and is easy to explore by car or bike. There is a regular bus from the harbour to Ano Sýros, the main resorts and outlying villages.

Kíni Bay and the town's harbour

Kíni ❷
Κίνι

9 km (6 miles) NW of Ermoúpoli.
🏠 300. 🚌 🚆 Delfíni 3 km
(2 miles) N.

The fishing village of Kíni is set in a horseshoe-shaped bay with two good sandy beaches. Kíni is a popular meeting place for watching the sunset over an ouzo, and it has some excellent fish tavernas.

North, over the headland, is the award-winning **Delfíni** beach – the largest on Sýros and popular with naturists.

Between Ermoúpoli and Kíni, set in pine-covered hills, is the red-domed convent of **Agía Varvára**. With spectacular views to the west, the

The red-tiles roofs of Agía Varvára convent near Kíni

Orthodox convent was once a girls' orphanage. The nuns run a weaving school and their knitwear and woven goods are on sale at the convent. The frescoes in the church depict the saint's martyrdom.

Environs
Boat services run from Kíni to some of the island's remote northern beaches. **Grámmata Bay** is one of the most spectacular, a deep sheltered inlet with golden sands where sea lilies grow in autumn. Some of the rocks here have a Hellenistic inscription carved on them, seeking protection for ships from sinking.

A boat trip around the tip of the island past Cape Diapóri to the east coast takes you to **Sykamiá** beach. Here there is a cave where the Syriot philosopher Pherekydes is thought to have lived during the summer months. A physicist and astronomer, Pherekydes pioneered philosophical thought in the mid-6th century BC, and was the inventor of the heliotrope, an early sundial. From Sykamiá you

can see the remains of the Bronze-Age citadel of **Kastrí** with its six towers perched on a steep rock.

Galissás ❸
Γαλησσάς

7 km (4 miles) W of Ermoúpoli.
🏠 500. 🚌 🚆 Armeós beach 1 km
(0.5 miles) N.

Lively Galissás has the most sheltered beach on the island, fringed by tamarisk trees and, across the headland to the north, **Armeós** beach is a haven for nudists. Galissás has both the island's campsites, making it popular with backpackers. In high season it can be a noisy place to stay, and is often full of bikers. To the south of the bay lies **Agía Pákou**, which is the site of the Classical city of Galissás.

Huge **Foínikas** bay, 3 km (2 miles) further south, was originally settled by the Phoenicians, and now houses more than 1,000 people. Foínikas is a popular resort with a pier and moorings for yachts and fishing boats.

Sweeping Foínikas bay on the southwest coast of Sýros

Poseidonía ❹
Ποσειδωνία

12 km (7 miles) SW of Ermoúpoli.
🏠 700. 🚌 🚆 Agathopés 1 km
(0.5 miles) S.

Poseidonía, or Dellagrázia, is one of the largest tourist sites on the island, with cosmopolitan hotels and restaurants. The island's first main road

An Italianate mansion in Poseidonía

was built in 1855 from Ermoú-poli through Poseidonía to Foínikas. The affluent village contains some Italianate mansions, which are the country retreats of wealthy islanders. A short walk to the southwest, quieter **Agathopés** is one of the island's best beaches with safe waters protected by an islet opposite. Mégas Gialós, 3 km (2 miles) away on the west coast, is a pretty beach shaded by tamarisk trees.

Vári **⑤**
Βάρη

8 km (5 miles) S of Ermoúpoli.
🚶 1,150. 🚌 🚐 Vári.

Quaint, sheltered Vári has become a major resort, but it still has traditional houses. On the Chontrá peninsula, east of the beach, is the site of the island's oldest prehistoric settlement (4000–3000 BC).

Kéa
Κέα

🚶 1,600. 🚢 🚐 Korissía.
ℹ 22880 21100. 🚤 Gialiskári 6 km (4 miles) NW of Ioulís

Kéa was first inhabited in 3000 BC and later settled by Phoenicians and Cretans. In Classical times it had four cities: Ioulís, Korissía, Poiíessa and Karthaía. The remains of Karthaía can be seen on the headland opposite Kýthnos. It is a favourite spot for rich Athenians due to its proximity to Attica. Mountainous, with fertile valleys, Kéa has been known since ancient times for its wine, honey and almonds.

Ioulís

The capital, Ioulís, or Ioulída, with its red terracotta-tiled roofs and winding alleyways, is perched on a hillside 5 km (3 miles) above Korissía. Ioulís is situated on the Mountain of the Mills. The town is a maze of tunnel-like alleys, and has a spectacular Neo-Classical **town hall** (1902) topped with statues of Apollo and Athena. On the west side are ancient bas-relief sculptures and in the entrance a sculpture of a woman and child found at ancient Karthaía.

The Kástro quarter is reached through a white archway, which stands on the site of the ancient acropolis. The Venetians, under the leadership of Domenico Micheli, built their castle in 1210 with stones from the ancient walls and original Temple of Apollo. There are panoramic views from here. The **Archaeological Museum**

is based in a fine Neo-Classical house. Its displays include an interesting collection of Minoan finds from Agía Eiríni; artifacts from the four ancient cities; Cycladic figurines and ceramics; and a copy of the stunning, marble, 6th-century BC *koúros* of Kéa. The smiling 6th-century BC **Lion of Kéa** is carved into the rock 400 m (1,300 ft) north of the town.

🏛 Archaeological Museum
Tel 22880 22079. ⏰ Tue–Sun.
🚫 main public hols. 🎫

Around the Island

The port of **Korissía** can be packed with Greek families on holiday breaks; as can **Vourkári**, an attractive and popular resort further north on the island that is famous for its fish tavernas.

The archaeological site of **Agía Eiríni** is topped by the chapel of the same name. The Bronze-Age settlement was destroyed by an earthquake in 1450 BC, and was excavated from 1960 to 1968. First occupied at the end of the Neolithic period, around 3000 BC, the town was fortified twice in the Bronze Age and there are still remains of the great wall with a gate, a tower and traces of streets. Many of the finds are displayed in the Archaeological Museum in Ioulís. The most spectacular monument on Kéa is the Hellenistic tower at Moní Agías Marínas, 5 km (3 miles) southwest of Ioulís.

A Hellenistic tower at Moní Agías Marínas on Kéa

Kýthnos
Κύθνος

🏛 1,500. 🚢 🚌 Mérichas.
ℹ 22810 31201.

Barren Kýthnos attracts more Greek visitors than foreign tourists, although it is a popular anchorage for flotilla holidays. Its dramatic, rugged interior and the sparsity of visitors make it an ideal location for walkers.

The local clay was traditionally used for pottery and ceramics, but is also used to make the red roofing tiles that characterize all the island's villages.

Known locally as Thermiá because of the island's hot springs, Kýthnos attracts visitors to the thermal spa at Loutrá. Since the closure of the iron mines in the 1940s, the islanders have lived off fishing, farming and basket-weaving. To celebrate festivals, such as the major pre-Lenten carnival, the islanders often wear traditional costumes.

Chora
Also known as Messariá, the capital is a charming mix of red roofs and Cycladic cube-shaped houses. Also worth visiting is the church of **Agios Sávvas**, founded in 1613 by the Venetian Cozzadini family whose coat of arms it bears. The oldest church is **Agía Triáda** (Holy Trinity), a domed, single-aisle basilica.

Interior of the church of Panagía Kanála in Kanála town on Kýthnos

Around the Island
The road network is limited, but buses connect the port of Mérichas with Kanála in the south and Loutrá in the north. The remaining areas of the island are mostly within a walkable distance of these points. **Mérichas**, on the west coast, has a small marina and tree-fringed beach, lined with small hotels and tavernas. Just to the north, the sandy beach of **Martinákia** is popular with families. Further along the coast are the lovely beaches at **Episkopí** and **Apókrisi**, overlooked by **Vryókastro**, the Hellenistic ruins of ancient Kýthnos.

Potter at work in Dryopída

You can walk to **Dryopída**, a good hour south of Chóra, down the ancient cobbled way with dramatic views. The town was named after the ancient Dryopes tribe whose king, Kýthnos, gave the island its name. The charming red-roofed village is divided into two districts by the river valley: Péra Roúga is lush with crops, while Galatás was once a centre for ceramics, but only one pottery remains.

At **Kanála**, 5 km (3 miles) to the south, holiday homes have sprung up by the church of Panagía Kanála, dedicated to the Virgin Mary, the island's patron saint. Set in attractive shaded picnic grounds, the church houses Kýthnos's most venerated icon of the Virgin. It is probably by master iconographer, Skordílis, as Kýthnos was a centre for icon-painting in the 17th century. Kanála beach has views to Sérifos and Sýros and there are good beaches nearby.

Loutrá is a straggling resort on the northeast coast with windswept beaches. Its spa waters are saturated with iron, and since ancient times the springs of Kákavos and Agioi Anárgyroi have been used as a cure for ailments ranging from gout, rheumatism and eczema to gynaecological problems. The Xenía Hotel, situated next door to the excellent Hydrotherapy Centre, has late 19th-century marble baths inside. A Mesolithic settlement to the north, dating from 7500–6000 BC, is the oldest in the Cyclades.

Sérifos
Σέριφος

🏛 1,100. 🚢 🚌 Livádi.
ℹ 22810 51210.

In mythology, the infant Perseus and his mother Danae were washed up on the shores of rocky Sérifos, known as "the barren one". Once rich in iron and copper mines, the island has bare hills

The red-roofed village of Dryopída on Kýthnos

For hotels and restaurants in this region see pp313–7 and pp336–8

The whitewashed village of Chóra on Sérifos

with small fertile valleys, and long sandy beaches.

Ferries dock at **Livádi** on the southeast coast. The town is situated on a sandy, tree-fringed bay backed by hotels and tavernas. Follow the stone steps up from Livádi, or use the sporadic bus service to reach the dazzling white **Chóra** high above on the steep hillside. It is topped by the ruins of a 15th-century Venetian kástro. Many of its medieval cube-shaped houses, some incorporating stone from the castle, have been renovated as holiday homes by Greek artists and architects. It is an attractive town with chapels and windmills perched precariously, offering breath-taking views of the island.

Near to the northern inland village of Galaní, the fortified **Moní Taxiarchón** (Archangel), built in 1500, is run by a single monk. The monastery contains fine 18th-century frescoes by Skordílis and some valuable Byzantine manuscripts.

Sífnos
Σίφνος

🏠 1,950. 🚤 🚢 Kamáres.
ℹ 22840 31977. **www**.sifnos.gr

Famous for its pottery, poets and chefs, Sífnos has become the most popular destination in the western Cyclades. Visitors in their thousands flock to the island in summer lured by its charming villages, terraced countryside dotted with ancient towers, Venetian dovecotes and long sandy beaches. In ancient times Sífnos was renowned for its gold mines. The islanders paid yearly homage to the Delphic sanctuary of Apollo with a solid gold egg. One year they cheated and sent a gilded rock instead, incurring Apollo's curse. The gold mines were flooded, the island ruined and from then on was known as *sífnos*, meaning empty.

Apollonia
The capital is set above Kamáres port and is a Cycladic labyrinth of white houses, flowers and belfries. It is named after the 7th-century BC Temple of Apollo, which overlooked the town, now the site of the 18th-century church of the Panagía Ouranofóra. The Museum of Popular Arts and Folklore in the main square has a good collection of local pottery and embroideries.

🏛 **Museum of Popular Arts and Folklore**
Plateía Iróön. 🕐 Apr–Oct:
9am–10pm daily. 🈂

Around the Island
Sífnos is a small, hilly island, popular with walkers. Buses from Kamáres port connect it with Apollonía and Kástro, on the east coast. **Artemónas** is Apollonía's twin village, the second largest on Sífnos, with impressive Venetian houses sporting distinctive chimneys. The 17th-century church, Agios Geórgios tou Aféndi, contains several fine icons from the period. The church of Panagía Kónchi, with its cluster of domes, was built on the site of a temple of Artemis.

Kástro, 3 km (2 miles) east of Artemónas, overlooks the sea, the backs of its houses forming massive outer walls *(see pp22–3)*. Some buildings in the narrow, buttressed alleys bear Venetian coats of arms. There are ruins of a Classical acropolis in the village. The **Archaeological Museum** has a collection of Archaic and Hellenistic sculpture, and Geometric and Byzantine pottery.

A fountain in Kástro, Sífnos

The port of **Kamáres** is a straggling resort, with waterside cafés and tavernas. The north of the harbour was once lined with pottery shops making Sífnos's distinctive blue and brown ceramics, but only two remain. Taxi boats go from Kamáres to the pretty pottery hamlet of **Vathý**, in the south. An hour's walk to the east is the busy resort of **Platýs Gialós**, with its long sandy beach. This is also connected by bus to Apollonía and Kamáres.

🏛 **Archaeological Museum**
Kástro. **Tel** 22840 31022.
🕐 Tue–Sun. 🈸 main public hols.

A chapel with steps leading down to a small quay at Platýs Gialós, Sífnos

Páros
Πάρος

Fertile, thyme-scented Páros is the third largest Cycladic island. Since antiquity it has been famous for its white marble, which ensured the island's prosperity from the early Cycladic age through to Roman times. In the 13th century Páros was ruled by the Venetian Dukes of Náxos, then by the Turks from 1537 until the Greek War of Independence *(see pp38–9)*. Páros is the hub of the Cycladic ferry system and is busy in high season. Buffeted by strong winds in July and August, it is a wind-surfer's paradise. There are several resorts, but it retains its charm with hill-villages, vineyards and olive groves.

The famous windmill beside Paroikiá's busy port

Paroikiá ❶
Παροικιά

🏠 *3,000.* ⛴ 🚌 *harbour.*
ℹ *22840 21673.* ⊙ *Apr–Oct.*
🚤 *Kriós 3 km (2 miles) N.*

The port of Paroikiá, or Chóra, owes its foundations to the marble trade. Standing on the site of a leading early Cycladic city, it became a major Roman marble centre. Traces of Byzantine and Venetian rule remain, although earthquakes have caused much damage.

Today it prospers as a resort town, with its quayside wind-mill and commercialized water-front crammed with ticket agencies, cafés and bars. The area behind the harbour is an enchanting Cycladic town, with narrow paved alleys, archways dating from medieval times and white houses overhung with cascading jasmine.

🔒 Ekatontapylianí
W *Paroikiá.* **Tel** *22840 21243.* ⊙ *daily.*
The Ekatontapylianí (Church of a Hundred Doors) in the west of town is the oldest in Greece in continuous use and

a major Byzantine monument. Its official name is the Dormition of the Virgin.

According to legend, the church was founded by St Helen, mother of Constantine, the first Christian Byzantine emperor. After having a vision here showing the path to the True Cross, she vowed to build a church on the site but died before fulfilling her promise. In the 6th century AD the Emperor Justinian carried out her wish, commissioning the architect Ignatius to design a cathedral. He was the apprentice of Isidore of Miletus, master builder of Agía Sofía in Constantinople. The result was so impressive that Isidore, consumed with jealousy, pushed his pupil off the roof. Ignatius grabbed his master's foot and they both fell to their deaths. The pair are immortalized in stone in the north of the court-yard in front of the church.

An ornate chandelier in the interior of Ekatontapylianí

Ekatontapylianí is made up of three interlocking buildings. It is meant to have 99 doors and windows. According to legend, when the 100th door is found, Constantinople (Istanbul) will return to the Greeks. Many earthquakes have forced much reconstruction, and the main church building was restyled in the 10th century in the shape of a Greek cross. The sanctuary columns date from the pre-Christian era and the marble screen, capitals and iconostasis are of Byzantine origin.

On the carved wooden iconostasis is an icon of the Virgin, worshipped for its healing virtues. Nearby a foot-print, set in stone, is claimed

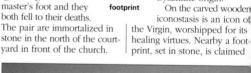

Theoktísti's footprint

Fishing boats, Paroikiá harbour

For key to map see back flap

KEY

Náxos, Kos,
Kálymnos,
Ikaría

Lágeri

Náousa

Moní Longovardas

Síros, Tínos
Mykonos

Kamínia · Kriós · Tris Ekklisíes

Íos, Piraeus

PAROIKIA · Maráthi quarries

Moní Christoú
tou Dásous · Léfkes · Marmara

Kéfalos

Mólos

Pródromos · Moní Agíou
Márpissa · Antoníou

Petaloúdes · Pounta · Piso Livádi

Antíparos town · Dryós · Chrysí Aktí

P A R O S

ANTIPAROS · Santoríni · Alykí

0 kilometres 5

0 miles 2

SIGHTS AT A GLANCE

Léfkes ❹
Náousa ❸
Paroikiá ❶
Petaloúdes ❻
Píso Livádi ❺
Tris Ekklisíes ❷

VISITORS' CHECKLIST

🏠 10,300. ✈ Alykí.
🚌 🚢 Paroikiá. ℹ Paroikiá
(22840 21673). 🎭 Fish & Wine
Festival at Náousa: 6 Aug; Festival
of the Dormition of the Virgin at
Paroikiá: 15 Aug; Agía Theoktísti
Saint's Day: 9 Nov.

⚓ **Kástro**
Built in 1260 on the site of
the ancient acropolis, the
Venetian kástro lies on a
small hill at the end of
the main street of the
town. The Venetians
used the marble remains
from the Classical
temples of Apollo and
Demeter to construct the
surviving eastern fortification
of the kástro. The
ancient columns
have also been
partially used to
form the walls of
neighbouring
houses. Next to the
site of the Temple
of Apollo stands
the 300-year-old
blue-domed church
of **Agía Eléni
and Agios
Konstantínos**.

*A Greco-Roman frieze
in the Archaeological
Museum*

Environs
Taxi boats cross the bay from
Paroikiá to the popular sands
of Kamínia beach and Kriós,
both sheltered from the pre-
vailing wind. The ruins of an
Archaic sanctuary of Delian
Apollo stand on the hill above.

a historical record
of the artistic
achievements of
ancient Greece up
to 264 BC. It is
carved on a marble
tablet and was
discovered in the
kástro walls during
the 17th century.
Also on display are finds
from the Temple of Apollo
including a 5th-century BC
Winged Victory, a mosaic
depicting Herakles hunting
and a frieze of Archilochus,
the 7th-century BC poet and
soldier from Páros.

to be that of Agía Theoktísti,
the island's patron saint. The
Greeks fit their feet into the
print to bring them luck. Also
displayed is her severed hand.
    From the back of the church
a door leads to the chapel of
Agios Nikólaos, an adapted 4th-
century BC Roman building.
It has a double row of Doric
columns, a marble throne and
a 17th-century iconostasis.
Next door, the 11th-century
baptistry has a marble font
with a frieze of Greek crosses.
Ekatontapylianí has no
belltower and instead the bells
are hung from a tree outside.

📷 **Archaeological Museum**
W Paroikiá. **Tel** 22840 21231. ◻
Tue–Sun. ◼ main public hols. 🎫 📷
The museum can be found
behind Ekatontapylianí. One
of its main exhibits is part of
the priceless Parian Chronicle,

THE LEGEND OF AGIA THEOKTISTI

Páros's patron saint, Theoktísti, was a young woman
captured by pirates in the 9th century. She escaped to Páros
and lived alone in the woods for 35 years, leading a pious
and frugal life. Found by a hunter, she asked him to bring
her some Communion bread. When he returned with the
bread she lay down and died. Realizing she was a saint, he
cut off her hand to take as a relic but found he could not
leave Páros until he reunited her hand with her body.

Around Páros Island

Páros is an easy island to explore, with an excellent bus service linking the three main towns: the capital Paroikía, the trendy fishing village resort of Náousa in the north and the central mountain town of Léfkes. There are plenty of cars and bikes for hire to get to the beaches and villages off the beaten track, and boat excursions and caïques to tour the remoter shores.

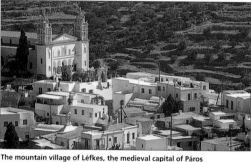

The mountain village of Léfkes, the medieval capital of Páros

Trís Ekklisíes ❷
Τρεις Εκκλησίες

3 km (2 miles) NE of Paroikiá.

North of Paroikiá the road to Náousa passes the remains of three 17th-century churches, Trís Ekklisíes, adapted from an original 7th-century basilica. That was in turn built from the marble of a 4th-century BC *heróon*, or hero's shrine, tomb of the Parian poet Archilochus.

In the mountains further north, the remote, 17th-century **Moní Longovárdas** is a hive of activity. The monks make wine and books and work in the fields, and the abbot is famous for his icon-painting. Visitors are, however, discouraged and women are banned.

Main door at Moní Longovárdas

Náousa ❸
Νάουσα

12 km (7 miles) NE of Paroikiá.
2,100. Lageri 5 km (3 miles) NE.

With its brightly painted fishing boats and winding white alleyways, Náousa has become a cosmopolitan destination for the jetset, with expensive boutiques and relaxed bars. It is the island's second largest town and the place to sit and watch the rich and the beautiful parade chic designer clothes along the waterfront.

The colourful harbour has a unique breakwater in the half-submerged ruin of a Venetian castle which has slowly been sinking with the coastline.

Every year, on the evening of 23 August, 100 torch-lit fishing caïques assemble to re-enact the battle of 1536 between the islanders and the pirate Barbarossa, ending with celebrations of music and dancing.

Léfkes ❹
Λεύκες

10 km (6 miles) SE of Paroikiá.
850.

The mountain road to Léfkes, the island's highest village, passes the abandoned marble quarries at Maráthi, last worked for Napoleon's tomb. It is possible to explore the ancient tunnels with a torch.

Léfkes, named after the local poplar trees, was the capital under Ottoman rule. It is a charming, unspoiled village with medieval houses, a labyrinth of alleys, *kafeneía* in shaded squares and restaurants with terraces overlooking the green valley below. Shops stock local weaving and ceramic handicrafts and the town has a tiny Folk Museum.

Folk Museum
Tel 22840 52284. ☐ Apr–Oct: daily; Nov–Mar: key at town hall.

Environs
From the windmills overlooking Léfkes, a Byzantine marble pathway leads 3 km (2 miles) southeast to **Pródromos**, an old fortified farming village. Walk a further 15 minutes past olive groves to reach **Mármara** village with its marble-paved streets. The pretty hamlet of **Márpissa** lies about 1.5 km (1 mile) south.

On Kéfalos hill, 2 km (1 mile) east of Márpissa, are the ruins of a 15th-century Venetian fortress and the 16th-century **Moní Agíou Antoníou**. The monastery is built from Classical remains and has a 17th-century fresco of the *Second Coming*.

Caïques at the attractive fishing harbour at Náousa

The convent of Moní Christoú tou Dásous near Petaloúdes

Píso Livádi **⑤**
Πίσω Λιβάδι

15 km (9 miles) SE of Paroikiá. 🏠 50. 🚌 to Márpissa. ⛴ Poúnta 1 km (0.5 mile) S.

Situated below Léfkes on the east coast of the island, the fishing village of Píso Livádi, with its sheltered sandy beach, has grown into a lively small resort. It was once the port for Páros's hill-villages and the island's marble quarries; today there are services operated over to nearby Agía Anna *(see p230)* on Náxos island. The small harbour has a wide range of bars and tavernas with a disco nearby and occasional local activities and entertainments.

The beautiful and fashionable beach at Poúnta

Environs
Mólos, 6 km (4 miles) north, has a long sandy beach with dunes, tavernas and a windsurfing centre. Just to the south lies **Poúnta** (not to be confused with the village of Poúnta on the west coast), one of the best and most fashionable beaches in the Cyclades with a trendy laid-back beach bar. The island's most famous east-coast

beach, 3 km (2 miles) south, is **Chrysí Aktí** (Golden Beach). With 700 m (2,300 ft) of golden sand it is perfect for families. It is also a well-known centre for watersports and has hosted the world windsurfing championships.

Dryós, 2 km (1 mile) further southwest, is an expanding resort but at its heart is a pretty village with a duck-pond, tavernas, a small harbour with a pebbly beach and a string of sandy coves.

Petaloúdes **⑥**
Πεταλούδες

6 km (4 miles) SW of Paroikiá. 🚌 🔲 *1 Jun–20 Sep: daily.* 🖼

Petaloúdes, or the Valley of the Butterflies, on the slopes of Psychopianá, is easily reached from Paroikiá. This lush green oasis is home to swarms of Jersey tiger moths, from May to August, which flutter from the foliage when disturbed. There are mule treks along the donkey paths that cross the valley. About 2 km (1 mile) north of Petaloúdes, the 18th-century convent of **Moní Christoú tou Dásous**, Christ of the Woods, is worth the walk, although women only are allowed into the sanctuary. Páros's second patron saint, Agios Arsénios, teacher and abbot, is also buried here.

Outlying Islands
The island of **Antíparos** used to be joined to Páros by a causeway. These days a small ferry links the two from the west-coast resort of Poúnta and there are also caïque trips from Paroikiá. Antíparos town has a relaxed and stylish café society, good for escaping from the Páros crowds. Activity centres around the quay and the Venetian **kástro** area. The kástro is a good example of a 15th-century fortress town, designed with inner courtyards and narrow streets to impede pirate attacks *(see pp22–3)*. The village also has two 17th-century churches, Agios Nikólaos and Evangelismós.

The island has fine beaches, but the star attraction is the massive **Cave of Antíparos**, with a breathtaking array of stalactites and stalagmites, discovered during Alexander the Great's reign. In summer, boats run to the cave from Antíparos town and Poúnta on Páros. From where the boat docks, it is a half-hour walk up the hill of Agios Ioánnis to the cave mouth, then a dramatic 70 m (230 ft) descent into the cavern. Lord Byron and other visitors have carved their names on the walls. In 1673 the French ambassador, the Marquis de Nointel, held a Christmas Mass here for 500 friends. The church outside, Agios Ioánnis Spiliótis, was built in 1774.

Bougainvillea on a house in Antíparos town

Náxos
Νάξος

The largest of the Cyclades, Náxos was first settled in 3000 BC. A major centre of the Cycladic civilization (*see pp28–9*), it was one of the first islands to use marble. Náxos fell to the Venetians in 1207, and the numerous fortified towers (*pýrgoi*) were built, still evident across the island today. Its landscape is rich with citrus orchards and olive groves, and it is famous in myth as the place where Theseus abandoned the Cretan princess Ariadne.

Mosaic from the Archaeological Museum in Náxos town

The Portára gateway from the unfinished Temple of Apollo

Náxos Town ❶
Χώρα

🏠 15,000. 🚢 🚌 Harbourfront 🅸 Harbourfront (22850 25201).

North of the port and reached by a causeway is the huge marble Portára gateway on the islet of Palátia, which dominates the harbour of Náxos town, or Chóra. Built in 522 BC, it was to be the entrance to the unfinished Temple of Apollo.

The town is made up of four distinct areas. The harbour bustles with its cafés and fishermen at work. To the south is Neá Chóra, or Agios Geórgios, a concrete mass of hotels, apartments and restaurants. Above the harbour, the old town divides into the Venetian Kástro, once home of the Catholic nobility, and the medieval Bourg, where the Greeks lived.

The twisting alleys of the Bourg market area are lined with restaurants and gift shops. The Orthodox cathedral in the Bourg, the fine 18th-century

Mitrópoli Zoödóchou Pigís, has an iconostasis, painted by Dimítrios Válvis of the Cretan School in 1786.

Uphill lies the imposing medieval north gate of the fortified Kástro, built in 1207 by Marco Sanudo. Only two of the original seven gate-towers remain. Little is left of the 13th-century outer walls, but the inner walls still stand, protecting 19 impressive houses. These bear the coats of arms of the Venetian nobles who lived there, and many of the present-day residents are descended from these families. Their remains are housed in the 13th-century Catholic **cathedral**, in the Kástro, beneath marble slabs dating back to 1619.

During the Turkish occupation, Náxos was famous for its schools. The magnificent Palace of Sanoúdo, dating from 1627, which incorporates part of the Venetian fortifications, housed the French school. The most famous pupil was Cretan novelist Níkos Kazantzákis (*see p276*) who wrote *Zorba the Greek*.

Angel from the Roman Catholic cathedral

The building now houses the **Archaeological Museum**, which has one of the best collections of Cycladic marble figurines (*see p211*) in the Greek islands, as well as some beautiful Roman mosaics.

🏛 **Archaeological Museum**
Palace of Sanoúdo. **Tel** 22850 22725. ⏰ Tue–Sun. ⬤ main public hols.

Environs
A causeway leads to the **Grótta** area, north of Náxos town, named after its numerous sea caves. To the south the lagoon-like bay of **Agios Geórgios** is the main holiday resort, with golden sands and shallow water. The best beaches are out of town along the west coast. **Agia Anna** is a pleasant small resort with silver sands and watersports. For more solitude, head south 3 km (2 miles) over the dunes to **Pláka**, the best beach on the island and mainly nudist. Further south the pure white sands of **Mikrí Vígla**, and **Kastráki**, named after a ruined Mycenaean fortress, are exceptionally good for both swimming and watersports.

The remote and beautiful Pláka beach south of Náxos town

Around Náxos Island

Inland, Náxos is a dramatic patchwork of rich gardens, vineyards, orchards and villages. These are backed by wild crags and dotted with Venetian watchtowers and a wealth of historical sites. Although there are organized tours from Náxos town and a good local bus service, a hired car is advisable to explore the island fully. The Tragaía region is, however, a walker's paradise.

Moní village in the Tragaía valley, surrounded by olive groves

VISITORS' CHECKLIST

🏠 20,000. ✈ 2 km (1 mile) S Náxos town. 🚌 Náxos town. 🚤
ℹ Náxos town (22850 25201).
🎭 Agios Nikódimos Folk Festival, Náxos town: 14 Jul; Dionysiac Festival, Náxos town: 1st week of Aug; Diorvoia Festival: Jul–Aug.

Bellonias tower, first of the fortified mansions on Náxos. The chapel of Agios Ioánnis Gýroulas in Ano Sagrí, south of Glinádo, is built over the ruins of a temple of Demeter.

Tragaía Valley ❸
Κοιλάδα Τραγαίας

15 km (9 miles) SE of Náxos town. 🚌

From Ano Sagrí the road twists to the Tragaía valley. The first village in the valley, **Chalkí**, is the most picturesque with its Venetian architecture and the old Byzantine Fragópoulos tower in its centre.

From Chalkí a road leads up to Moní, home of the most unusual church on Náxos, **Panagía Drosianí**. Dating from the 6th century, its domes are made from field stones.

Filóti is a traditional village, the largest in the region. It sits on the slopes of Mount Zas, which, at 1,000 m (3,300 ft), is the highest in the Cyclades.

Mélanes Valley ❷
Κοιλάδα Μελάνων

10 km (6 miles) S of Náxos town. 🚌 to Kinídaros.

The road south of Náxos town passes through the Livádi valley, the heart of ancient marble country, to the Mélanes villages. In **Kournochóri**, the first village, is the Venetian Della Rocca tower. At **Mýloi**, near the ancient marble quarry at Fleriό, lie two 6th-century BC *koúroi*, huge marble statues. One, 8 m (26 ft) long, lies in a private garden, open to visitors. The other, 5.5 m (18 ft) long, lies in a nearby field.

Environs
Southeast of Náxos town is **Glinádo**, home to the Venetian

SIGHTS AT A GLANCE

Apeíranthos ❹
Apóllon ❻
Komiakí ❺
Mélanes Valley ❷
Náxos town ❶
Tragaía Valley ❸

Koúros in a private garden in Mýloi in the Mélanes valley

KEY

For key to map see back flap

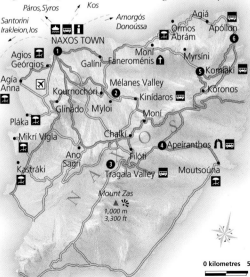

0 kilometres 5

0 miles 3

Terraced fields outside the village of Komiakí

Apeíranthos ❹
Απείρανθος

25 km (16 miles) SE of Náxos town.
👥 1,500. 🚌

Apeíranthos was colonised in the 17th and 18th centuries by Cretan refugees fleeing Turkish oppression and coming to work in the nearby emery mine. It is the island's most atmospheric village, with marble-paved streets and 14th-century towers (*pýrgoi*) built by the Venetian Crispi family. Locals still wear traditional costume, women weave on looms and farmers sell their wares from donkeys.

The small **Archaeological Museum** has a collection of proto-Cycladic marble plaques depicting scenes from daily life as well as Neolithic finds. There is also a small **Geological Museum** on the second floor of the village school. Below the village is the port of **Moutsoúna** where ships were once loaded with emery before the industry's decline. The fine beach is now lined with holiday villas.

🏛 **Archaeological Museum**
Off main road. **Tel** 22850 61725.
⬜ daily. ⬛ main public hols. ♿

🏛 **Geological Museum**
Village school. **Tel** 22850 61724.
⬜ daily. ⬛ main public hols. 🖼

Komiakí ❺
Κωμιακή

42 km (26 miles) E of Náxos town.
👥 500. 🚌

Approaching from Kóronos the road becomes a tortuous succession of hairpin bends before finally arriving in pretty Komiakí (also known as Koronída). This is the highest village on Náxos and a former home of the emery miners. It is covered with vines and is known for being the place where the local *kítro* liqueur originated. There are wonderful views over the surrounding terraced vineyards. The village is the start of one of the finest walks on Náxos. The walk takes you down into the lush valley and the charming oasis hamlet of **Myrsíni**.

Apóllon ❻
Απόλλων

49 km (30 miles) NE of Náxos town.
👥 100. 🚌

Originally a fishing village that is slowly turning into a resort, Apóllon gets busy in the summer with coach trips of people coming to visit the fish tavernas and the huge *koúros* found here. Steps lead up the hillside above the village to ancient marble quarries where the vast unfinished statue has lain abandoned since 600 BC. The bearded marble figure, which is believed to represent the god Apollo, is 10.5 m (35 ft) long and weighs 30 tonnes. There is also a lively festival in the village for St John the Baptist on 28 August.

Environs

At Agiá, 10 km (6 miles) west of Apóllon, stands the **Cocco Pýrgos**, built by the Venetian Cocco clan at the beginning of their rule of northern Náxos in 1770. *Pýrgoi* are fortified watchtowers that were built during the Venetian occupation of Náxos. Further along the north coast road lies the idyllic beach at **Órmos Abrám** with a good family-run taverna.

Dating from 1606, the abandoned **Moní Faneroménis** is 13 km (8 miles) south on the road winding down the west coast from Apóllon. Slightly further south towards Galíni, a road leads up to the most famous *Pýrgos*, the **High Tower** of the Cocco clan. It was built in 1660 in a commanding position overlooking a valley. During the 17th century a family feud between the Orthodox Cocco and the Catholic Barozzi families

The harbour at Moutsoúna, Náxos

The huge *koúros* in Apóllon's ancient quarries

roke out as a result of an nsult. The feud led to the oombardment of the High Tower when a Barozzi woman oersuaded her husband, who vas a Maltese privateer, to oesiege it. The Cocco clan nanaged to hold out but the vendetta continued to rage or another 20 years until a marriage eventually united the two families.

A Venetian fortified watchtower, or *pýrgos*, west of Apóllon

Outlying Islands
Between Náxos and Amorgós lie **Donoússa, Koufoníssi, Iraklía** and **Schinoússa**, the "Back Islands". They all have rooms to rent, a post and tourist office, but no banks. Iraklía, the largest, boasts impressive stalactites in the Cave of Aï-Giánni as well as Cycladic remains. Koufoníssi consists of two islands, Ano (upper), the most developed of the Back Islands, with good sandy beaches, and the uninhabited Káto (lower). Schinoússa has wild beaches and great walking over cobbled mule tracks. Donoússa, the most northerly of the chain, is more isolated and food can be scarce. A settlement from the Geometric era was excavated on the island, but most of its visitors come for the fine sandy beaches at Kéntros and Livádi.

Amorgós
Αμοργός

🏠 1,800. ⛴ Katápola & Aigiáli.
🚌 Katápola & Aigiáli harbours.
ℹ️ Katápola quay (22850 71278).
🎉 Ormos Aigiális 12 km (7 miles) NE of Amorgós town.

Dramatically rugged, the small island of Amorgós is narrow and long with a few beaches. Inhabited from as early as 3300 BC, its peak was during the Cycladic civilization, when there were three cities: Minoa, Arkesini and Egiali. In 1885 a find of ceramics and marble was taken to the Archaeological Museum in Athens (*see p286*).

Chóra
The capital, Chóra, or Amorgós town, is a dazzling clutch of whitewashed houses with windmills standing nearby. Above the town is **Apáno Kástro**, a Venetian fortress, which was built by Geremia Ghisi in 1290. Chóra also boasts the smallest church in Greece, the tiny **Agios Fanoúrios**.

Environs
Star attraction on the island is the spectacular Byzantine **Moní Panagías Chozoviótissas**, below Chóra on the east coast. The stark white monastery clings to the 180-m (590-ft) cliffs. It is a huge fortress, built into the rock, housing the miraculous icon of the Virgin Mary. Founded in 1088 by the Byzantine Emperor Alexios I Comnenos, the monastery has a library with a collection of ancient manuscripts.

Around the Island
The best way to get around the island is by boat or walking, although there is a limited bus service. The main port of **Katápola** in the southwest is set in a horseshoe-shaped bay with tavernas, *pensions*, fishing boats and a small shingly beach. The harbour area links three villages: Katápola in the middle where the ferries dock, quieter **Xylokeratídi** to the north and **Rachídi** on the hillside above. A track leads from Katápola to the hilltop ruins of the ancient city of **Minoa**. All that remains are the Cyclopean walls, the gymnasium and the foundations of the Temple of Apollo.

The northern port of **Ormos Aigiális** is the island's main resort, popular for its sandy beach. It is worth following the mule paths north to the hill-villages of **Tholária**, which has vaulted Roman *tholos* tombs, and **Lagáda**, one of the prettiest villages on the island, with a stepped main street painted with daisies.

The cliff-top Moní Chozoviótissas

The white walls and blue-domed churches of Ios town

Ios
`Ίος`

🏛 1,654. 🚢 Gialós. 🚌 Ios town. 🛈 Ano Chóra, Ios town (22860 91505). 🚤 Mylopótas 2 km (1 mile) E of Ios town.

In ancient times Ios was covered in oak woods, later used for shipbuilding. The Ionians built cities at the port of Gialós and at Ios town, later to be used as Venetian strongholds. Ios is also known as the burial place of Homer, and 15 May is the Omíria, or Homer festival. A local speciality is its cheese, *myzíthra*, similar to a soft cream cheese.

Ios is renowned for its nightlife and as a result is a magnet for the young. However, it remains a beautiful island. Its mountainous coastline has over 400 chapels and some of the finest sands in the Cyclades.

Ios town, also known as the Village, is a dazzling mix of white houses and blue-domed churches fast being swamped by discos and bars. There are ruins of the Venetian fortress, built in 1400 by Marco Crispi, remains of ancient walls, and 12 windmills above the town.

The port of **Gialós**, or Ormos, has a busy harbour, with yachts and fishing boats, good fish tavernas and quieter accommodation than Ios town. The beach here is windy, although a 20-minute walk west leads to the sandy

Windmill above Ios town

cove at Koumpará. A bus service runs from here to Ios town and the superb **Mylopótas** beach which has two campsites. Excursion boats run from Gialós to the beach at **Manganári** bay, in the south and **Psáthi** bay in the east.

On the northeast coast the beach at **Agía Theodóti** is overlooked by the medieval ruins of Palaiókastro fortress. A festival is held at nearby **Moní Agías Theodótis** on 8 September to mark the islanders' victory over medieval pirates. You can see the door the pirates broke through only to be scalded to death by boiling oil.

Homer's tomb is supposedly in the north at **Plakotós**, an ancient Ionian town which has slipped down the cliffs over the ages. Homer died on the island after his ship was forced to dock en route to Athens. The tomb entrance, ruined houses and the remains of the Hellenistic **Psarópyrgos tower** can be seen today.

Síkinos
Σίκινος

🏛 300. 🚢 Aloprónoia. 🚌 Síkinos town. 🛈 Kástro, Síkinos town (22860 51222). 🚤 Agios Geórgios 7 km (3 miles) NE of Síkinos town.

Síkinos is quiet, very Greek and one of the most ruggedly beautiful islands in the Cyclades. Known in Classical Greece as Oinoe (wine island), it has remained a traditional backwater throughout history. Fishing and farming are the main occupations of the 300 or so islanders and, although there are some holiday homes, there is little mass tourism.

Síkinos town is divided into twin villages: Kástro and the pretty and unspoilt Chóra perched high up on a ridge overlooking the sea. Kástro is a maze of lanes and *kafeneía* At the entrance to the village is Plateía Kástrou where the wall of 18th-century stone mansion formed a bastion of defence. The church of the Pantánassa forms the focal point and among the ruined houses is a huge marble portico.

The partly ruined Moní Zoödóchou Pigís, fortified against pirate raids, looms down from the crag above Chóra and has icons by the 18th-century master Skordílis.

In medieval Chóra there is a private **Folk Museum**, which is in the family home of an American expatriate. It has an olive press and a wide range of local domestic and agricultural artifacts.

From Chóra a path leads past the ruined ancient Cyclopean walls southwest to **Moní Episkopís**, a good hour's trek. With Doric

The golden sands of Mylopótas beach, Ios

columns and inscriptions it is thought to be a 3rd-century AD mausoleum, converted in the 7th century to the Byzantine church of Koímisis Theotókou. A monastery was added in the 17th century, but is now disused.

On the east coast 3 km (2 miles) southeast of Síkinos town, the port of **Aloprónoia**, also known as Skála, has a few small cafés that double as shops, a modern hotel complex and a wide sandy beach that is safe for children.

Folk Museum

no Chorió, Síkinos town.
May–Sep: daily.

The sleepy port of Aloprónoia

Folégandros
Φολέγανδρος

650. Karavostásis.
Chóra (22860 41249).
Agáli 2 km (1 mile) W of Folégandros town.

Bleak and arid, Folégandros is one of the smallest inhabited islands in the Cyclades. It aptly takes its name from the Phoenician for rocky. Traditionally a place of exile, this remote island passed quietly under the Aegean's various rulers, suffering only from the threat of pirate attack. Popular with photographers and artists for its sheer cliffs, terraced fields and striking Chóra, it can be busy in peak season, but is still a good place for walkers, with a wild

Koímisis tis Theotókou in Folégandros town

beauty and unspoiled beaches. **Folégandros town** or Chóra, perched 300 m (985 ft) above the sea to avoid pirates, is spectacular. It divides into the fortified Kástro quarter (see p 22) and Chóra, or main village. Kástro, built in the 13th century by Marco Sanudo, Duke of Náxos, is reached through an arcade. The tall stone houses back on to the sea, forming a stronghold along the ridge of the cliff with a sheer drop below. Within its maze of crazy-paved alleys full of geraniums are the distinctive two-storey cube houses with brightly painted wooden balconies.

In Chóra village life centres on four squares with craft shops and lively tavernas and bars. The path from the central bus stop leads to the church of Koímisis tis Theotókou, (Assumption of the Virgin Mary). It was built after a silver icon was miraculously saved by an islander from medieval pirates who drowned in a storm. Forming part of the ancient town walls, it is thought to have once been the site of a Classical temple of Artemis.

Ferries dock at **Karavostási** on the east coast, a tiny harbour with a tree-fringed pebble beach, restaurants, hotels and rooms. There is a bus to Chóra, and **Livádi** beach is a short walk from the port. In season there are excursions available to the western beaches at **Agáli, Agios**

Nikólaos and **Latináki**, as well as to the island's most popular sight, the **Chrysospiliá** or Golden Cave. Named after the golden shade of its stalactites and stalagmites, the grotto lies just below sea level in the northeast cliffs.

Ano Meriá, 5 km (3 miles) to the west of Folégandros town, is a string of farming hamlets on either side of the road, surrounded by terraced fields. There are wonderful sunset views from here and on a clear day it is possible to see Crete in the distance. There is a good **Ecology and Folk Museum** with a display of farming implements, and reconstructions of traditional peasant life. On 27 July a major local festival is held for Agios Panteleímon.

From Ano Meriá steep paths weave down to the remote beaches at **Agios Geórgios** bay and **Vígla**.

Ecology and Folk Museum

Ano Meriá. **Tel** 22860 41370.
Jul–mid-Sep: 5–8pm.

Traditional houses in Kástro, Folégandros town

Mílos
Μήλος

Volcanic Mílos is the most dramatic of the Cyclades with its extraordinary rock formations, hot springs and white villages perched on multicoloured cliffs. Under the Minoans and Mycenaeans the island became rich from trading obsidian. However, the Athenians brutally captured and colonized Mílos in the 4th century BC. Festooned with pirates, the island was ruled by the Crispi dynasty during the Middle Ages and was claimed by the Turks in 1580. Minerals are now the main source of the island's wealth, although tourism is growing.

Museum in Athens *(see p286)*. There are also finds from the neighbouring island of Kímolos. The **History and Folk Museum** is housed in a 19th-century mansion in the centre of Pláka. It has costumes, four-poster beds and handicrafts.

The *Lady of Phylakopi* in the Archaeological Museum

Steps lead to the ruined **kástro** which was built by the Venetians on a volcanic plug 280 m (920 ft) above sea level. Only the houses that formed the outer walls of the fortress remain.

Above the kástro, the church of Mésa Panagía was bombed during World War II. It was rebuilt and renamed **Panagía Schiniótissa** (Our Lady of the Bushes) after an icon of the Virgin Mary appeared in a bush where the old church used to stand.

Just below, the church of **Panagía Thalassítra** (Our Lady of the Sea), built in 1728, has icons of Christ, the Virgin Mary and Agios Eleuthérios.

The massive stone blocks of the Cyclopean walls that formed the city's East Gate in 450 BC remain, while 15 m (50 ft) west there are marble

View across the houses of Pláka in the mid-morning sun

Pláka

On a clifftop 4 km (2.5 miles) above the port of Adámas, Pláka is a pretty mix of churches and white cube houses. These blend into the suburb of Trypití which is topped by windmills.

It is believed that Pláka is sited on the acropolis of ancient Mílos, built by the Dorians between 1100 and 800 BC. The town was then destroyed by the Athenians and later settled by the Romans.

The principal sight is the **Archaeological Museum**, its entrance hall dominated by a plaster copy of the *Venus de Milo*, found on Mílos. The collection includes Neolithic finds, particularly obsidian, Mycenaean pottery, painted ceramics, and terracotta

animals from 3500 BC, found at the ancient city of Philakopí. The most famous of the ceramics is the *Lady of Phylakopi*, an early Cycladic goddess decorated in Minoan style. However, the Hellenistic 4th-century BC statue of Poseidon and the *koûros* of Mílos (560 BC) are now in the National Archaeological

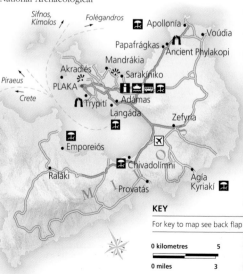

KEY

For key to map see back flap

0 kilometres 5

0 miles 3

he twin rocks, known as The Bears, on the approach to Adámas

VISITORS' CHECKLIST

4,500. ✈ 7 km (4 miles) SE of Adámas. ⛴ Adámas. 🚌 Adámas. 🛈 Harbourfront, Adámas (22870 21370 ext 112). ⛵ Nautical week: end Jun–beg Jul; Panagía at Zefyría: 15 Aug.

elics and a Christian baptismal ont from a Byzantine basilica. Roman amphitheatre nearby still used for performances.

Archaeological Museum
Main square. **Tel** 22870 28026. ◯ ue–Sun. ● 1 May. 🖼

History and Folk Museum
láka. **Tel** 22870 21292. ◯ Tue–Sat & Sun am. ● main public hols. 🚻 🖼

side the Christian Catacombs

nvirons
n the nearby town of Trypití re well-preserved 1st-century D **Christian Catacombs**. arved into the hillside, the massive complex of galleries as tombs in arched niches, ach one containing up to even bodies. The catacomb etwork is 184 m (605 ft) long, ith 291 tombs. Archaeologists elieve that as many as 8,000 odies were interred here.

From the catacombs, a track eads to the place where the enus de Milo was discovered, ow marked by a plaque. It as found on 8 April 1820, by farmer, Geórgios Kentrótas. e uncovered a cave in the orner of his field with half of e ancient marble statue side. The other half was und by a visiting French fficer and both halves were

bought as a gift for Louis XVIII, on 1 March 1821. The statue is now on show in the Louvre, Paris. The missing arms are thought to have been lost in the struggle for possession.

Christian Catacombs
Trypití, 2 km (1 mile) SE of Pláka. **Tel** 22870 21625. ◯ Tue–Sun.

Around the Island
The rugged island is scattered with volcanic relics and long stretches of beach. The vast Bay of Mílos, the site of the volcano's central vent, is one of the finest natural harbours in the Mediterranean, and has some of Mílos's best sights.

West of Adámas, the small and sandy **Langáda** beach is popular with families. On the way to the beach are the municipal baths with their warm mineral waters.

South of Adámas, the Bay of Mílos has a succession of attractive beaches, including **Chivadolímni**, backed by a turquoise saltwater lake. On the south coast is the lovely beach of Agía Kyriakí, near the village of Provatás.

Situated on the northeast tip of the island is **Apollonía**, a popular resort with a tree-fringed beach. Water taxis leave here for the island of **Kímolos**, named after the chalk (kimolía) mined there.

Once an important centre of civilization, little remains now of **Ancient Phylakopi**, just southwest of Apollonía. You can make out the old Mycenaean city walls, ruined houses and grave sites, but a large part of the city has been submerged beneath the sea.

GEOLOGY OF MILOS
Due to its volcanic origins, Mílos is rich in minerals and has some spectacular rock formations. Boat tours from Adámas go to the eerie pumice moonscape of Sarakíniko, formed two to three million years ago, the lava formations known as the "organ-pipes" of Glaronísia (offshore near Philakopí), and the sulphurous blue water at Papáfragkas. Geothermal action has provided a wealth of hot springs; in some areas, such as off the Mávra Gkrémna cliffs, the sea can reach 100° C (212° F) only 30 cm (12 inches) below the surface.

Mineral mine at Voúdia, still in operation

The white pumice landscape at Sarakíniko

The sulphurous blue water at Papáfragkas

Santoríni
Σαντορίνη

Early Cycladic figurine

Colonized by the Minoans in 3000 BC, this volcanic island erupted in 1450 BC, forming Santoríni's crescent shape. The island is widely believed to be a candidate for the lost kingdom of Atlantis. Named Thíra by the Dorians when they settled here in the 8th century BC, it was renamed Santoríni, after St Irene, by the Venetians who conquered the island in the 13th century. Despite tourism, Santoríni remains a stunning island with its white villages clinging to volcanic cliffs above black sand beaches.

One of the many cliffside bars in Firá, with views over the caldera

Firá ❶
Φηρά

🏙 1,550. 🚢 🚌 50 m (165 ft) S of main square. 🛈 22860 22231.
🚌 Monólithos 5 km (2.5 miles) E.

Firá, or Thíra, overlooking the caldera and the island of Néa Kaméni, is the island's capital. It was founded in the late 18th century when islanders moved from the Venetian citadel of Skáros, near present day Imerovígli, to the clifftop plains for easier access to the sea.

Devastated by an earthquake in 1956, Firá has been rebuilt, terraced into the volcanic cliffs with domed churches and barrel-roofed cave houses (skaftá). The terraces are packed with hotels, bars and restaurants in good positions along the lip of the caldera to

enjoy the magnificent views, especially at sunset. The tiny port of Skála Firón is 270 m (885 ft) below Firá, connected by cable car or by mule up the 580 steps. Firá is largely pedestrianized with winding

cobbled alleys. The town's main square, Plateía Theotokopoúlou, is the bus terminal and hub of the road network. All the roads running north from here and the harbour eventually merge in Plateía Firostefáni. The most spec-

Firá's whitewashed buildings lining the clifft[op]

SIGHTS AT A GLANCE

Akrotíri ❹
Ancient Thíra ❸
Firá ❶
Oía ❷

KEY

For key to map see back flap

0 kilometres ⊢————⊣ 5
0 miles ⊢————⊣ 3

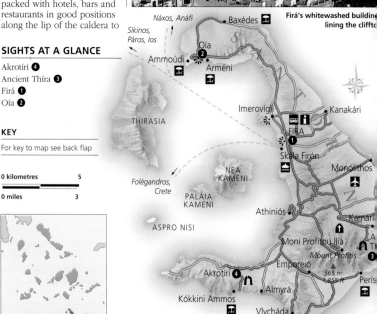

Náxos, Anáfi
Sikinos, Páros, Ios
Baxédes 🚉
Ammoúdi • Oía ❷
Ammoúdi • Arméni 🚉

THIRASIA

Imerovígli ☀ 🏨🛈 FIRA
FIRÁ ❶

Kanakári

Skala Firón 🚢

Monólithos ✈

NEA KAMENI

Folégandros, Crete

PALAIA KAMENI

Athiniós ⚓

Kamári

ASPRO NISI

Moní Profítou Ilía 🛈
Mount Profítis
Emporeió · 565 m
1,855 ft

Ancient Thíra ❸

Akrotíri ❹ 🕳 • Almyrá
Kókkini Ammos 🚉
Vlycháda 🚉

Perís 🚉

cular street, Agíou
iná, runs south
long the edge of
e caldera to the
8th-century church
f **Agíos Minás**.
7ith its distinctive
lue dome and its
vhite belltower, it
as become the
ymbol of Santoríni.
he **Archaeological
Museum** houses
nds from Akrotíri
(*see p241*) and the
ncient city of Mésa
ounó (*see p240*),
cluding early Cycladic
gurines found in local
umice mines. The **New
rcheological Museum** con-
ins the colourful Firá frescoes
rginally thought to be from
e mythical city of Atlantis.
  Housed in a beautiful 17th-
entury mansion, the **Mégaro
hísi Museum**, in the northern
art of the town, holds manu-
cripts from the 16th to 19th
enturies, maps, paintings, and
hotographs of Firá before
nd after the earthquake.
  Despite the 1956 earthquake
ou can still see vestiges of
'irá's architectural glory from
ne 17th and 18th centuries,
n Nomikoú and Erythroú
tavroú where several
nansions have been restored.
  The pretty ochre chapel of
Agios Stylianós, clinging to
ne edge of the cliff, is worth
visit on the way to the Fran-
ika, or Frankish quarter, with
s maze of arcaded streets. To
he south, the Orthodox
athedral is dedicated to the
'papantí (the Presentation of
Christ in the Temple). Built
n 1827, it is an imposing
ochre building with two

Detail of bright orange volcanic cliff in Firá

belltowers and murals by the
artist Christóforos Asimís.
The belltower of the **Dómos**
dominates the north of town
on Agíou Ioánnou. Though
severely damaged in the earthquake, much of its Baroque
interior has now been restored.

VISITORS' CHECKLIST

🏠 12,500. ✈ 5 km (3 miles) SE
of Firá. ⛴ Skála Firón. 🚌
ℹ Firá (22860 23021). 🎵
Classical Music, Firá: Aug & Sep.

🏛 **Archaeological Museum**
Opposite cable car. **Tel** 22860
22217. ⏱ 8:30am–3pm Tue–Sun.
⛔ main public hols. 🎫 📷

🏛 **Mégaro Ghísi Museum**
Near cable car. **Tel** 22860 22244.
⏱ May–late Oct: daily. 🎫

🏛 **New Archaeological
Museum**
Near Firá central square. **Tel** 22860
23217. ⏱ 8:30am–3pm Tue–Sun.
⛔ main public hols. 🎫

GEOLOGICAL HISTORY OF SANTORINI

Santoríni is one of several ancient volcanoes lying on the
southern Aegean volcanic arc. During the Minoan era, around
1450 BC, there was a huge eruption which began Santoríni's
transformation to how we see it today.

1 *Santoríni was a
circular volcanic
island before the
massive eruption that
blew out its middle.*

The volcano was active
for centuries, building up
to the 1450 BC explosion.

Clouds containing molten rock
spread over 30 km (19 miles).

**Crater of 22 sq km
(8.5 sq miles)**

2 *The eruption left
a huge crater, or
caldera. The rush of water
into the void created a
tidal wave, or tsunami, which
devastated Minoan Crete.*

A huge volume of lava
was ejected, burying
Akrotíri (*see p237*).

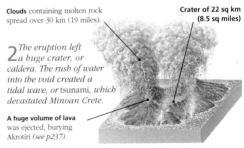

**Néa Kaméni and its active
volcanic cone**

Thirasía

**Volcano walls up
to 300 m (985 ft)
high**

3 *The islands of Néa
Kaméni and
Palaiá Kaméni, visible
today, emerged after
more recent volcanic
activity in 197 BC
and 1707. They are
still volcanically
active.*

Aspro Nisí

Palaiá Kaméni

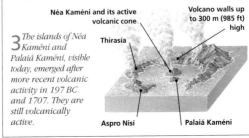

A donkey ride up the steps from
Skála Firón to Firá

Around Santoríni Island

Santoríni has much to offer apart from the frequently photographed attractions of Firá. There are some charming inland villages, and excellent beaches at Kamári and Aríssa with their long stretches of black sand. You can also visit some of Santoríni's wineries, or take a ferry or boat to the smaller islands. There are good bus services but a car or bike will allow you more freedom to explore. Major sites such as Ancient Thíra and Akrotíri have frequent bus or organized tour services.

A blue and ochre painted housefront in Oía

Oía ➋

Οία

11 km (7 miles) NW of Firá.
🏛 400. 🚌

At the northern tip of the island, the beautiful town of Oía is famous for its spectacular sunsets. A popular island excursion is to have dinner in one of the many restaurants at the edge of the abyss as the sun sinks behind the caldera. According to legend, the atmospheric town is haunted and home to vampires.

Reached by one of the most tortuous roads in the Cyclades, Oía is the island's third port and was an important and wealthy commercial centre before it was badly damaged in the 1956 earthquake.

Today Oía is designated a traditional settlement, having been carefully reconstructed after the earthquake. Its white and pastel-coloured houses with red pebble walls cling to the cliff face with the famous *skaftá* cave houses and blue-domed churches. Some of the Neo-Classical mansions built by shipowners can still be seen. A marble-paved pathway skirts the edge of the caldera to Firá. Staircases lead down to Arméni and the nearby fishing harbour at **Ammoúdi** with its floating pumice stones and red pebble beach. The tradition of boat-building continues at Arméni's small ferry dock at the base of the cliff, although the port is now mainly used by tourist boats departing daily for the small island of Thirasía.

Ancient Thíra, situated at the end of the Mésa Vounó peninsula

Ancient Thíra ➌

Αρχαία Θήρα

11 km (7 miles) SW of Firá.
🚌 to Kamári. ⬜ 8am–2pm Tue–Sun. ⬛ main public hols. 🚏 Aríssa 200 m (600 ft) below.

Commanding the rocky head-land of Mésa Vounó, 370 m (1,215 ft) up on the southeast coast, the ruins of the Dorian town of Ancient Thíra are still visible. Recolonized after the great eruption (*see p239*), the ruins stand on terraces over-looking the sea.

Excavated by the German archaeologist Hiller von Gortringen in the 1860s, most of the ruins date from the Ptolemies who built temples to the Egyptian gods in the 4th and 3rd centuries BC. There are also Hellenistic and Roman remains. The 7th-century Santoríni vases that were discovered here are now housed in Firá's Archaeological Museum (*see p239*).

A path through the site passes an early Christian basilica, remains of private houses, some with mosaics, the agora (or market) and a theatre, with a sheer view down to the sea. On the far west is a 3rd-century BC sanctuary cut into the rock, founded by Artemídoros of Perge, an admiral of the Ptolemaic fleet. It features relief carvings of an eagle, a lion, a dolphin and a phallus symbolizing the gods Zeus, Apollo, Poseidon and Priapus. To the east, on the Terrace of Celebrations, you can find

Rock carving in Ancient Thíra

Ammoúdi fishing village overlooked by Oía on the clifftop above

he view from ancient Thíra down to Kamári

raffiti which dates back as
ar as 800 BC. The messages
raise the competitors and
ancers of the *gymnopediés* –
estivals in which boys danced
aked and sang hymns to
pollo, or competed in feats
f physical strength.

nvirons
he headland of Mésa Vounó,
hich rises to the peak of
Mount Profítis, juts out into
he sea between the popular
eaches of Kamári and Veríssa.
Kamári is situated below
ncient Thíra to the north, and
s the island's main resort. The
each is a mix of stone and
lack volcanic sand, and is
acked by bars, tavernas and
partments. **Veríssa** has 8 km (5
niles) of black volcanic sand,
wide range of watersports
nd a campsite. A modern
hurch stands on the site of the
Byzantine chapel of Irene, after
whom the island is named.

Akrotíri ❹
Ακρωτήρι

2 km (7 miles) SW of Firá.
350. Kókkini Ammos 1 km
0.5 miles) S.

Akrotíri was once a Minoan
utpost on the southwest tip
f the island and is one of the
nost inspiring archaeological
ites in the Cyclades. After an
ruption in 1866, French
rchaeologists discovered
Minoan pots at Akrotíri,

though it was
Professor Spyrídon
Marinátos who,
digging in 1967,
unearthed the com-
plete city; it was
wonderfully
preserved after
some 3,500 years
of burial under
tonnes of volcanic
ash. The highlight
was the discovery
of frescoes which
are now displayed
at the National
Archaeological
Museum in Athens
(see p286).
Marinátos was
killed in a fall on
the site in 1974
and his grave is
beside his life's
work. Covered by a modern
roof, the excavations include
late 16th-century BC houses
on the Telchínes road, two
and three storeys high, many
still containing huge *pithoi*, or
ceramic storage jars. The
lanes were covered in ash
and it was here that the well-
known fresco of the two boys
boxing was uncovered.
Further along there is a mill
and a pottery. A flyover-style
bridge enables you to see the
town's layout including a
storeroom for *pithoi* which
held grain, flour and oil. The
three-storey House
of the Ladies is
named after the
fresco of two
voluptuous dark
women. The
Triangle Square has
large houses that

were originally decorated with
frescoes of fisherboys and
ships, now removed to Firá's
New Archaeological Museum.
The city's drainage system
demonstrates how sophis-
ticated and advanced the
civilization was. No human
or animal remains or treasure
were ever found, suggesting
that the inhabitants were
probably warned by tremors
before the catastrophe and
fled in good time.

Outlying Islands
From Athiniós, 12 km (7 miles)
south of Firá, excursion boats
run to the neighbouring islands.
The nearest are **Palaiá Kaméni**
and **Néa Kaméni**, known as
the Burnt Islands. You can
take a hot mud bath in the
springs off Palaiá Kaméni and
walk up the volcanic cone
and crater of Néa Kaméni.
Thirasía has a few tavernas
and hotels. Its main town, the
picturesque Manolás, has fine
views across the caldera to
Firá. Remote **Anáfi** is the most
southerly of the Cyclades and
shares the history of the other
islands in the group. It is a
peaceful retreat with good
beaches. There are a few
ancient ruins but nothing remains of the sanctuaries of
Apollo and Artemis that
once stood here.

Storage jars found at Akrotíri

FRESCOES OF AKROTÍRI
Painted around 1500 BC, these
Minoan-style murals are similar
to those found at Knosós *(see
pp 272–5)*. The best known
are *The Young Fisherman*,
depicting a youth holding blue
and yellow fish, and *The Young
Boxers*, showing two young
sparring partners with long
black hair and almond-shaped
eyes. Preserved by lava, the
frescoes have kept their colour
and are displayed on a rotating
basis at the New Archaeologica
Museum in Firá *(see p239)*.

CRETE

CHANIA · RETHYMNO · IRAKLEIO · LASITHI

*T*he island of Crete is dominated by harsh, soaring mountains whose uncompromising impregnability is etched deep into the Cretan psyche. For centuries, cut off by these mountains and isolated by sea, the character of the island people has been proudly independent. Many conquerors have come and gone but the Cretan passion for individuality and freedom has never been extinguished.

For nearly 3,000 years the ruins of an ancient Minoan civilization lay buried and forgotten beneath the coastal plains of Crete. It was not until the early 20th century that the remains of great Minoan palaces at Knosós, Phaestos, Mália and Zákros were unearthed. Their magnificence demonstrates the level of sophistication and artistic imagination of the Minoan civilization, now considered the wellspring of European culture.

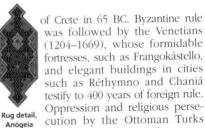

Rug detail, Anógeia

Historically, the island and its people have endured occupation by foreign powers and the hardships of religious persecution. The Romans brought their administrative expertise to the island, and the ancient city-state of Górtys became capital of the Roman province of Crete in 65 BC. Byzantine rule was followed by the Venetians (1204–1669), whose formidable fortresses, such as Frangokástello, and elegant buildings in cities such as Réthymno and Chaniá testify to 400 years of foreign rule. Oppression and religious persecution by the Ottoman Turks (1669–1898) encouraged a strong independence movement. By 1913, led by Elefthérios Venizélos (1864–1936), Crete had become a province of Greece. The island was again occupied by German forces during World War II despite valiant resistance.

Today, mountains, sparkling seas and ancient history combine with the Cretans' relaxed nature to make the island an idyllic holiday destination.

A local in Réthymno wearing traditional Cretan boots and headdress

◁ A palm-fringed estuary meets the sea at Préveli Beach on Crete's southern coast

The Flora and Fauna of Crete

Crete's wildlife is as varied as its landscape. In spring, flowers cover the coastal strip and appear inland in the patchwork of olive groves, meadows and orchards. Stony, arid *phrygana* habitat is widespread and pockets of native evergreen forests still persist in remote gorges. Freshwater marshes act as magnets for waterbirds, while Crete's position between North Africa and the Greek mainland makes it a key staging post for migrant birds in spring and autumn. Its comparative isolation has meant that several unique species of plant have evolved.

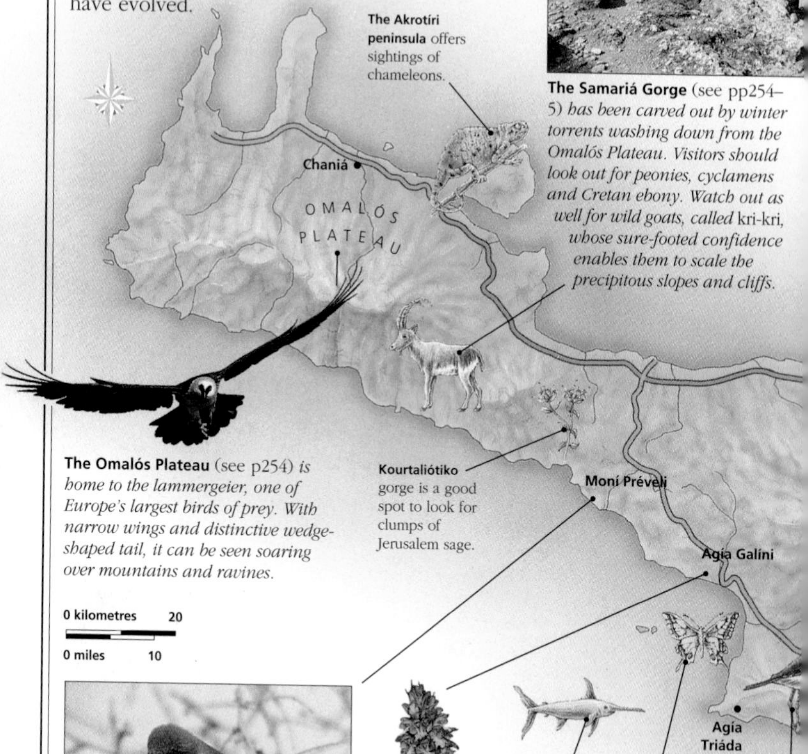

The Akrotíri peninsula offers sightings of chameleons.

The Samariá Gorge (see pp254–5) *has been carved out by winter torrents washing down from the Omalós Plateau. Visitors should look out for peonies, cyclamens and Cretan ebony. Watch out as well for wild goats, called kri-kri, whose sure-footed confidence enables them to scale the precipitous slopes and cliffs.*

Chaniá

OMALÓS PLATEAU

The Omalós Plateau (see p254) *is home to the lammergeier, one of Europe's largest birds of prey. With narrow wings and distinctive wedge-shaped tail, it can be seen soaring over mountains and ravines.*

Kourtaliótiko gorge is a good spot to look for clumps of Jerusalem sage.

Moní Préveli

Agía Galíni

0 kilometres   20

0 miles   10

Marlin and swordfish are the largest fish in the seas around Crete.

Agía Triáda

The Gulf of Mesará has a rough, grassy shoreline that is home to butterflies like the swallowtail.

Agía Triáda's wetlands are the haunt of black-winged stilts.

Moní Préveli (see p260) *is visited by the migrant Ruppell's warbler between May and August. With his bold black and white head markings and beady red eyes, the male is a striking bird.*

Agía Galíni (see p263) *is an excellent spot for spring flowers, and in particular the striking giant orchid. It stands more than 60 cm (24 inches) tall and can bloom as early as February or early March.*

The colourful yellow bee orchid   **The catchfly with its sticky stems**   **Cretan ebony, endemic to Crete**

WILD FLOWERS ON CRETE

Botanists visit Crete in their thousands each year to enjoy the spectacular display of wild flowers. They are at their best, and in greatest profusion, from February to May. By late June, with the sun at its highest in the sky, many have withered and turned brown. Most of those that undergo this transformation survive the summer as underground bulbs or tubers.

WILDLIFE TOUR OPERATORS

Honeyguide Wildlife Holidays
36 Thunder Lane, Norwich, Norfolk NR7 0PX. *Tel* 01603 300552.
www.honeyguide.co.uk

Naturetrek
Cheriton Mill, Cheriton, Alresford, Hampshire SO24 0NG. *Tel* 01962 733051. www.naturetrek.co.uk

Pure Crete
Bolney Place, Cowfold Road, Haywards Heath, RH17 5QT.
Tel 0845 070 1571.
www.purecrete.com

Wildlife Travel
The Manor House, Broad St, Great Cambourne, Cambridge CB3 6DH.
Tel 01954 713575. www.wildlife-travel.co.uk

Mál.ia (see p277) *is one of the many coastal resorts on Crete that provide a temporary home for migrant waders in spring and autumn. This wood sandpiper will stay and feed for a day or so around the margins of pools and marshes.*

Mount Díkti's slopes are covered in wild flowers in spring, including Cretan bee orchids.

Dolphins can be spotted from northern headlands.

Eloúnta has saltpans that are much favoured by avocets.

Siteía's *precipitous cliffs* (see p280) *are the habitat for Cretan ebony, a shrub unique to the island, which produces pinkish-purple spikes of flowers in the spring.*

Siteía

Lasíthi's fields are feeding grounds for colourful hoopoes.

Ierápetra

Geckos can be found on stone walls beside many roads in eastern Crete.

Agios Nikólaos is a stopping-off place for migrants such as wagtails.

Ierápetra (see p279) *attracts the migrant woodchat shrike in summer. Woodchats feed on insects and small lizards, which they sometimes impale on thorns to make them easier to eat.*

Zákros (see p281), *with its high cliffs, is where you find Eleonora's falcons performing aerobatic displays in summer.*

Exploring Crete

The most southerly of the Greek islands, Crete boasts clear blue seas, sandy beaches and glorious sunshine. Its north coast bustles with thriving resorts as well as historic towns such as Réthymno and Chaniá. Its rugged southern coast, in particular the southwest, is less developed. Four great mountain ranges stretch from east to west, forming the spine of the 250-km (155-mile) long island. A hiker's paradise, they offer magnificent scenery and some spectacular gorges. The island's capital, Irákleio, is famous for its Archaeological Museum and is also a good base for exploring the greatest of Crete's Minoan palaces, Knosós.

Card players in the vine-canopied streets of Réthymno's old town

View of the harbour, Sfakiá

SEE ALSO

- **Where to Stay** pp317–20
- **Where to Eat** pp338–40
- **Travel Information** pp366–9

GETTING AROUND

The provincial capitals of Chaniá, Réthymno, Irákleio and Agios Nikólaos act as the main transport hub for each region. Crete's bus service is quite well developed, with regular buses running along the north coast road. For touring the island a car is the most convenient mode of transport, though taxi fares are reasonable. Mountain roads between villages are now largely paved.

Large domed mosque inside Réthymno's Venetian Fortétsa

IGHTS AT A GLANCE

LOCATOR MAP

The north entrance to the Palace of Knosós

Piraeus Santoríni

Día

Kárpathos

Kárpathos,
Kásos

Mílos

Dragonada

Kato
Gouves
CHERSONISOS
㉔   Sisi

Agios Georgios

Agios Georgios

EIO ㉑

㉒ **KNOSOS**

㉕ **MALIA**

Spinalónga

Itanos

VAÏ
BEACH ㉞

ssos

HANES ㉓ Myrtiá

Agna

Selena Oros

Tzermiádo

Lató

SITEIA ㉜

㉝
**MONI
TOPLOU** Palaikastro

ari

Dikttan
Cave

㉖ **LASITHI
PLATEAU** ㉙ **KRITSA**

Psari Madara
2148m

Oros Dikti

Móchlos

Hamezi

Tefelion

Marta

GOURNIA ㉛

Hrisopigi

Voila

㉟
ZAKROS

éka Protoria

Koútsouras

nas Oros

Akhendrias

Nea Arvi Sidonia

㉚
IERAPETRA

Prasonision

Paranimfi

Alikapunta

0 kilometres   10

Chrysí

Koufonissi

0 miles   10

ELOUNTA ㉗

AGIOS
NIKOLAOS ㉘

Mirtos

A pelican in the picturesque harbour at Siteía

KEY

▬▬	Motorway
▬▬	Main road
═══	Minor road
▬▬	Scenic route
– –	Track
⋯⋯	High-season, direct ferry route
△	Summit

The magnificent beach of Falásarna with its long stretch of sand and turquoise waters

Kastélli Kissámou ❶
Καστέλλι Κισσάμου

Chaniá. 🏘 3,000. 🚌 ⛴ 🚆 Kastélli Kissámou.

The small, unassuming town of Kastélli Kissámou, also known simply as Kastélli, sits at the eastern base of the virtually uninhabited Gramvoúsa Peninsula, once a stronghold of pirates. While not a tourist-oriented town, it has a scattering of hotels and restaurants along its pebbly shore. In the town square there is a fine **Archaeological Museum** housing some spectacular Roman mosaics excavated in the area. The town is also a good base from which to explore the west coast of Crete. Boat trips run to the tip of the **Gramvoúsa Peninsula**, where there are some isolated and beautiful sandy beaches.

🏛 Archaeological Museum
Platía Tzanakáki (near the bus station). ◷ 8:30am–3pm Tue–Sun.

Environs
Some 7 km (4 miles) south of Kastélli, the ruins of the ancient city of **Polyrínia** are scattered above the village of Ano Palaiókastro (also known as Polyrínia). Dating from the 6th century BC, the fortified city-state was developed by the Romans and later the Byzantines and Venetians. The present church of **Enenínta ennéa Martýron** (Ninety-Nine Martyrs), built in 1894, stands on the site of a large Hellenistic building.
On the west coast of the Gramvoúsa Peninsula, 16 km

(10 miles) west of Kastélli, a winding road descends to the spectacular and isolated beach at **Falásarna**. Once the site of a Hellenistic city-state of that name, earthquakes have obliterated almost all trace of the once-thriving harbour and town. Today a few small guesthouses and tavernas are scattered along the northern end of the beach.
About 20 km (12 miles) east of Kastélli lies the picturesque fishing village of **Kolympári**. Head 1 km (0.5 miles) north of Kolympári for the impressive 17th-century **Moní Panagías Goniás**, with a magnificent seaside setting and a fine collection of 17th-century icons. Every year on 29 August (Feast of St John the Baptist), hundreds of pilgrims make the three-hour walk up the peninsula to the church of **Agios Ioánnis** to witness the mass baptism of boys named John (Ioánnis).

Palaióchora ❷
Παλαιόχωρα

Chaniá. 🏘 1,800. ⛴ 🚌 🚆 Elafónisos 14 km (9 miles) W.

First discovered in the 1960s by the hippie community, Palaióchora has become a haven for backpackers and package holiday-makers. This small port began life as a castle built by the Venetians in 1279. Today the remains of the fort, destroyed by pirate attacks in 1539, stand guard on a little headland dividing the village's two excellent beaches. To the west is a wide sandy beach with a windsurfing school, while to the east is a rocky but sheltered beach.

Environs
Winding up through the Lefká Ori (White Mountains), a network of roads passes through a stunning landscape

Moní Chrysoskalítissas near Palaióchora

THE BATTLE OF CRETE (1941)

Following the occupation of Greece in World War II, German forces invaded Crete. Thousands of German troops were parachuted into the Chaniá district, where they seized Máleme airport on 20 May 1941. The Battle of Crete raged

fiercely for ten days, with high casualties on both sides. Allied troops retreated through the Lefká Ori (White Mountains) to the south where, with the help of locals, they were evacuated from the island. Four years of German occupation followed, during which time implacable local resistance kept up the pressure on the invaders, until their final surrender in 1945.

German parachutists in Crete, 1941

Chaniá ❹

See pp252–3.

Akrotíri Peninsula ❺
Χερσόνησος Ακρωτηρίου

6 km (3.5 miles) NW of Chaniá. ⛴ Soúda. 🚌 Chaniá & Soúda. ✈ Stavrós 14 km (9 miles) N of Chaniá. Maráthi 10 km (6 miles) E of Chaniá.

Flat by Cretan standards, the Akrotíri Peninsula lies between Réthymno (see pp258–9) and Chaniá (see pp252–3). At its base, on top of Profítis Ilías hill, is a shrine to Crete's national hero, Elefthérios Venizélos (see p43). His tomb is a place of pilgrimage, for it was here that Cretan rebels raised the Greek flag in 1897 in defiance of the Great Powers.

There are several monasteries in the northeastern hills of the peninsula. **Moní Agías Triádas**, which has an impressive multidomed church, is 17th century, while **Moní Gouvernétou** dates back to the early Venetian occupation. Monks still inhabit both. Nearby, but accessible only on foot, the abandoned **Moní Katholikoú**, is partly carved out of the rock. Situated at the neck of the peninsula is a military base and the **Commonwealth War Cemetery**, burial ground of over 1,500 British, Australian and New Zealand soldiers killed in the Battle of Crete.

🏛 **Commonwealth War Cemetery**
4 km (2.5 miles) SE of Chaniá
⬜ daily.

of terraced hills and mountain villages, noted for their Byzantine churches. The closest of these is **Anýdri**, 5 km (3 miles) east of Palaióchora, with the 14th-century double-naved church of **Agios Geórgios** containing frescoes by Ioánnis Pagoménos (John the Frozen) from 1323.

In summer, a daily boat service runs to **Elafonísi**, a lagoon-like beach of golden sand and brilliant blue water. From here, a 5-km (3-mile) walk north takes you to **Moní Chrysoskalítissas** (Golden Step), named for the 90 steps leading up to its church, one of which is said to appear golden, at least in the eyes of the virtuous. It can also be reached by road 28 km (17 miles) south of Kastélli Kissámou. From Palaióchora, boat trips make the rough, 64-km (40-mile) crossing to **Gávdos** island, Europe's southernmost point.

Soúgia ❸
Σούγια

Chaniá. 🏘 270. ⛴ 🚌 ✈ Soúgia; Lissós 3 km (1.5 miles) W.

Once isolated from the rest of the world at the mouth of the Agía Eiríni Gorge, the hamlet of Soúgia is now linked with Chaniá and the north coast by a good road.

Still growing as a resort, the village has rooms to rent, and a few tavernas and bars. The beach is long and pebbly. It is overlooked by the village church which is built on top of a Byzantine structure, whose mosaic floors have been largely removed.

Fresco by Ioánnis Pagoménos, Agios Geórgios

Environs
Just over an hour's walk west of Soúgia, the ancient city-state of **Lissós** was a flourishing commercial centre in Hellenistic and Roman times. Among the remains are two fine 13th-century Christian basilicas, a 3rd-century BC Asklepieion (temple of healing) and a sanctuary. The route to Lissós leads up through the **Agía Eiríni Gorge**. Popular with experienced hikers, plans are under way to develop the gorge along the lines of the Samariá Gorge.

Goats grazing on the Akrotiri Peninsula

Chaniá 4

Χανιά

Olive oil tin, Chaniá covered market

Set against a spectacular backdrop of majestic mountains and aquamarine seas, Chaniá is one of the island's most appealing cities and a good base from which to explore western Crete. Its stately Neo-Classical mansions and massive Venetian fortifications testify to the city's turbulent and diverse past. Once the Minoan settlement of ancient Kydonia, Chaniá has been fought over and controlled by Romans, Byzantines, Venetians, Genoese, Turks and Egyptians. Following unification with Greece in 1913, the island saw yet another invasion during World War II – this time by the German army in 1941, when the Battle of Crete raged around Chaniá *(see p251)*.

The Mosque of the Janissaries

The Venetian Fort Firkás overlooking Chaniá's outer harbour

The Harbour

Most of the city's interesting sights are to be found in the old Venetian quarter, around the harbour and surrounding alleyways. At the northwest point of the outer harbour, the **Naval Museum**'s collection of model ships and other maritime artifacts is displayed in the well-restored Venetian Fort Firkás – also the setting for theatre and evenings of traditional dance in summer.

On the other side of the outer harbour, the **Mosque of the Janissaries** dates back to the arrival of the Turks in 1645 and is the oldest Ottoman building on the island. It was damaged during World War II and reconstructed soon after. Behind the mosque rises the hilltop quarter of Kastélli, the oldest part of the city, where the Minoan settlement of **Kydonia** is undergoing exca-vation. The site, closed to the

public but clearly visible from the road, is approached along Líthinon, a street lined with ornate Venetian doorways. Many of the finds from the site are on display in Chaniá's Archaeological Museum, including a collection of clay tablets inscribed with Minoan Linear A script.

By the inner harbour stand the now derelict 16th-century Venetian arsenals, where ships were once stored and repaired. The Venetian lighthouse, at the end of the sea wall, offers superb views over Chaniá.

🎥 Naval Museum
Fort Firkás, Aktí Kountourióti.
Tel 28210 91875. ☐ *daily.*
● *main public hols.* 🎫

Around the Covered Market

Connected to the harbour by Chálidon, this turn-of-the-century covered market sells local fruit and vegetables and Cretan souvenirs. Alongside the market, the bustling Skrýdlot, or Stivanádika, has shops selling leather goods, including traditional Cretan boots and made-to-measure

A tranquil view of Chaniá's old harbour at dawn

For hotels and restaurants in this region see pp317–20 and pp338–40

The atmospheric backstreets of the old Splántzia quarter

VISITORS' CHECKLIST

Chaniá. 🏘 *50,000.* ✈ *16 km (10 miles) E of Chaniá.* ⚓ *Soúda bay.* 🚌 *Kydonías (long distance), Plateía Agorás (around Chaniá).* ℹ️ *Kriári 40 (28210 92943).* 🛒 *Mon–Thu, Sat (veg & clothes).* 🎭 *Nautical Week (end June).* 🏖 *Agía Marína 9 km (6 miles) W; Plataniás 11 km (7 miles) W.*

sandals. The nearby **Archaeological Museum** is housed in the church of San Francesco and displays artifacts from western Crete including pottery, sculpture, mosaics and coins. Across a small square next to the museum is the 19th-century cathedral of **Agía Triáda**. Also nearby is the restored 15th-century **Etz Hayyim Synagogue**, which was used by Chania's Jewish population until the German occupation of 1941–45 when they were deported to death camps.

Dionysos and Ariadne mosaic, Chaniá Archaeological Museum

📷 Archaeological Museum
Chálidon 21. **Tel** 28210 90334. ⏱ 8:30am–3pm Tue–Sun. 🚫 main public hols. 📷 ♿

✡ Etz Hayyim Synagogue
Parados Kondylaki **Tel** 28210 86286. ⏱ 8am–6pm Mon–Fri.

The Splántzia Quarter
Northeast of the market, the Splántzia quarter is a picturesque area of the old town, where houses with wooden balconies overhang the cobbled backstreets. The tree-lined square known as **Plateía 1821** commemorates a rebellion against the occupying Turks, during which an Orthodox bishop was hanged. Overlooking the square stands the Venetian church of **Agios**

Nikólaos with and, nearby, are the 16th-century church of **Agioi Anárgyroi**, with its beautiful icons and paintings, and the church of **San Rocco** which was built in 1630.

Outside the City Walls
South of the covered market along Tzanakáki are the **Public Gardens**. They were laid out in the 19th century by a Turkish *pasha* (governor). The gardens include a modest zoo which houses a few animals, including the *kri-kri* (the Cretan wild goat). The gardens also offer a children's play area, a café and an open-air auditorium, which is often used for local ceremonies and cultural performances. The nearby **Historical Museum and Archives** is housed in a Neo-Classical building, and is devoted to the Cretan pre-occupation with rebellions and invasions. Its exhibits include photographs and letters of the famous statesman Elefthérios Venizélos (1864–1936), as well as many other historical records.

📷 Historical Museum and Archives
Sfakianáki 20. **Tel** 28210 52606. ⏱ Mon–Fri. 🚫 main public hols.

Environs
A series of sandy beaches stretches west from Chaniá all the way to the agricultural town of Tavronítis, 21 km (13 miles) away. A short walk west of Chaniá, the sandy beach of Agioi Apóstoloi is quieter and less developed than the city beaches.

Further west, the well-tended **German War Cemetery** stands witness to the airborne landing at Máleme of the German army in 1941 *(see p251)* . Built into the side of a hill, the peaceful setting is home to over 4,000 graves whose simple stone markers look out over the Mediterranean. A small pavilion by the entrance to the cemetery houses a display commemorating the event.

✝ German War Cemetery
19 km (12 miles) W of Chaniá. **Tel** 28210 62296. ⏱ daily.

The sandy beach of Agioi Apóstoloi, a short walk west of Chaniá

Samariá Gorge ❻
Φαράγγι της Σαμαριάς

The most spectacular landscape in Crete lies along the Samariá Gorge, the longest ravine in Europe. When the gorge was established as a national park in 1962, the inhabitants of pastoral Samariá village moved elsewhere, leaving behind the tiny chapels seen today. Starting from the Xylóskalo, 44 km (27 miles) south of Chaniá, a well-trodden trail leads down a tortuous 18-km (11-mile) course to the seaside village of Agía Rouméli. The walk takes from five to seven hours. Water fountains can be found en route and sturdy shoes should be worn.

Paeonia clusii,
Samariá Gorge

Facing east across the spectacular Samariá Gorge

Omalós Plateau

★ **Xylóskalo (Wooden Stairs)**
The Samariá Gorge is reached via the Xylóskalo, a zigzag path with wooden handrails which drops a staggering 1,000 m (3,280 ft) in the first 2 km (1 mile) of the walk.

Agios Nikólaos
This tiny chapel nestles under the shade of pines and cypresses near the bottom of the Xylóskalo.

THE KRI-KRI (CRETAN WILD GOAT)

Found in only a few areas of Crete, notably the Samariá Gorge, the Cretan wild goat is thought to be a truly wild relative of the all-too-numerous feral goats that are found throughout the Mediterranean region, as well as in other parts of the world. A protected species, the Cretan wild goat is nimble and sure-footed on rugged terrain, attributes that help guard against attacks by other predators. Mature adults have attractively marked coats and horns with three rings along their length.

A kri-kri on rocky terrain

0 kilometres		2
0 miles	1	

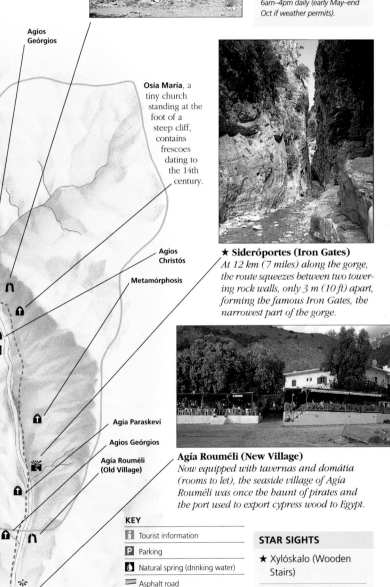

★ **Samariá Village**
Once inhabited, the village was abandoned in 1962 when the gorge was designated as a national park.

Agios Geórgios

Osía María, a tiny church standing at the foot of a steep cliff, contains frescoes dating to the 14th century.

Agios Christós

Metamórphosis

Agía Paraskeví

Agios Geórgios

Agía Rouméli (Old Village)

★ **Sideróportes (Iron Gates)**
At 12 km (7 miles) along the gorge, the route squeezes between two towering rock walls, only 3 m (10 ft) apart, forming the famous Iron Gates, the narrowest part of the gorge.

Agía Rouméli (New Village)
Now equipped with tavernas and domátia (rooms to let), the seaside village of Agía Rouméli was once the haunt of pirates and the port used to export cypress wood to Egypt.

VISITORS' CHECKLIST

44 km (27 miles) S of Chaniá, Chaniá. to Xylóskalo. Agía Rouméli to Sfakiá (via Loutró); to Palaiochóra (via Soúgia); last boat back leaves at 5pm. **Gorge** May–mid-Oct: 6am–4pm daily (early May–end Oct if weather permits).

KEY

- 🛈 Tourist information
- 🅿 Parking
- 💧 Natural spring (drinking water)
- Asphalt road
- Park boundary
- -- Path
- ❃ Viewpoint

STAR SIGHTS

★ Xylóskalo (Wooden Stairs)

★ Samariá Village

★ Sideróportes (Iron Gates)

Réthymno ❼

Ρέθυμνο

Once the Greco-Roman town of Rithymna, the site of today's Réthymno has been occupied since Minoan times. The city flourished under Venetian rule during the 16th century, developing into a literary and artistic centre, and becoming a haven for scholars fleeing Constantinople. Despite modern development and tourism, the city today has retained much of its charm and remains the intellectual capital of Crete. The old quarter is rich in elegant, well-preserved Venetian and Ottoman architecture. The huge Venetian Fortétsa, built in the 16th century to defend the island against the increasing attacks by pirates, overlooks the picturesque harbour with its charming 13th-century lighthouse.

The 17th-century Nerantzés Mosque

Exploring Réthymno

Réthymno's bustling harbour-front serves as one great outdoor cafeteria, catering almost exclusively for tourists. It is skirted along most of its length by a good, sandy beach, but at its western end lies a small inner harbour. A restored 13th-century **lighthouse** stands on its breakwater.

The **Fortétsa** dominates the town, above the inner harbour. Designed by Pallavicini in the 1570s, it was built to defend the port against pirate attacks (Barbarossa had devastated the town in 1538) and the threat of expansionist Turks. The ramparts are still largely intact. Within them, a mosque, a small church and parts of the governor's quarters can still be seen, though most are now in ruins. During the summer there are open-air concerts.

Traditional weaving in the Historical and Folk Art Museum

Directly opposite the main entrance to the Fortétsa, the **Archaeological Museum** occupies a converted Turkish bastion. Its collection is set out chronologically from Neolithic through Minoan to Roman times and includes artifacts from cemeteries, sanctuaries and caves in the region. Highlights include the late Minoan burial caskets *(larnakes)* and grave goods.

The old town clusters behind the Fortétsa, characterized by a maze of narrow vine-canopied streets and its Venetian and Ottoman houses with wrought-iron balconies. Off Plateía Títou Peocháki is the **Nerantzés Mosque**. This is the best-preserved mosque in the city. Built as a church by the Venetians, it was converted in 1657 into a mosque by the Turks. It now serves as the city's concert hall.

On Palaiológou, the 17th-century Venetian **Rimóndi Fountain**, with lion-headed spouts, stands alongside busy cafés and shops selling fresh produce. The elegant 16th-century Venetian **Lótzia** (Loggia) can also be seen here.

The small **Historical and Folk Art Museum** is housed in a Venetian mansion. On display here are local crafts, including some brilliantly coloured weaving, pottery, lace and jewellery.

⚑ Fortétsa
Katecháki. *Tel* 28310 28101. ◯ May–Oct: daily. ◉ main public hols. ▨

◖◗ Archaeological Museum
Cheimárras. *Tel* 28310 54668. ◯ Tue–Sun. ◉ main public hols. ▨

▥ Lótzia
Palaiológou & Arkadíou. *Tel* 28310 53270. ◯ Mar–Sep: Mon–Fri. ◉ main public hols. ▨

◖◗ Historical and Folk Art Museum
Vernárdou 30. *Tel* 28310 23398. ◯ Apr–Oct: Mon–Sat. ◉ main public hols. ▨

Tavernas and bars along Réthymno's waterfront, the focus of the town's activity

◁ Fishing boats lining the picturesque Venetian harbour of Réthymno

The magnificent shell of Frangokástello set against a dramatic backdrop

Environs

East of Réthymno, towards
Pánormos, the resort developments flow one into another,
while west of the city a 20-km
(12-mile) stretch of relatively
uncrowded beach culminates
in the village of **Georgioúpoli**.
Despite wholesale tourist
development, this small
community still retains some
of its traditional atmos-
phere. Massive eucalyptus
trees line the streets and
a picturesque, turtle-
inhabited river flows
placidly down to the sea.
Lake Kournás, 5 km
(3 miles) inland from
Georgioúpoli, is set in
a hollow among the
steeply rising hills.
Pedaloes, windsurfs
and canoes can be
hired at the lake and
a few shady tavernas
offer refreshments.

In Arménoi, on the main
Réthymno–Agía Galíni road,
there is an extensive late
Minoan cemetery where a
large number of graves have
been excavated, some with
imposingly long entrances.
Among the contents unearthed
are bronze weapons, vases
and burial caskets *(larnakes)*,
now on view in the archaeological museums of Chaniá
(see p253) and Réthymno.

Minoan Cemetery
9 km (6 miles) S of Réthymno.
Tue–Sun. main public hols.

Sfakiá 8
Σφακιά

Chaniá. 400. Sweetwater
3 km (2 miles) W of Loutró.

Overlooking the Libyan
Sea at the mouth of the
breathtaking Impros Gorge,
Sfakiá (also known as Chóra
Sfakíon) enjoys a commanding
position as the last coastal

community of any size until
Palaióchora *(see pp250–51)*.
Cut off from the outside
world until recently, it is little wonder that historically the local Sfakiot
clansmen enjoy their
reputation for rugged self-
sufficiency and indivi-
dualism, albeit accom-
panied by the notorious
feuding. The village
today is largely devoted
to tourism and makes a
good stepping-off point
for the southwest coast.

A Sfakiot in
traditional dress

Environs
West of Sfakiá, almost
impregnable mountains plum-
met into the Libyan Sea, allowing space for just a couple of
tiny settlements accessible
only by boat or on foot along
the E4 coastal path. The closest of these is **Loutró**, a charming and remote spot whose
sheltered cove, curving beach
and little white houses with
blue shutters fulfil every

The quiet bay and whitewashed
houses of Loutró

traveller's fantasy of a "real"
Greek village. In summer a
dozen tavernas and houses
provide rooms and meals for
tourists. Small boats are
available to take tourists to
nearby Gávdos island and the
breathtaking bay around
Sweetwater beach.

Frangokástello 9
Φραγκοκάστελλο

14 km (9 miles) E of Sfakiá, Chaniá.
daily.

Built by the Venetians as a
bulwark against pirates and
unruly Sfakiots in 1371, little
remains of the interior of
Frangokástello. However, its
curtain walls are well preserved
and from above the south
entrance, the Venetian Lion of
St Mark looks out to sea.

Ioánnis Daskalogiánnis, the
Sfakiot leader, surrendered
here in 1770 and was flayed
alive in Irákleio by his Turkish
captors. Fifty years later
Chatzimichális Daliánis, a
Greek freedom fighter, wrested
the fort from the Turks and
tried to hold it with an army of
just 385 men. Hopelessly
outnumbered, he and all his
followers were massacred by
the pitiless Turks. Legend has it
that at the end of May at dawn,
their solemn shadows can be
seen climbing up to the castle.

Directly below the fortress
is a sandy beach whose waters
are shallow and warm, an ideal
spot for families with young
children. A scattering of hotels
and tavernas cater for holiday-
makers and passing motorists.

Boats lining the small harbour at Plakiás

Plakiás ⑩
Πλακιάς

Réthymno. 🏠 100. 🚌 🚗 *Damnóni 3 km (2 miles) E.*

Once a simple fishing harbour serving the villages of Mýrthios and Selliá, Plakiás has grown into a full-scale resort with all the usual facilities. Its grey sandy beach is nearly 2 km (1 mile) long. Sited at the mouth of the Kotsyfoú Gorge, and with good road connections, Plakiás makes an excellent base for exploring the region.

Environs
A 5-minute drive, or a scenic walk around the headland, leads east to the beach of **Damnóni**. Tiny coves beyond it offer good swimming. Holiday apartments are being built on the adjoining hill. Quiet **Soúda** beach lies 3 km (2 miles) west of Plakiás.

Moní Préveli ⑪
Μονή Πρέβελη

14 km (9 miles) E of Plakiás, Réthymno. **Tel** *28320 31246.* 🚌 ◯ *daily.* 🏛 *museum only.* ♿

Accessible by road through the Kourtaliótiko Gorge, the working monastery of Préveli stands in an isolated but beautiful spot overlooking the sea. It played a prominent role in the evacuation of Allied forces from nearby beaches during World War II *(see p251)*.

The buildings cluster around a large central courtyard dating from 1731. There is a 19th-century church and a small museum displaying religious artifacts, including silver candlesticks and some highly

decorative robes. Further inland, the original 16th-century **Moní Agíou Ioánnou** (now known as Káto Préveli) was founded by Abbot Préveli and abandoned in the 17th century in favour of the more strategic position of the present monastery. About 1 km (0.5 mile) east of Moní Préveli, a steep path leads to **Préveli** beach (also known as Kourtaliótiko or Palm Beach), a crystal-clear, palm-fringed oasis.

Moní Arkadíou ⑫
Μονή Αρκαδίου

24 km (15 miles) SE of Réthymno, Réthymno. 🚌 *to Réthymno.* ◯ *daily.* 🏛 *museum only.* ♿

The 5th-century monastery of Arkadíou stands at the top of a winding gorge, at the edge of a fertile region of fruit trees and cypresses. Largely

Venetian façade of the church at Moní Arkadíou

rebuilt at the end of the 16th century, the most impressive of its buildings is the double-naved church with an ornate Venetian façade which dates back to 1587.

The monastery provided a safe haven for its followers in times of religious persecution by local Muslims. On 9 November 1866, when its buildings were crowded with hundreds of refugees, it came under attack by the Ottoman army. Choosing death over surrender, the Cretans torched the gunpowder storeroom, killing Christian and Muslim alike. The ensuing carnage created instant martyrs for freedom whose sacrifice is not forgotten. A sculpture outside the monastery depicts the only surviving girl and the abbot who lit the gunpowder. Today, a small museum displays sacramental vessels, icons, prayer books, vestments and tributes to the martyrs.

Environs
At Archaía Eléftherna, 10 km (6 miles) northeast of Moní Arkadíou, lie the ruins of the ancient city-state of **Eléftherna**. Founded in 700 BC, all that remains is a tower on a rocky ridge and a derelict Hellenistic bridge in the valley below. Northeast of Eléftherna the village of **Margarítes** is well known for its pottery.

The isolated buildings of Moní Préveli, nestled into the rocks

Tour of the Amári Valley ⓭

Dominated by the peaks of Mount Idi to its east, the Amári Valley offers staggering views over the region's rock-strewn peaks, broad green valleys and dramatic gorges. Twisting but well-paved roads link the many small agricultural communities of the Amári where, even today, moustachioed men in knee-high

Detail from the church of the Panagía at Méronas

boots and baggy trousers *(vrákes)* can be seen outside the local tavernas. The area is dotted with shrines, churches and monasteries harbouring Byzantine frescoes and icons. Traditionally an area of Cretan resistance, many of the Amári villages were destroyed during World War II.

Olive groves in the Amári Valley

Thrónos ①
The beautifully frescoed church of the Panagía at Thrónos dates back to the 14th century and has traces of 4th-century Christian mosaics. A key is available from the nearby taverna.

Méronas ⑧
At the centre of Méronas is the Venetian-style church of the Panagía with its early 14th-century frescoes.

Moní Asomáton ②
The Venetian buildings of **Moní Asomáton**, now an agricultural college, stand in a lush oasis of palm, plane and eucalyptus trees.

Amári ③
Sweeping views of Mount Idi can be seen from the Venetian clock tower in the centre of Amári. Just outside the village, the church of Agía Anna shelters the island's oldest frescoes, dated 1225.

Gerakári ⑦
Gerakári is famous for its fresh and bottled cherries and cherry brandy.

Vizári ④
Just west of the village of Vizári are the ruins of an early Christian basilica dating from the 6th century.

Kardáki ⑥
The 13th-century ruined church of Agios Ioánnis Theológos stands by the roadside north of Kardáki.

RETHYMNO • Agía Foteiní • Opsiglas • Monastiráki • Platánia • Fourfourás • Vrýses • SPILI • Platís • AGIAGALINI • Ádodoúlou

Ano Méros ⑤
A large marble war memorial just outside Ano Méros depicts a woman hewing out the names of World War II Resistance heroes.

TIPS FOR DRIVERS

Length: 92 km (57 miles).
Stopping-off points: There are local tavernas in every village en route. The taverna at Ano Méros offers spectacular views over the valley. Opposite the ruined church outside Kardáki is a shaded area and water fountain, an ideal stop in the heat of summer (see also p370).

KEY
— Tour route
= Other roads
⚡ Viewpoint

0 kilometres 5
0 miles 2

For additional map symbols *see back flap*

Mount Idi
Ψηλορείτης

Réthymno. ▦ to Anógeia & Kamáres.

At 2,456 m (8,080 ft) the soaring peaks of Mount Idi (or Psiloreítis) are the crowning glory of the massive Psiloreítis range. The highest mountain in Crete, it is home to many sanctuaries including the famous Idaian Cave.

From Anógeia, a paved road leads to the **Nída Plateau**, a journey of 23 km (14 miles) through rocky terrain, punctuated by the occasional stone shepherd's hut. Here a lone taverna caters to visitors en route to the **Idaian Cave**, a further 20-minute hike up the hill. This huge cavern, where Zeus was reared, has yielded artifacts, including some remarkable bronze shields, dating from c.700 BC. Some of the artifacts can be seen in the Irákleio Archaeological Museum (see pp270–71). From the plateau, marked trails lead up to the peak of **Mount Idi**, while a short distance away a tiny ski resort operates at weekends, conditions permitting, from December to March.

On the mountain's southern face, a 3-hour scramble from Kamáres village leads to the **Kamáres Cave**. Here the famous Minoan pottery known as Kamáres ware was discovered and examples are now on display in the Irákleio Archaeological Museum.

CRETAN CAVES AND THE MYTH OF ZEUS

The island of Crete is home to 4,700 caves and potholes of which some 2,000 have been explored. Since Neolithic times, caves have been used as cult centres by successive religions and have yielded many archaeological treasures. Bound up with ancient Cretan mythology, the Diktian (see p277) and Idaian caves are two of the island's most visited. According to legend, Rhea gave birth to the infant god Zeus in the Diktian Cave where he was protected by *kourítes* (warriors) and nurtured by a goat. He was then concealed and raised in the Idaian Cave to protect him from his father, Kronos, who had swallowed his other offspring after a warning that he would be dethroned by one of his sons. The Idaian Cave was an important pilgrimage centre during Classical times.

Stalagmites in the Diktian Cave (see p273), Lasíthi

Anógeia ⓯
Ανώγεια

Réthymno. ▨ 2,300. ▦

High up in the Psiloreítis mountain range, the small village of Anógeia dates back to the 13th century. The village has suffered a turbulent past, having been destroyed by the Turks in 1821 and 1826, and then completely rebuilt after destruction by the German army in 1944.

Modern Anógeia runs along a rocky ridge, with its own square and **war memorial –**

a bronze statue of a Cretan hero in traditional dress. Inscribed on the memorial are the most significant dates in Crete's recent past: 1821, Greek Independence; 1866, slaughter of Christian refugees at Moní Arkadíou (see p260); 1944, liberation from German occupation. Tavernas, shops and banks are also situated in this part of town.

The old village tumbles down the steep slopes into a warren of narrow stepped alleys, ultimately converging on a little square of stalls and tavernas. Here, a marble bust

The Nída Plateau between Anógeia village and the Idaian Cave, Mount Idi

For hotels and restaurants in this region see pp317–20 and pp338–40

Woman selling locally made rugs and lace in Anógeia

f local politician Vasíleios
Skoulás stands next to a less
formal woodcarving of his
friend Venizélos *(see p43)*, by
local artist Manólis Skoulás.

The stalls in the old part of
the village abound in locally
made embroidery, lace and
brightly coloured rugs, forming
one of Crete's main centres
for woven and embroidered
goods. Nearby tavernas serve
grilled goats' meat and other
Cretan specialities. Music en-
thusiasts can pay their respects
at the shrine of Níkos Xyloúris,
a 1970s folk singer who died
at an early age and whose
little whitewashed house
overlooks the main square.

Agía Galíni 16
Αγία Γαλήνη

Réthymno. 1,040. Agía
Galíni.

Formerly a fishing village
situated at the southern end of
the Amári Valley, Agía Galíni is
today a full-blown tourist
resort. The original village,
now only a handful of old
houses and narrow streets, is
dwarfed by the mass of holiday apartments stretching up
the coast. The harbourfront
is alive with busy tavernas
snuggled between the water
and cliffs. Just beyond the
harbour, the small sandy beach
is popular with sunbathers.

Environs
Taxi boat trips sail daily from
Agía Galíni's harbour to the
neighbouring beaches of
Agios Geórgios and **Agios
Pávlos** and, further still, to

Préveli beach at Moní Préveli
(see p260). There are also
daily excursions to the
Paximádia islands where
there are good sandy beaches.

Agía Triáda 17
Αγία Τριάδα

3 km (2 miles) W of Phaestos, Irákleio.
to Phaestos. *Tel 28920 91564.*
daily. main public hols.
Kómo 10 km (6 miles) SW; Mátala
15 km (9 miles) SW.

The Minoan villa of Agía
Triáda was excavated by the
Italians from 1902 to 1914. An
L-shaped structure, it was built
around 1700 BC, the time of
the Second Palace period *(see
p275)*, over earlier houses. Its
private apartments and public
reception rooms are located
in the angle of the L, overlooking a road that may have
led to the sea. Gypsum facing
and magnificent frescoes used
to adorn the walls of these
rooms. Rich Minoan treasures,
including the carved stone
Harvester Vase, Boxer Rhyton
(jug) and Chieftain Cup, were

all found in this area and are
on display at the Irákleio
Archaeological Museum *(see
pp270–71)*. Evidence of the
villa's importance is provided
by a find of clay seals and rare
tablets bearing the undeci-
phered Minoan Linear A script.

Following the villa's destruction by fire in around 1400
BC, a Mycenaean megaron
(hall) was built on the site. The
ruined settlement to the north,
with its unique porticoed row
of shops, dates mostly from
this period, as does the magnificent painted sarcophagus
that was found in the cemetery to the north. The paint-
work on the sarcophagus
depicts a burial procession; it
can be seen in the Irákleio
Archaeological Museum.

Agía Triáda archaeological site

Environs
At the village of Vóroi, 6 km
(4 miles) north-east of Agía
Triáda, is the fascinating
Museum of Cretan Ethnology.
Displayed here is a collection
of tools and materials used in
the everyday life of rural Crete
up to the early 20th century.

**Museum of Cretan
Ethnology**
Tel 28920 91110. daily.
main public hols.

Agía Galíni resort, nestled into the rocks at the foot of the Amári Valley

Mátala's town beach flanked by sandstone cliffs

Mátala ⑱
Μάταλα

Irákleio. ⬙ 132. ▦ ▦ Kalamáki 5 km
(3 miles) N; Léntas 24 km (15 miles) SE.

Clustered around an idyllic sweeping bay, Mátala remained a small fishing hamlet until the tourist boom of the 1960s, when it was transformed into a pulsating resort. Hotels, bars and restaurants abound in the lively town centre and development here is steadily on the increase.

Despite present appearances, Mátala has not passed untouched by history. Homer described Menelaos, husband of Helen of Troy *(see p54)*, being shipwrecked here on his way home from Troy. During Hellenistic times, around 220 BC, Mátala served as the port for the ancient city-state of Górtys. The resort's pitted sandstone cliffs, looming dramatically over the town beach, were originally carved out for use as tombs in the Roman era. Later they were extended as cave dwellings for early Christians, shepherds and recently hippies.

Environs
The area around Mátala has some beautiful beaches including the bay of **Kaloí Liménes** to the southeast. This was said to have been the landing place of St Paul the Apostle on his way to Egypt. To the north, a sandy track leads to **Kommós**, one of the best sandy beaches on the south coast. In this magnificent setting lay the Minoan settlement of Kommós, thought to have been a major port serving Phaestos *(see pp266–7)*. The extensive site is currently under excavation.

Boat excursions run daily from Mátala to the Paximádia islands in the bay and to palm-fringed Préveli beach *(see p260)* further west. There are also several bus tours to the important archaeological sites of Phaestos, Agía Triáda *(see p263)* and Górtys.

Phaestos ⑲

See pp266–7.

Górtys ⑳
Γόρτυς

Irákleio. **Tel** 28920 31144. ▦
◯ 8am–7pm daily. ● main public hols. ▦ ⬙

A settlement from Minoan through to Christian times, the ancient city-state of Górtys began to flourish under Dorian rule during the 6th century BC. Following its defeat of Phaestos in the 2nd century BC, Górtys became the most important city on Crete. Its pre-eminence was sealed following the Roman invasion of 65 BC, when Górtys was appointed capital of the newly created Roman province of Crete and Cyrene (modern-day Libya). Górtys continued to flourish under Byzantine rule, strategically sited at the point where a tributary of the ancient river Lethe (today's Mitropolianós) flowed into the fertile Messará Plain, with coastal ports to the west and south. It was not until the late 7th century AD that the great city was destroyed by Arab invaders. Today, the most visited ruins of this extensive site lie to the north of the main road.

Statue at the ancient site of Górtys

Section of the Law Code of Górtys, housed in the odeion, Górtys

THE LAW CODE OF GORTYS
The most extensive set of early written laws in the Greek world was found at the archaeological site of ancient Górtys and dates from c.500 BC. Each stone slab of the Górtys Code contains 12 columns of inscriptions in a Doric Cretan dialect. There is a total of 600 lines which read alternately from left to right and from right to left (a style known as *boustrophedon*, literally "as the ox-plough turns"). The laws were on display to the public and related to domestic matters including marriage, divorce, adoption, the obligations and rights of slaves, and the sale and division of property.

The bema (area behind altar) of Agios Títos basilica, Górtys

Exploring the Ruins

A car park, ticket booth and café are located near the entrance to the site. Immediately beyond stand the remains of the 6th-century basilica of **Agios Títos**, once an impressive, three-aisled edifice whose floorplan is still clearly visible. In its heyday it was the premier Christian church of Crete, traditionally held to be the burial place of St Titus, first bishop and patron saint of Crete, who was sent by St Paul to convert the heathens. Behind the basilica is an area thought to be a Greek **agora** (market place). Beyond this stand the semicircular tiered benches of the Roman **odeion**, originally used for concerts and now home to the famous stone slabs inscribed with the Law Code of Górtys.

Behind the odeion, a path leads up to the **acropolis** hill above Górtys, where a post-Minoan settlement was built

around 1000 BC. Parts of the fortifications still remain. On the east slope of the hill are the foundations of the 7th-century BC **Temple of Athena**. A statue and other votive objects found at a sacrificial altar lower down are in Irákleio Historical Museum *(see p268)*.

To the south of the main road, an extensive area of Roman Górtys remains only partially excavated. Standing in a grove of old olive trees is the 7th-century BC **Temple of Pythian Apollo**, to which a monumental altar was added in Hellenistic times. The temple was converted into a Christian basilica in the 2nd century AD and remained important until AD 600, when it was superseded by the basilica of Agios Títos. At the far end of the site are the ruins of the 1st-century AD **praetorium**, the grand palace of the Roman provincial governor.

Environs

East of Górtys, in the nearby village of **Agioi Déka**, is the 13th-century Byzantine church of the same name. It was built on the spot where ten early Christian Cretans were martyred in AD 250 for their opposition to the Roman Emperor Decius. In the nave of the church is an icon portraying the ten marytrs.

13th-century icon of the ten martyrs, Agioi Déka church

North of Górtys, a scenic drive heads to the mountain village of **Zarós**, a surprisingly green oasis famous for its clear spring water. From here, a clearly marked trail leads north through the spectacular **Zarós Gorge**. About 3 km (2 miles) northwest of Zarós lies **Moní Vrontisíou**. The monastery's icons by Michaíl Damaskinós (c.1530–91), a famous painter of the Cretan School, are now on display in the Museum of Religious Art in Irákleio *(see p268)*.

The ruins of the *praetorium*, the once-grand palace complex of the governor of the province, Górtys

Phaestos ⑲

Το Ανάκτορο της Φαιστού

Spectacularly situated on a ridge overlooking the fertile Messará Plain, Phaestos was one of the most important Minoan palaces in Crete. Excavations by the Italian archaeologist Frederico Halbherr, in 1900, unearthed two palaces. Remains of the first palace, constructed around 1900 BC and destroyed by an earthquake in 1700 BC, are still visible. However, most of the present ruins are of the second palace which was severely damaged around 1450 BC, possibly by a tidal wave. The city-state was finally destroyed by Górtys *(see pp264–5)* in the 2nd century BC. Today, the superimposed ruins of both palaces make interpretation of the site difficult.

View of the Messará Plain from the north court

The archives room consists of a series of mudbrick chests. It was here that the famous Phaestos disc was discovered.

The peristyle hall, a colonnaded courtyard, bears traces of an earlier structure dating from the Prepalatial period (3500–1900 BC).

North court

First Palace shrine complex

★ **Grand Staircase**
This monumental staircase, which leads up to a propylon (porch) and colonnaded lightwell, was the main entrance to the palace.

THE PHAESTOS DISC

This round clay disc, 16 cm (6 inches) in diameter, was discovered at Phaestos in 1903. Inscribed on both sides with pictorial symbols that spiral from the circumference into the centre, no one has yet been able to decipher its meaning or identify its origins, though it is possibly a sacred hymn. The disc is one of the most important exhibits at the Irákleio Archaeological Museum *(see pp270–71)*.

West Courtyard and Theatre Ar
The ruins of the west court date to c.1900 BC, the First Palace period. T seats on its north side were used fo viewing rituals and ceremonies.

Royal Apartments
Now fenced, these rooms were the most elaborate, consisting of the Queen's Megaron or chamber (left), the King's Megaron, a lavatory and a lustral basin (covered pool).

VISITORS' CHECKLIST

65 km (40 miles) SW of Irákleio.
Tel 28920 42315. 🚌 🔲 Nov–May: 8:30am–3pm daily; Jun–Oct: 8:30am–7pm daily. 🔴 1 Jan, 25 Mar, Good Fri am, Easter Sun, 1 May, 25, 26 Dec. 🈺 📷

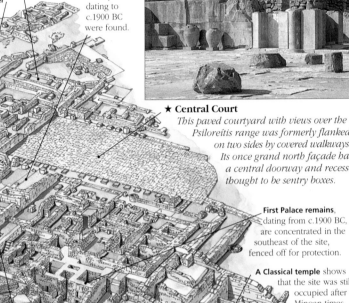

Workshops

Northeast quarter

The main hall is where clay seals dating to c.1900 BC were found.

★ Central Court
This paved courtyard with views over the Psiloreítis range was formerly flanked on two sides by covered walkways. Its once grand north façade has a central doorway and recesses thought to be sentry boxes.

First Palace remains, dating from c.1900 BC, are concentrated in the southeast of the site, fenced off for protection.

A Classical temple shows that the site was still occupied after Minoan times.

STAR SIGHTS

★ Grand Staircase

★ Central Court

Storerooms

Storage Pits
Dating from around 1900 BC, these circular walled pits were used for storing the palace's grain.

RECONSTRUCTION OF SECOND PALACE

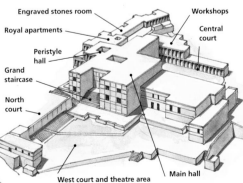

Engraved stones room

Royal apartments

Peristyle hall

Grand staircase

North court

Workshops

Central court

Main hall

West court and theatre area

Irákleio ㉑
Ηράκλειο

A settlement since the Neolithic era, Irákleio served as the port for Knosós in Roman times. Under Venetian rule in the 13th century, it became known as Candia, the capital of the Aegean territories. Today the sprawl of traffic-jammed streets and concrete apartment buildings detracts from Irákleio's appeal. Yet, despite first impressions, the island's capital harbours a wealth of Venetian architecture, including the city walls and fortress. Its Archaeological Museum houses the world's greatest collection of Minoan art, and the city provides easy access to the Palace of Knosós *(see pp272–5)*.

Façade of the Venetian church of Agios Títos

Exploring Irákleio
At the heart of Irákleio is Plateía Eleftheríou Venizélou, a pedestrian zone with cafés and shops grouped around the ornate 17th-century **Morosini fountain**. Facing the square, the restored church of **Agios Márkos** was built by the Venetians in 1239 and is now used as a venue for concerts and exhibitions. From here, 25 Avgoústou (25 August Street) leads north to the Venetian harbour. On this street, the elegantly restored 17th-century **Loggia** was a meeting place for the island's nobility and now serves as Irákleio's city hall. Beyond the Loggia, in a small square set back from the road, is the refurbished 16th-century church of **Agios Títos**, dedicated to the island's patron saint. On the other side of 25 Avgoustou, the tiny **El Greco Park** is named after Crete's most famous painter.

At the northern end of 25 Avgoustou, the old harbour is dominated by the Venetian **fortress**, whose dauntingly massive structure successfully repulsed prolonged assaults by the invading Turks in the 17th century. Named the *Rocca al Mare* (Fort on the Sea) by

Lion of St Mark detail, fortress

the Venetians and *Koulés* by the Turks, it was erected by the Venetians between 1523 and 1540. Opposite the fortress are the arcades of the 16th-century Venetian **Arsenali** where ships were built and repaired. West along the waterfront, the **Historical Museum** traces the history of Crete since early Christian times. Its displays include Byzantine icons, friezes, sculptures, and archives of the Battle of Crete *(see p251)*. Pride of place is given to the only El Greco painting in Crete, *The Landscape of the Gods-Trodden Mount Sinai* (c.1570).

A short walk two blocks southwest of Plateía Venizélou, on Plateía Agías Aikaterínis, is the 16th-century Venetian church of Agía Aikateríni of Sinai. Once a monastic foundation famous as a centre of art and learning, it now houses the **Museum of Religious Art**, a magnificent collection of

EL GRECO
Domínikos Theotokópoulos (alias El Greco) was born in Crete in 1545. His art was rooted in the Cretan School of Painting, an influence that permeates his highly individualistic use of dramatic colour and elongated human forms. In Italy, El Greco became a disciple of Titian before moving to Spain. He died in 1614, and his works can to be seen in major collections around the world. Ironically, only one exists in Crete, at Irákleio's Historical Museum.

El Greco's *The Landscape of the Gods-Trodden Mount Sinai* (c. 1570), Historical Museum

Byzantine icons, frescoes and manuscripts. The most significant exhibits are six icon by Michaíl Damaskinós, a 16th-century Cretan artist wh learnt his craft here at Agía Aikateríni and taught El Grecc Next door, the 19th-century cathedral of **Agios Minás** towers over the square.

To the east, the street marke in 1866 Street leads south to Plateía Kornárou. Here, coffe is served from a charming con verted Turkish pump-house, next to which a headless Roman statue graces the

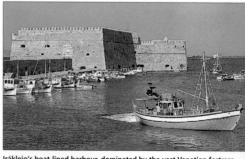

Irákleio's boat-lined harbour, dominated by the vast Venetian fortress

The Bembo drinking fountain, Plateía Kornárou

6th-century **Bembo fountain**. East, along Avérof, Plateía Eleftherías (Freedom Square) is dominated by a statue of Elefthérios Venizélos (1864–1936), the politician central to Crete's union with Greece. Off the square, the pedestrianized Daidálou is good for shops and restaurants. Just to the north is the **Irákleio Archaeological Museum** (see pp270–71) and main tourist office.

South of town, beyond the old city walls, the small **Museum of Natural History** deals with the natural environment of the Aegean. Exhibits include fossils and live animals.

🏛 **Loggia**
25 Avgoústou. **Tel** 2810 399399.
⬜ Mon–Sat. ⬛ main public hols.

⚓ **Fortress**
Venetian harbour. ⬜ Tue–Sun.
⬛ main public hols.

🏛 **Historical Museum**
Lysimáchou Kalokairinoú 7. **Tel** 2810 283219. ⬜ Mon–Sat.
⬛ main public hols.

🏛 **Museum of Religious Art**
Agía Aikateríni of Sinai, Plateía Agías Aikaterínis. **Tel** 2810 288825. ⬜ Mon–Sat. ⬛ main public hols. ♿

🏛 **Museum of Natural History**
Sofokli Venizélou. **Tel** 2810 285740. ⬜ Sun–Fri.

VISITORS' CHECKLIST

Irákleio. 👥 116,000. ✈ 5 km (3 miles) E. ⛴ E of Venetian harbour. 🚌 Leofóros Papadimitríou (for Réthymno, Chaniá, Agios Nikólaos and Ierápetra); Plateía Kóraka (for Mátala). 🛈 Xanthoudídou 1 (2810 246299, dtkritis@otenet.gr). ⬛ Sat. 🎭 Summer Festival: Jul–Sep. 🏖 Amoudára 10 km (6 miles) W.

Environs
Travelling west by the main Irákleio–Réthymno road, a turn-off to Anógeia (see pp262–3) climbs to the village of **Tylissos**, where the remains of three Minoan villas were found in 1902. West of Irákleio, the road leads to the village of **Fódele**, claimed to be the birthplace of El Greco. His house lies above the Byzantine church to the northwest. The **CretAquarium**, 15km (9 miles) east from Irákleio, exhibits around 200 species of fish and invertebrates.

🐟 **CretAquarium**
Near Gournes. ⬜ Jun–Sep: 9am–9pm daily; Oct–May: 9am–7pm daily. ♿

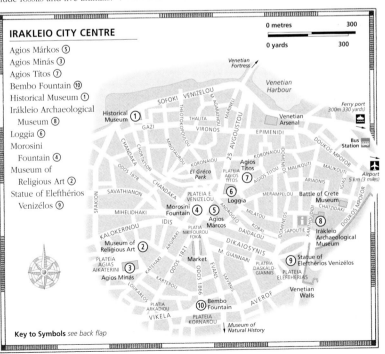

IRAKLEIO CITY CENTRE

Agios Márkos ⑤
Agios Minás ③
Agios Títos ⑦
Bembo Fountain ⑩
Historical Museum ①
Irákleio Archaeological Museum ⑧
Loggia ⑥
Morosini Fountain ④
Museum of Religious Art ②
Statue of Elefthérios Venizélos ⑨

0 metres 300
0 yards 300

Key to Symbols see back flap

Irákleio Archaeological Museum
Αρχαιολογικό Μουσείο Ηρακλείου

The Irakleio Archaeological Museum houses the world's most important collection of Minoan artifacts, giving a unique insight into a highly sophisticated civilization that existed on Crete over 3,000 years ago. On display are exhibits from all over Crete amassed since 1883, including the famous Minoan frescoes from Knosós *(see pp272–5)* and the Phaestos Disc*(see p266)*. Finely carved stone vessels, exquisite jewellery, Minoan double axes, and other artifacts make up only part of the museum's vast collection. The museum is partially closed for renovation, but major exhibits can be seen in a temporary display.

Gold Bee Pendant
Found in the Chrysólakkos cemetery at Mália (see p277), this exquisite gold pendant of two bees joined together dates from the 17th century BC.

★ **Bull's Head Rhyton**
This 16th-century BC vessel (see p63) was used for the pouring of ritual wines. Found at Knosós, it is carved from steatite, a black stone, with inset rock crystal eyes and a mother-of-pearl snout.

★ **Phaestos Disc**
Made of clay, the disc was found at the Palace of Phaestos in 1903.

Ground floor

STAR SIGHTS

★ The Hall of the Frescoes

★ Phaestos Disc

★ Bull's Head Rhyton

★ Snake Goddesses

Octopus Vase
This fine late Minoan vase from Palaí-kastro (see p281) is decorated with images from the sea.

Stairs to first floor

Minoan vase with double axe motif

THE MINOAN DOUBLE AXE

The Minoan double axe served both as a common tool used by carpenters, masons and shipbuilders, and as an extremely powerful sacred symbol thought to have been a cult object connected with the Mother Goddess. The famous Labyrinth at Knosós *(see pp272–5)* is believed to have been the "dwelling place of the double axe", the word *labrys* being the ancient Greek name for double axe. Evidence of the importance of the axe for the Minoans is clear from the many vases, *larnakes* (clay coffins), seals, frescoes and pillars that were inscribed or painted with the ceremonial double axe, including the walls of the Palace of Knosós. The ceremonial axe is often depicted between sacred horns or in the hands of a priest. Votive axes (ritual offerings) were highly decorated and made of gold, silver, copper or bronze. A stylized version of the double axe also features in early Linear A and B scripts.

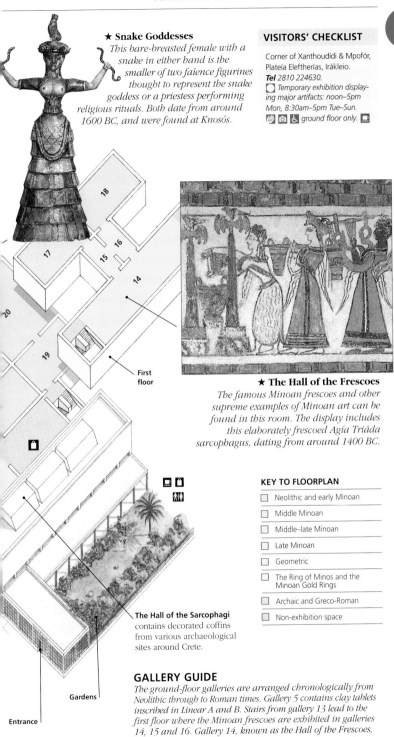

★ Snake Goddesses
This bare-breasted female with a snake in either hand is the smaller of two faïence figurines thought to represent the snake goddess or a priestess performing religious rituals. Both date from around 1600 BC, and were found at Knosós.

VISITORS' CHECKLIST

Corner of Xanthoudídi & Mpofór, Plateía Eleftherías, Irákleio.
Tel *2810 224630.*
☐ *Temporary exhibition display-ing major artifacts: noon–5pm Mon, 8:30am–5pm Tue–Sun.*
🖼 📷 ♿ *ground floor only.* ☐

18

17   16
      15
         14

20

19

First floor

★ The Hall of the Frescoes
The famous Minoan frescoes and other supreme examples of Minoan art can be found in this room. The display includes this elaborately frescoed Agía Triáda sarcophagus, dating from around 1400 BC.

KEY TO FLOORPLAN
- ☐ Neolithic and early Minoan
- ☐ Middle Minoan
- ☐ Middle–late Minoan
- ☐ Late Minoan
- ☐ Geometric
- ☐ The Ring of Minos and the Minoan Gold Rings
- ☐ Archaic and Greco-Roman
- ☐ Non-exhibition space

The Hall of the Sarcophagi contains decorated coffins from various archaeological sites around Crete.

GALLERY GUIDE
The ground-floor galleries are arranged chronologically from Neolithic through to Roman times. Gallery 5 contains clay tablets inscribed in Linear A and B. Stairs from gallery 13 lead to the first floor where the Minoan frescoes are exhibited in galleries 14, 15 and 16. Gallery 14, known as the Hall of the Frescoes, houses a model of the Palace of Knosós.

Gardens

Entrance

The Palace of Knosós ㉒

Ανάκτορο της Κνωσού

Built around 1900 BC, the first palace of Knosós was destroyed by an earthquake in about 1700 BC and was soon completely rebuilt. The restored ruins visible today are almost entirely from this second palace. The focal point of the site is its vast north–south aligned Central Court, off which lie many of the palace's most important areas (see pp274–5). The original frescoes are in the Archaeological Museum of Irákleio (see pp270–71).

View across the Central Court towards the northeast

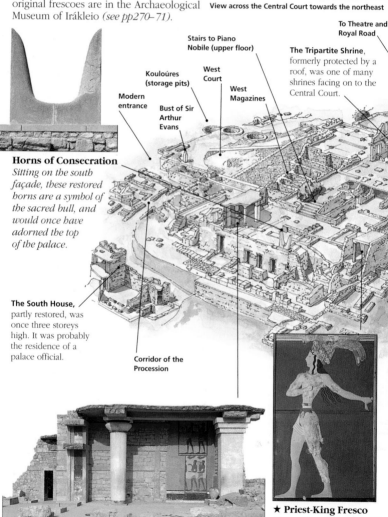

Stairs to Piano Nobile (upper floor)

Kouloúres (storage pits)

West Court

West Magazines

Modern entrance

Bust of Sir Arthur Evans

To Theatre and Royal Road

The Tripartite Shrine, formerly protected by a roof, was one of many shrines facing on to the Central Court.

Horns of Consecration
Sitting on the south façade, these restored horns are a symbol of the sacred bull, and would once have adorned the top of the palace.

The South House, partly restored, was once three storeys high. It was probably the residence of a palace official.

Corridor of the Procession

South Propylon
Entrance to the palace was through this monumental, pillared gateway, decorated with a replica of the Cup-Bearer figure, a detail from the Procession fresco.

★ **Priest-King Fresco**
This replica of the Priest-King fresco, also known as the Prince of the Lilies, is a detail from the Procession fresco and depicts a figure wearing a crown of lilies and feathers

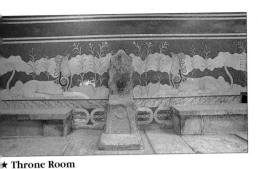

★ Throne Room
With its adjoining antechamber and lustral basin, the Throne Room is believed to have served as a shrine. The original stone throne, thought to be that of a priestess, is guarded by a restored fresco of griffins, sacred symbols in Minoan times.

VISITORS' CHECKLIST

5 km (3 miles) S of Irákleio, Irákleio. **Tel** 2810 231940. 🚌
◯ Nov–mid-May: 8:30am–3pm daily; mid-May–Oct: 8:30am–7pm daily. ⬤ 1 Jan, 25 Mar, Good Fri am, Easter Sun, 1 May, 25, 26 Dec. 🎫 📷 📹 📄

★ Giant Pithoi
Over 100 giant pithoi (storage jars) were unearthed at Knosós. The jars were used to store palace supplies.

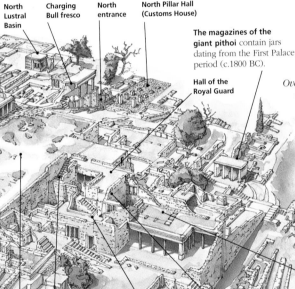

North Lustral Basin

Charging Bull fresco

North entrance

North Pillar Hall (Customs House)

The magazines of the giant pithoi contain jars dating from the First Palace period (c.1800 BC).

Hall of the Royal Guard

King's Megaron (Hall of the Double Axes)

Central Court

Grand Staircase

Queen's Megaron

★ Royal Apartments
These rooms include the King's Megaron, also known as the Hall of the Double Axes; the Queen's Megaron, which is decorated with a copy of the famous dolphin fresco and has an en suite bathroom; and the Grand Staircase.

STAR SIGHTS

* ★ Priest-King Fresco
* ★ Throne Room
* ★ Giant Pithoi
* ★ Royal Apartments

Exploring the Palace of Knosós

Unlike other Minoan sites, the Palace of Knosós was imaginatively restored by Sir Arthur Evans between 1900 and 1929. While his interpretations are the subject of academic controversy, his reconstructions of the second palace do give the visitor an impression of life in Minoan Crete that cannot so easily be gained from the other palaces on the island.

Restored clay bath tub adjacent to the Queen's Megaron

AROUND THE SOUTH PROPYLON

The palace complex is entered via the **West Court**, the original ceremonial entrance now marked by a bust of Sir Arthur Evans. To the left are three circular pits known as *kouloúres*, which probably served as granaries. Ahead, along the length of the west façade, are the **West Magazines**. These contained numerous large storage jars *(pithoi)*, and, along with the granaries, give an impression of how important the control of resources and storage was as a basis for the power of the palace.

At the far right-hand corner of the West Court the west entrance leads to the **Corridor of the Procession**. Now cut short by erosion of the hillside, the corridor's frescoes, depicting a series of gift-bearers, seem to reflect the ceremony that accompanied state and religious events at the palace. This is further revealed in the frescoes of the **South Propylon**, to which one branch of the corridor led. From the South Propylon,

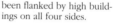

Shield motif, Knosós

steps lead up to the reconstructed **Piano Nobile**, the name given by Sir Arthur Evans to the probable location of the grand state apartments and reception halls. Stone vases found in this part of the palace were used for ritual purposes and indicate the centrality of religion to palace life. The close link between secular and sacred power is also reinforced by the **Throne Room**, where ritual bathing in a lustral basin (sunken bath) is thought to have taken place. Steps lead from the Throne Room to the once paved **Central Court**. Now open to the elements, this would have once been flanked by high buildings on all four sides.

THE ROYAL APARTMENTS

On the east side of the Central Court lie rooms of such size and elegance that they have been identified as the Royal Apartments. The apartments are built into the side of the hill and accessed by the **Grand Staircase**, one of the most impressive surviving architectural features of the

palace. The flights of gypsum stairs descend to a colonnaded courtyard, providing a source of light to the lower storeys. These light-wells were a typical feature of Minoan architecture.

A drainage system was provided for the toilet beside the **Queen's Megaron**, which enjoyed the luxury of an *en suite* bathroom complete with clay bathtub. Corridors and rooms alike in this area were decorated with frescoes of floral and animal motifs. The walls of the **Hall of the Royal Guard**, a heavily guarded landing leading to the Royal Apartments, were decorated with a shield motif. The **King's Megaron**, also known as the Hall of the Double Axes, takes its name from the fine double-axe symbols incised into its stone walls. The largest of the rooms in the Royal Apartments, the King's Megaron could have been divided by multiple doors, giving it great flexibility of space. Remains of what may have been a plaster throne were found here, suggesting that the room was also used for some state functions.

NORTH AND WEST OF THE CENTRAL COURT

The north entrance of the Central Court was adorned with remarkable figurative decoration. Today, a replica of the *Charging Bull* fresco can be seen on site. The north

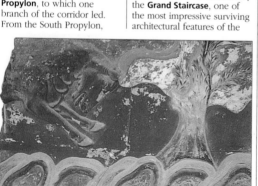

Replica of the celebrated *Charging Bull* fresco

entrance leads to the **North Pillar Hall**, named as the Customs House by Sir Arthur Evans who believed merchandise was inspected here. The hall is an addition of the Second Palace period (c.1700 BC). Immediately to the west is a room with restored steps leading into a pool, known as the **North Lustral Basin**. Traces of burning and finds of oil jars suggest that those coming to the palace were purified and annointed here. Further west is the **Theatre**, a stepped court whose position at the end of the Royal Road suggests that rituals connected with the reception of visitors

The stepped court of the theatre

may have occurred here. The **Royal Road**, which leads away from the Palace to the Minoan town of Knosós, was lined with houses. Just off the Royal Road lies the so-called **Little Palace**. This building has been excavated, but is not open to the public. It is architecturally very similar to the main palace and was destroyed at the same time.

THE HISTORY OF KNOSOS

The capital of Minoan Crete, Knosós was the largest and most sophisticated of the palaces on the island. It contained over 1,000 rooms and enjoyed the comforts of an elaborate drainage system, flushing toilets and paved roads. In legend, Knosós was believed to be the setting of an underground labyrinth designed to imprison the Minotaur. This half-man, half-bull was born of King Minos's wife, Pasiphaë, and slain by Theseus. This reconstruction shows the second palace as it might have looked in about 1700 BC.

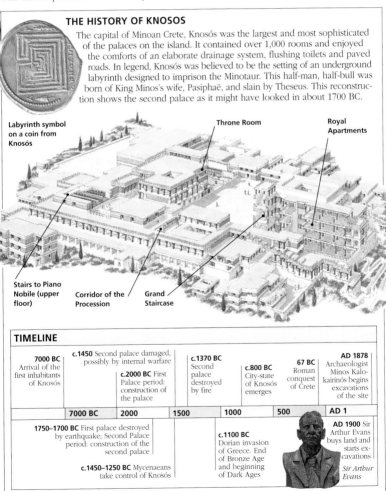

Labyrinth symbol on a coin from Knosós

Throne Room

Royal Apartments

Stairs to Piano Nobile (upper floor)

Corridor of the Procession

Grand Staircase

TIMELINE

7000 BC Arrival of the first inhabitants of Knosós	**c.1450** Second palace damaged, possibly by internal warfare	**c.1370 BC** Second palace destroyed by fire	**c.800 BC** City-state of Knosós emerges	**67 BC** Roman conquest of Crete	**AD 1878** Archaeologist Minos Kalokairinós begins excavations of the site
	c.2000 BC First Palace period: construction of the palace				
7000 BC	**2000**	**1500**	**1000**	**500**	**AD 1**
1750–1700 BC First palace destroyed by earthquake; Second Palace period: construction of the second palace			**c.1100 BC** Dorian invasion of Greece. End of Bronze Age and beginning of Dark Ages		**AD 1900** Sir Arthur Evans buys land and starts excavations
c.1450–1250 BC Mycenaeans take control of Knosós					*Sir Arthur Evans*

The modern seafront of Chersónisos, the busiest of Crete's package-holiday resorts

Archánes ㉓
Αρχάνες

Irákleio. 🏠 4,000. 🚌 ℹ️ 2810
246299 (Irákleio office).

A way from Crete's coastal
holiday resorts, Archánes is a
down-to-earth farming centre,
where olive groves and small
vineyards chequer the rolling
landscape. Lying at the foot of
the sacred **Mount Gioúchtas**
(burial place of Zeus according to local tradition),
Archánes was a thriving and
important settlement in
Minoan times.

In 1964, the remains of a
Minoan **palace** were found in
the town of Tourkogeitoniá. A
short walk out of town, on
Fourní hill to the north, lies
an extensive **Minoan cemetery**. Among the treasures
unearthed here was the tomb
of a princess with mirror and
gold diadem in place, as well
as exquisitely engraved signet
rings. Some of these are now
on display at the **Archaeological Museum** of Archánes.

🏚 **Minoan cemetery**
Fourní hill. ◯ Mon, Wed–Sun.
● main public hols.

🏛 **Archaeological Museum**
Kalochristianáki. ◯ Mon,
Wed–Sun. ● main public hols. ♿

Environs
On the north slope of Mount
Gioúchtas is the site of a
Minoan sanctuary at **Anemos-
piliá**. Excavations unearthed
a shocking scene of human
sacrifice here, seemingly
interrupted by an earthquake
around 1700 BC which killed
all four participants. Though
little remains to be seen
today, there are sensational
views of Mount Idi *(see p262)*.

The **Kazantzákis Museum**
at Myrtiá displays memorabilia of the author of *Zorba the
Greek*. Nearby in Scalani, the
Boutari Winery offers good
guided tours and tastings.

🏛 **Kazantzákis Museum**
Myrtiá, 14 km (9 miles) E of
Archánes. **Tel** 2810 742451.
◯ Mar–Oct: daily; Nov–Feb: Sun.
● main public hols. 🖼

Chersónisos ㉔
Χερσόνησος

Irákleio. 🏠 4,050. 🚌 🚕
Chersónisos.

A flourishing and busy port
from Classical to early
Byzantine times, Chersónisos
(strictly Liménas Chersonísou)
is today the centre of the
package-holiday business.
Amid the plethora of tavernas,
souvenir shops and discos the
harbour still retains faint
intimations of the old
Chersónisos. Along the waterfront a pyramid-shaped
Roman **fountain** with fish
mosaics dates from the 2nd–
3rd century AD. Some
remains of the **Roman harbour**, now mostly submerged,
can also be seen here.

On the coast at the eastern
edge of town, traditional Cretan life is recreated at the **Cretan Open-Air Museum** or
"Lychnostátis", where exhibits
include a windmill, a stone
house and a gallery. The
Museum of Rural Life, housed
in a 19th-century olive oil
mill, displays a range of traditional farming tools used
before the introduction of
modern technology. To cool
off, the **Aqua Splash Water
Park** is a playground of pools,
waterslides and waterfalls.

🏛 **Cretan Open-Air Museum**
Lychnostátis. **Tel** 28970 23660.
◯ Apr–Oct: Tue–Sun.
● main public hols. 🖼 ♿

🏛 **Museum of Rural Life**
Piskopianá. **Tel** 28970 23303.
◯ Apr–Oct: daily. 🖼 ♿

💦 **Aqua Splash Water Park**
5 km (3 miles) S of National
Highway. **Tel** 28970 24950.
◯ May–Oct: daily. 🖼 ♿

NIKOS KAZANTZAKIS
From the village of Myrtiá, Níkos Kazantzákis (1883–1957)
was Crete's greatest writer. Dedicated to the Cretan struggle
for freedom from Turkish rule, he wrote poems, philosophical
essays, plays and novels including *Zorba the Greek* and

*The Last Temptation of
Christ* (both made into
films). Excommunicated
by the Orthodox church,
the epitaph on his grave
in Irákleio consists of
his own words: "I hope
for nothing. I fear
nothing. I am free."

**Poster of the 1960s film
version of Zorba the Greek**

Mália 25
Μάλια

6 km (22 miles) E of Irákleio.
2,700. Stalida 3 km
(2 miles) NW.

The Mália of package-holiday fame bustles noisily with sun-seekers hellbent on enjoying the crowded beaches by day and the cacophony of competing discos by night.

In marked contrast, the less visited Minoan **Palace of Mália** lies in quiet ruins along the coastal plain to the east. The first palace was built in 1900 BC but, like all the other major palaces, it suffered destruction in 1700 BC and again in 1450 BC *(see p275)*. The site incorporates many features characteristic of other Minoan palaces – the great central court with its sacrificial altar, royal apartments, lustral basins (water pools) and light-wells (court-yards). In a small sanctuary in the west wing of the palace, the Minoan religious symbol of the double axe *(labrys)* can be seen inscribed on twin pillars.

Giant *pithos* in the Palace of Mália

Beyond the palace, remains thought to be of a town are currently under excavation while further north lies the burial site of **Chrysólakkos** (pit of gold). Important treasures were recovered here, including the famous gold bee pendant displayed in the Irákleio Archaeological Museum *(see pp270–71)*.

The chequered landscape of the agricultural plateau of Lasíthi

🏛 Palace of Mália
3 km (2 miles) E of Mália. **Tel** 28970 31597. ⬜ Tue–Sun. ⬤ 28 Oct, main public hols.

Environs
The fast developing village of **Sisi** is situated 6.5 km (4 miles) east of Mália. Continuing east-wards, stunning views mark the descent to Mílatos. From here a well-signposted trail leads to the **Mílatos Cave** where a shrine and glass-fronted casket of bones are a memorial to those massacred here by the Turks in 1823 during the Greek War of Independence.

Lasíthi Plateau 26
Ωροπέδιο Λασιθίου

Díkti mountains, Irákleio. to Tzermiádo.

High up in the formidable Díkti mountains, the bowl-shaped plain of Lasíthi was for centuries shut off from the outside world. A row of stone

windmills at the Séli Ampélou Pass marks the main entry to the plateau, a flat agricultural area lying 800 m (2,600 ft) above sea level and encircled by mountains. Fruit, potatoes, and cereals are the main crops here, thanks to the fertile allu-vial soil washed down from the mountains. A few cloth-sailed windmills are still used today to pump irrigation water.

Along the perimeter of the plain are several villages, the largest of which is **Tzermiádo** with good tourist facilities. A path from Tzermiádo to the **Trápeza Cave** (also known as Króneion Cave) is signposted from the village centre. At the west end of the village a rough road (just over an hour's walk) leads up to the archaeological site of **Karfí**, the last retreat of Minoan civilization. On the southern edge of the plain, the village of **Agios Geórgios** has a small **Folk Museum** set in two old village houses and displaying a collection of embroidery, paintings and Kazantzákis memorabilia.

The highlight of a visit to Lasíthi is the climb to the **Diktian Cave** at Psychró, birthplace of Zeus *(see p262)*. A wealth of artifacts have been unearthed here including votive offerings, double axes and bronze statuettes, now in the Irákleio Archaeological Museum *(see pp270–71)*.

🏛 Folk Museum
Agios Geórgios. ⬜ Mar–Oct: daily.

🏛 Diktian Cave
Psychró. **Tel** 28440 31316. ⬜ daily. ⬤ 27 Sep, 28 Oct, public hols.

A small shrine in the multichambered Mílatos Cave

The fortified islet of Spinalónga off the coast of Eloúnta

Eloúnta ㉗
Ελούντα

Lasíthi. 🏠 1,500. 🚌 ⛴ Tue.
🚖 Eloúnta.

Once the site of the ancient city-state of Oloús, the town of Eloúnta was developed by the Venetians in 1579 as a fortified port. Today, the town is a well-established holiday resort idyllically situated on the Mirabéllou Bay. The town is blessed with attractive sandy coves and offers a good range of accommodation.

East of the village an isthmus joins the mainland to the long strip of land forming the Spinalónga peninsula. Here, remains of the Greco-Roman city-state of **Oloús**, with its temples of Zeus and Artemis, can be discerned just below the water's surface. To the north of the peninsula is the small island of **Spinalónga** where a forbidding 16th-century Venetian fortress now stands deserted. Having with-stood assault from the Turks for many years, its last function was as a leper colony until the mid-1950s. Today, boats regularly ferry tourists to the island from Eloúnta and elsewhere.

Environs
The small hamlet of **Pláka**, 5 km (3 miles) north of Eloúnta, makes for a pleasant retreat from the bustle of Eloúnta. Dine on fresh fish at the water-front, where boat trips are available to Spinalónga island.

Skull and wreath, Archaeological Museum, Agios Nikólaos

Agios Nikólaos ㉘
Αγιος Νικόλαος

Lasíthi. 🏠 10,000. 🚌 ⛴
🛈 Koundoúrou 21 (28410 22357).
🐓 Wed. 🚖 Almyrós 2 km (1.5 miles) E; Chavánia 3 km (2 miles) W.

One of the most delightful holiday centres in Crete, Agios Nikólaos boasts a superb setting on the Mirabéllou Bay. In Hellenistic times, according to inscriptions dating back to 193 BC, this was one of two flourishing cities called Lató: Lató pros Kamára (towards the arch) and Lató Etéra (Other Lato). Having declined in importance under Venetian rule, it was not until the 19th century that modern Agios Nikólaos began to develop.

Now a thriving resort, its centre is the harbour and, with a depth of 64 m (210 ft), the Almyrí Lake or Voulisméni. Overlooking the lake, the **Folk Museum** houses a colourful display of traditional Cretan crafts and domestic items. Just north of town, in the grounds of the Mínos Palace Hotel, is the tiny 10th–11th-century church of **Agios Nikólaos** after which the town is named.

Close to several important Minoan sites, the **Archaeo-logical Museum** at Agios Nikólaos possesses a treasure trove of artifacts from Lasíthi Province. Pieces housed here include carved stone vases, gold jewellery from the Minoan site of Móchlos near Gourniá and pottery, including the drinking vessel known as the Goddess of Mýrtos. One unique exhibit is the skull of a man thought to be an athlete, complete with a wreath made of gold laurel leaves and a silver coin for his fare across the mythical River Styx.

In summer, boat trips run to Spinalónga island and Agioi Pántes, an island refuge for the Cretan wild goat, the kri-kri (see p254).

🏛 **Folk Museum**
Koúndourou 23. **Tel** 28410 25093.
☐ May–Oct: Tue–Sun. ☐ main public hols. 🈲 ♿

🏛 **Archaeological Museum**
Palaiológou 68. **Tel** 28410 24943.
☐ Tue–Sun. ☐ main public hols. 🈲

The attractive inner harbour of Agios Nikólaos, with Lake Voulisméni in the foreground

For hotels and restaurants in this region see pp317–20 and pp338–40

Section of the *Paradise* fresco at Panagía Kerá in Kritsá

Kritsá ㉙

Κριτσά

Lasíthi. 🏛 *2,500.* 🚌 🏠 *Mon.* 🚉 *Ammoudára 11 km (7 miles) E; Ístro 15 km (9 miles) SE.*

Set at the foot of the Lasíthi mountains, Kritsá is a small village known throughout Crete for its famous Byzantine church. Also a popular centre for Cretan crafts, its main street is awash with lace, elaborately woven rugs and embroidered tablecloths during the summer months. From the cafés and tavernas along the main street, fine views of the valley leading down to the coast can be enjoyed. By November, Kritsá reverts back to life as a workaday Greek village.

East of Kritsá, situated just off the road among olive groves, the hallowed 13th-century church of **Panagía Kerá** contains some of the finest frescoes in Crete, dating from the 13th to mid-14th century. The building is triple-aisled with the central aisle being the oldest. Beautiful representations of the life of Christ and the Virgin Mary cover the interior.

Environs

North of Kritsá lie the ruins of a fortified city founded by the Dorians in the 7th century BC. **Lató Etéra** flourished until Classical times when its fortunes declined under Roman rule: it was superseded by the more easily reached port of Lató pros Kamára (today's Agios Nikólaos). Sitting perched on a saddle between two peaks, the site offers fine views of the Mirabéllou Bay. A paved road, with workshops and houses clustered on the right, climbs up to a central agora, or marketplace, with a cistern to collect rainwater and a shrine. On the north side of the agora, a staircase flanked by two towers leads to the place where the city's archives would once have been stored. To the south of the agora a temple and a theatre can be seen.

🏛 Lató

4 km (2 miles) N of Kritsá.
⬜ *Tue–Sun.* ⬛ *main public hols.*

Lace shop on Kritsá's main street

Ierápetra ㉚

Ιεράπετρα

Lasíthi. 🏛 *15,000.* 🚌 🛈 *Adrianoú (28420 22246).* 🏠 *Sat.* 🚉 *Agiá Fotiá 17 km (11 miles) E; Makrýs Gialós 30 km (19 miles) E.*

Situated on the southeast coast of Crete, Ierápetra boasts of its position as the most southerly city in Europe. A settlement since pre-Minoan times, trade and cultural connections with North Africa and the Middle East were an important basis of the city's existence. Sir Arthur Evans *(see p274)* declared it the "crossroads of Minoan and Achaian civilizations". Once a flourishing city with villas, temples, amphitheatres, and imposing buildings, the town today has an air of decline. Gone are all signs of its ancient history, thanks partly to past pillage and, more recently, to modern "development".

The entrance to the old harbour is guarded by an early 13th-century Venetian **fortress**. West of the fortress is the attractive Turkish quarter where a restored **mosque** and elegant Ottoman fountain can be seen. Also in this area, on Kougioumtzáki, is the 14th-century church of **Aféntis Christós** and, off Samouíl, **Napoleon's House**, where he is said to have spent a night en route to Egypt in 1798. Today it is not open to the public.

The small **Archaeological Museum** in the centre of town displays a collection of local artifacts that managed to survive marauders and various archaeological predators. The exhibits date from Minoan to Roman times and include *larnakes* (burial caskets), *pithoi* (storage jars), statues, bronze axes and stone carvings.

An almost unbroken line of sandy beaches stretches eastwards from Ierápetra, overlooked by the inevitable plethora of hotels and restaurants. From Ierápetra's harbour, a daily boat service runs to the idyllic white sands and cedar forests of the uninhabited **Chrysí** island.

🏛 Fortress
Old port. ⬜ *daily.* ⬛ *main public hols.* 🖼

🏛 Archaeological Museum
Adrianoú Koustoúla. **Tel** 28420 28721. ⬜ *Tue–Sun.* ⬛ *main public hols.* ♿

Mosque and Ottoman fountain in Ierápetra's old Turkish quarter

Gourniá archaeological site

Gourniá **③**
Γουρνιά

18.5 km (11 miles) E of Agios
Nikólaos, Lasíthi. ☐ Tue–Sun.
☐ main public hols. ☐ Istro 8 km
(5 miles) W.

The Minoan site of Gourniá
stands on a low hill overlooking the tranquil Mirabéllou
Bay. Excavated by the
American archaeologist Harriet
Boyd-Hawes between 1901
and 1904, Gourniá is the best-
preserved Minoan town in
Crete. A mini-palace (one-tenth
the size of Knosós) marks its
centre, surrounded by a labyrinth of narrow, stepped streets
and one-room dwellings. The
site was inhabited as early as
the 3rd millennium BC, though
what remains dates from the
Second Palace period, around
1700 BC (see p275). A fire,
caused by seismic activity in
around 1450 BC, destroyed
the settlement at Gourniá.

Environs
Along the National Highway, 2
km (1.5 miles) west of Gourniá,
an old concrete road turns left
up a spectacular 6-km (4-mile)
climb to **Moní Faneroménis**.
Here, the 15th-century chapel
of the **Panagía** has been built
into a deep cave and is the
repository for sacred (and
some say miraculous) icons.
East along the National
Highway, a left turning from
Sfáka leads down to the
delightful fishing village of
Móchlos. The small island
of Móchlos, once joined to
the mainland by a narrow
isthmus, is the site of a Minoan
settlement and cemetery.

Siteía **②**
Σητεία

Lasíthi. ☐ 7,500. ☐ ☐ ☐
☐ Tue. ☐ Siteía.

Snaking its way through the
mountains between Gourniá
and Siteía, the National Highway traverses some of the
most magnificent scenery
in Crete. Towards Siteía,
the landscape gives way to
barren hills and vineyards.
Although there is evidence
of a large Greco-Roman city in
the region, modern Siteía dates
from the 4th century AD. It
flourished under Byzantine
and early Venetian rule but its
fortunes took a downturn in
the 16th century as a result of
earthquakes and pirate attacks.
When rebuilding took place
in the 1870s Siteía began to
prosper once again.
Today, the production of
wine and olive oil is important to the town's economy
and the mid-August Sultana
Festival celebrates its success
as a sultana exporter.
At the centre of Siteía's old
quarter lies a picturesque

harbour, with tavernas and
cafés clustering around its
edges. Above the north end
of the harbour the restored
Venetian **fort** (now used as
an open-air theatre) is all that
remains of the once extensive
fortifications of the town. The
Kornaria Festival is a cultural
event held in the fort from the
beginning of July until mid-
August and is a great way for
visitors to learn about the
customs and traditions of
Siteía. Events include music
and dance events, theatre,
exhibitions and sports events.
On the southern outskirts
of town, the **Archaeological
Museum** displays artifacts
from the Siteía district. Exhibits
range from Neolithic to Roman
times and include an exquisite
Minoan ivory statuette known
as the *Palaíkastro Koúros*.
There are pottery finds from
all over the region including a
large collection of material
from Zákros Palace.

🏛 **Archaeological Museum**
Piskokefálou 3. *Tel* 28430 23917.
☐ Tue–Sun. ☐ main public hols.
☐ ☐

Siteía's old quarter on the hillside overlooking the tree-lined harbour

Moní Toploú 🏵
Μονή Τοπλού

6 km (10 miles) W of Siteía, Lasíthi.
Tel 28430 61226. 🚌 to Váï.
Site & Museum ⬜ daily. 📷
🚩 Itanos 7.5 km (4.5 miles) NE.

Founded in the 14th century,
Moní Toploú is now one of
the wealthiest and most influential monasteries in Crete.
The present buildings date
from Venetian times, when the
monastery was fortified against
pirate attacks. The Turkish
name "Toplou" refers to the
cannon installed here. During
World War II, Resistance radio
broadcasts were transmitted
from the monastery, an act for
which Abbot Siligknákis was
executed by German forces
near Chaniá.

Three levels of cells overlook
the inner courtyard, where a
small 14th-century church
contains frescoes and icons.
The most famous of these
is the *Lord, Thou
Art Great* icon,
completed in
1770 by the
artist Ioánnis
Kornáros. On
the façade of
the church,
an inscription
records the
arbitration
of Magnesia in 132
BC. This was an
order that settled a
dispute between the rival city-
states of Ierapytna (today's
Ierápetra) and Itanos, over the
control of the Temple of Zeus
Diktaios at Palaíkastro. The
inscription stone was used
originally as a tombstone. The
monastery's small museum
houses etchings and 15th- to
18th-century icons.

*Lord, Thou Art Great
icon by Ioánnis Kornáros,
Moní Toploú*

Väï Beach 🏵
Ιαραλία Βάι

28 km (17 miles) NE of Siteía,
Lasíthi. 🚌

The exotic Väï Beach is a
tropical paradise of dense
palm trees known to have
existed in Classical times and
reputedly unique in Europe.
This inviting sandy cove is
tremendously popular with

holiday-makers. Although
thoroughly commercialized,
with overpriced tavernas and
the constant arrival of tour
buses, great care is taken to
protect the palm trees.

Environs
In the desolate landscape 2 km
(1 mile) north of Väï, the ruins
of the ancient city-state of
Itanos stand on a hill
between two
sandy coves.
Minoan,
Greco-
Roman and
Byzantine
remains
have
been ex-
cavated
(the scant traces of
which can be seen
today), including a
Byzantine basilica and the
ruins of some Classical temples.
The agricultural town of
Palaíkastro, 10 km (6 miles)
south of Väï, is the centre of an
expanding olive business. At
the south end of Chióna beach,
2 km (1 mile) to the east, the
Minoan site of Palaíkastro is
presently under excavation.

Zákros archaeological site, situated behind the hamlet of Káto Zákros

Zákros 🏵
Ζάκρος

Káto Zákros, Lasíthi. **Tel** 28430
93338. 🚌 ⬜ Tue–Sun. ⬤ main
public hols. 📷 🚩 Káto Zákros;
Xerókampos 13 km (8 miles) S.

In 1961, Cretan archaeologist
Nikólaos Pláton discovered
the unplundered Minoan
palace of Zákros. The fourth
largest of the palaces, it was
built around 1700 BC and
destroyed in the island-wide
disaster of 1450 BC. Its ideal
location made it a centre of
trade with the Middle East.

The two-storied palace was
arranged around a central
courtyard, the east side of
which contained the royal
apartments. Remains of a
colonnaded cistern hall can still
be seen, and a stone-lined
well in which some perfectly
preserved 3,000-year-old olives
were found in 1964. The main
hall, workshops and storerooms are in the west wing.
Finds from the palace include
an exquisite rock crystal jug
and numerous vases, now in
the Irákleio Archaeological
Museum *(see p270–71)*.

Väï Beach with its calm waters and native palms

A SHORT STAY IN ATHENS

A vast, sprawling metropolis surrounded by rocky mountains, Athens covers 457 sq km (176 sq miles) and has a population of four million people. The city prides itself on being home to the 2,500-year-old temple of Athena – the Parthenon – as well as some superb museums. A stopover in Athens en route to the islands offers the ideal opportunity to visit the best sights in the city.

The birthplace of European civilization, Athens has been inhabited for 7,000 years, since the Neolithic era. Ancient Athens reached its high point in the 5th century BC, when Perikles commissioned many fine new buildings, including some of the temples on the Acropolis. Other relics from the Classical period can be seen in the Ancient Agora, a complex of public buildings dominated by the reconstructed Stoa of Attalos, a long, covered colonnade.

Evzone in Plateía Syntágmatos

There is little architectural evidence of the city's more recent history of occupation. With the exception of some fine Byzantine churches, particularly those in historic Pláka, one of the oldest areas of Athens, nothing of importance has survived from the years of Frankish, Venetian and Ottoman rule. In 1834, inspired by the Classical buildings of the Acropolis, King Otto declared Athens the new capital of Greece, and his Greek, German and Danish town-planners and architects created a modern city of Neo-Classical municipal buildings, wide boulevards and elegant squares around the ancient "Sacred Rock".

The rich cultural heritage of Athens can be appreciated in some magnificent museums, including the National Archaeological Museum, where an unrivalled collection beautifully illustrates the glories of ancient Greece. The National Gallery of Art includes well-known works by both Greek and European artists.

The nightlife in Athens is excellent, with tavernas, clubs and bars open until the early hours. Open-air cinemas and theatres, such as the Theatre of Herodes Atticus at the foot of the Acropolis, are popular in summer. There is music for every taste, from traditional Greek to pop, jazz and classical concerts. Shopping ranges from the flea market and antique and bric-a-brac shops in Monastiráki, to designer boutiques in Kolonáki. Pedestrianization of the city centre makes Athens a pleasant place to explore on foot.

View of the Acropolis from Filopáppos Hill

Lykavittós Hill rising above the spread of concrete apartment blocks and Byzantine churches, in Athens

Exploring Athens

Even with only an afternoon to spend in Athens, it is possible to visit a few of the main sights. The Acropolis is the most popular attraction, along with the Ancient Agora. The National Archaeological Museum houses many finds from these sites in its fine collection of ancient Greek art. The renovated Benáki Museum houses a glittering array of jewellery, costumes and ceramics from Greece and the Middle East, as well as many temporary exhibitions. Shopping provides an alternative to sightseeing, from the bric-a-brac in Pláka to the designer stores in Kolonáki. For information on getting around Athens, see pp292–5.

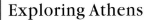

Avyssinías in Monastiráki
(see p286)

The Central Market
has a fine array of foods, herbs and spices.

Figure from the Museum of Cycladic Art
(see p291)

Mitrópoli is
Athens' cathedral. It towers over the tiny Byzantine Panagía Gorgoepíkoös (or Little Cathedral) next to it.

0 metres   250
0 yards   250

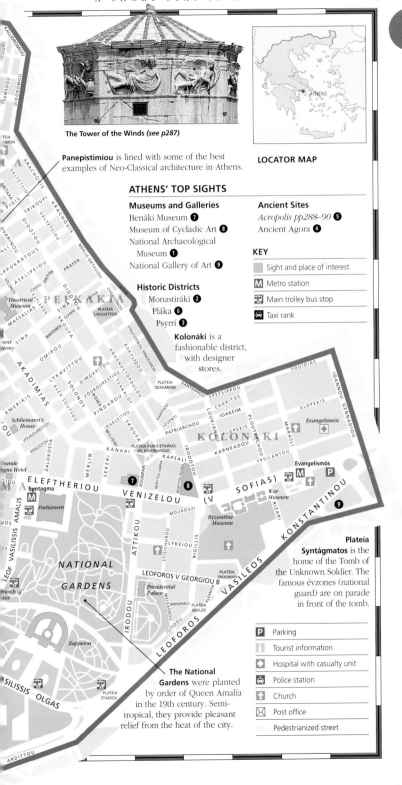

The Tower of the Winds *(see p287)*

Panepistimíou is lined with some of the best examples of Neo-Classical architecture in Athens.

LOCATOR MAP

ATHENS' TOP SIGHTS

Museums and Galleries
Benáki Museum **7**
Museum of Cycladic Art **8**
National Archaeological Museum **1**
National Gallery of Art **9**

Historic Districts
Monastiráki **2**
Pláka **6**
Psyrrí **3**

Ancient Sites
Acropolis pp288–90 **5**
Ancient Agora **4**

KEY

	Sight and place of interest
M	Metro station
	Main trolley bus stop
	Taxi rank

Kolonáki is a fashionable district, with designer stores.

Plateía Syntágmatos is the home of the Tomb of the Unknown Soldier. The famous évzones (national guard) are on parade in front of the tomb.

The National Gardens were planted by order of Queen Amalía in the 19th century. Semi-tropical, they provide pleasant relief from the heat of the city.

P	Parking
	Tourist information
	Hospital with casualty unit
	Police station
	Church
⊠	Post office
	Pedestrianized street

National Archaeological Museum ❶
Εθνικό Αρχαιολογικό Μουσείο

Patissíon 44, Exárcheia.
Tel 210 821 7717. Ⓜ *Omónoia, Viktória.* ◯ *1–7:30pm Mon, 8am–7:30pm Tue–Sun.* ♨ ◎ ☐ ☐

Shoppers browsing in Athens' lively Monastiráki market

When it was opened in 1891, this museum brought together a collection that had previously been stored all over the city. New wings were added in 1939, but during World War II this priceless collection was dispersed and buried underground to protect it from any possible damage. The museum reopened in 1946, but it has taken another 50 years of renovation and reorganization to finally do justice to its formidable collection. With its comprehensive assembly of pottery, sculpture and jewellery, it definitely deserves ranking as one of the finest museums in the world. It is a good idea to plan ahead and be selective when visiting the museum and not attempt to cover everything in one visit.

The museum's exhibits can be divided into five main collections: Neolithic and Cycladic, Mycenaean, Geometric and Archaic, Classical sculpture, Roman and Hellenistic sculpture and the pottery collections. There are also other smaller collections that are well worth seeing. These include the stunning Eléni Stathátou jewellery collection and the recently opened Egyptian rooms.

High points of the museum include the unique finds from the grave circle at Mycenae, in particular the gold *Mask of Agamemnon*. Also not to be missed are the Archaic *koúroi* statues and the unrivalled collection of Classical and Hellenistic statues. Two of the most important and finest of the bronzes are the *Horse with the Little Jockey* and the *Poseidon*.

Also housed here is one of the world's largest collections of ancient ceramics comprising elegant figure vases from the 6th and 5th centuries BC (*see pp62–3*) and some Geometric funerary vases that date back to 1000 BC. The collection is currently closed due to earthquake damage.

The *Mask of Agamemnon* in the National Archaeological Museum

Monastiráki ❷
Μοναστηράκι

Ⓜ *Monastiráki.* **Market** ◯ *daily.*

This area, named after the little monastery church in Plateía Monastirakíou, is synonymous with Athens' famous fleamarket. Located next to the Ancient Agora, it is bounded by Sari in the west and Aiólou in the east. The streets of Pandrósou, Ifaístou and Areos leading off Plateía Monastirakíou are full of shops, selling a range of goods from antiques, leather and silver to tourist trinkets.

The heart of the flea market is in Plateía Avyssinías, west of Plateía Monastirakíou, where every morning junk dealers arrive with pieces of furniture and various odds and ends. During the week and on Sunday mornings the shops and stalls are filled with antiques, second-hand books, rugs, leatherware, taverna chairs, army surplus gear and tools.

The market flourishes particularly along Adrianoú and in Plateía Agíou Filíppou. There are always numerous bargains to be had. Items particularly worth investing in include some of the colourful woven and embroidered cloths and an abundance of good silver jewellery.

Psyrrí ❸
Ψυρρί

Ⓜ *Monastiráki.*

For a taste of Athens as it was through most of its modern history, wander the warren of streets comprising the Psyrrí district. Bordered by the Central Market, Athinas and Ermou Streets, this neighbourhood is becoming the city's trendiest area. Many of the handsome Neo-Classical buildings have been renovated for art galleries and restaurants while theatres, wine bars and boutiques pop up daily. Tiny stores specialize in unique, handmade items like copper kitchenware, belt buckles, wickerwork and icons. At night the district's transformed, commercial buzz is replaced by the gentle pleasures of cafés, restaurants and wine bars. The food here is some of the most interesting in the city and prices are reasonable. This is very much an Athenian part of town.

Ancient Agora ❹
Αρχαία Αγορά

Main entrance at Adrianoú,
Monastiráki. **Tel** *210 321 0185.* Ⓜ
Thiseío, Monastiráki. **Museum and
site** ◯ *8am–7pm daily, noon–3pm
Good Fri.* ● *main public hols.* 🎫
◗ & *limited.*

The American School of
Archaeology commenced
excavations of the Ancient
Agora in the 1930s, and since
then a complex array of
public buildings and temples
has been revealed. The
democratically governed
Agora was the political and
religious heart of Ancient
Athens. Also the centre of
commercial and daily life, it
abounded with schools and
elegant stoas filled with
shops. The state prison was
here, as was the mint, which
was used to make the city's
coins inscribed with the
famous owl symbol. Even the
remains of an olive oil mill
have been found here.

The main building standing
today is the impressive two-
storey stoa of Attalos. This
was rebuilt between 1953
and 1956 on the original
foundations and using
ancient building materials.
Founded by King Attalos of
Pergamon (ruled 159–138
BC), it dominated the eastern
quarter of the Agora until it
was destroyed in AD 267. It
is used today as a museum,
exhibiting the finds from the
Agora. These include legal
finds, such as a *klepsydra* (a
water clock that was used for
timing plaintiffs' speeches),

The rooftop of the church of Agios Nikólaos Ragavás, Pláka

bronze ballots and items
from everyday life such as
some terracotta toys and
leather sandals. The best-
preserved ruins on the site
are the Odeion of Agrippa,
a covered theatre, and the
Hephaisteion, a temple to
Hephaistos, which is also
known as the Theseion.

Acropolis ❺

See pp288–9.

Pláka ❻
Πλάκα

Ⓜ *Monastiráki.* 🚌 *1, 2, 4, 5, 9, 10,
11, 12, 15, 18.*

The area of Pláka is the
historic heart of Athens. Even
though only a few buildings
date back further than the
Ottoman period, it remains
the oldest continuously
inhabited area in the city.

One probable explanation of
its name comes from the
word used by Albanian
soldiers in the service of the
Turks who settled here in the
16th century – *pliaka* (old)
was how they used to
describe the area. Despite the
constant swarm of tourists
and Athenians, who come to
eat in old-fashioned tavernas
or browse in the antique
and icon shops, Pláka
still retains the
atmosphere of a
traditional neigh-
bourhood. The
only choregic
monument still
intact in Athens is
the **Lysikrates
Monument** in
Plateía Lysikrátous.
Built to commem-
orate the victors at
the annual choral
and dramatic festival at the
Theatre of Dionysos, these
monuments take their name
from the sponsor (*choregos*)
of the winning team.

Many churches are worth a
visit: the 11th-century **Agios
Nikólaos Ragavás** has ancient
columns built into the walls.

The **Tower of the Winds**, in
the far west of Pláka, lies in
the grounds of the Roman
Agora. It was built by the
Syrian astronomer Andronikos
Kyrrestes around 100 BC as
a weather vane and water-
clock. On each of its marble
sides one of the eight mythological winds is depicted.

**Detail from a
terracotta
roof, Pláka**

🏛 **Tower of the Winds**
Plateía Aéridon. **Tel** *210 324 5220.*
◯ *daily.* ● *main public hols.* 🎫

The façade of the Hephaisteion in the Ancient Agora

Acropolis ⑤
Ακρόπολη

In the mid-5th century BC, Perikles persuaded the Athenians to begin a grand programme of new building work in Athens that has come to represent the political and cultural achievements of Greece. The work transformed the Acropolis with three contrasting temples and a monumental gateway. The Theatre of Dionysos on the south slope was developed further in the 4th century BC, and the Theatre of Herodes Atticus was added in the 2nd century AD.

The Acropolis with the Temple of Olympian Zeus in the foreground

★ **Porch of the Caryatids**
These statues of women were used in place of columns on the south porch of the Erechtheion. The originals, four of which can be seen in the Acropolis Museum, have been replaced by casts.

An olive tree now grows where Athena first planted her tree in a competition against Poseidon.

The Propylaia was built in 437–432 BC to form a new entrance to the Acropolis.

★ **Temple of Athena Nike**
This temple to Athena of Victory is on the west side of the Propylaia. It was built in 427–424 BC.

The Belué Gate was the first entrance to the Acropolis.

Pathway to Acropolis from ticket office

Theatre of Herodes Atticus
Also known as the Odeion of Herodes Atticus, this superb theatre was originally built in AD 161. It was restored in 1955 and is used today for outdoor concerts.

STAR SIGHTS

★ Parthenon

★ Porch of the Caryatids

★ Temple of Athena Nike

★ **Parthenon**
Although few sculptures are left on this famous temple to Athena, some can still be admired, such as this one from the east pediment (see p290).

VISITORS' CHECKLIST

Dionysíou Areopagítou (main entrance), Pláka. **Map** 6 D2. *Tel* 210 321 0219.
Ⓜ *Acropolis.* 🚌 *230, 231.*
◯ *Apr–Oct: 8am–7pm daily, Nov–Mar: 8am–2:30pm daily.* ⬤ *1 Jan, 25 Mar, Easter Sun, 1 May, 25, 26 Dec.*
📷 ◎ ✦ www.culture.gr

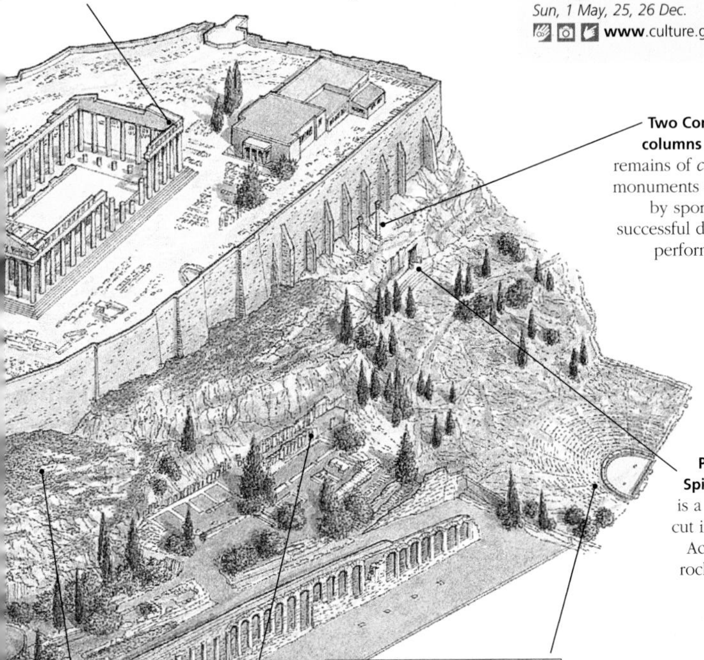

Two Corinthian columns are the remains of *choregic* monuments erected by sponsors of successful dramatic performances.

Panagía Spiliótissa is a chapel cut into the Acropolis rock itself.

Theatre of Dionysos
This figure of the comic satyr, Silenus, can be seen here. The theatre visible today was built by Lykourgos in 333–330 BC.

Shrine of Asklepios

Stoa of Eumenes

The Acropolis rock was an easily defended site. It has been in use for nearly 5,000 years.

TIMELINE

3000 BC First settlement on the Acropolis during Neolithic period	**AD 51** St Paul delivers sermon on Areopagos hill
	AD 267 Germanic Heruli tribe destroy Acropolis

480 BC All buildings of Archaic period destroyed by the Persians

St Paul

3000 BC	2000 BC	1000 BC	AD 1	AD 1000

1200 BC Cyclopean wall built to replace original ramparts

510 BC Delphic Oracle declares Acropolis a holy place of the gods, banning habitation by mortals

447–438 BC Construction of the Parthenon under Perikles

AD 1687 Parthenon damaged by Venetians

AD 1987 Restoration of the Erechtheion completed

Perikles (495–429 BC)

Exploring the Acropolis

Once through the Propylaia, the grand entrance to the site, the Parthenon exerts an overwhelming fascination. The other fine temples on "the Rock" include the Erechtheion and the Temple of Athena Nike. Since 1975, access to all the temple precincts has been banned. However, it is a miracle that anything remains at all. The ravages of war, the removal of treasures and pollution have all taken their irrevocable toll on the Acropolis.

A section from the north frieze of the Parthenon

The *Moschophoros* (or Calf-Bearer) in the New Acropolis Museum

⋂ The Parthenon

One of the world's most famous buildings, the Parthenon was commissioned by Perikles as part of his rebuilding plan. Work began in 447 BC when the sculptor Pheidias was entrusted with supervising the building of a magnificent new Doric temple to Athena, the patron goddess of the city. It was built on the site of earlier Archaic temples, and was designed primarily to house the *Parthenos*, Pheidias's impressive 12-m (39-ft) high cult statue of Athena covered in ivory and gold.

Taking nine years to complete, the temple was dedicated to the goddess during the Great Panathenaia festival of 438 BC. Designed and constructed in Pentelic marble by the architects Kallikrates and Iktinos, the complex architecture of the Parthenon replaces straight lines with slight curves. This is generally thought to have been done to prevent visual distortion or perhaps to increase the impression of grandeur. All the columns swell in the middle and all lean slightly inwards, while the foundation platform rises towards the centre.

For the pediments and the friezes which ran all the way round the temple, an army of sculptors and painters was employed. Agorakritos and Alkamenes, both pupils of Pheidias, are two of the sculptors who worked on the frieze, which represented the people and horses in the Panathenaic procession.

Despite much damage and alterations made to adapt to its various uses, which include a church, a mosque, and even an arsenal, the Parthenon remains a powerful symbol of the glories of ancient Greece. It is currently being restored.

🏛 New Acropolis Museum

Makrigiánni 2–4. **Tel** *210 923 9381.* ☐ *due to open late 2008; phone for opening hours.*

After decades of planning and delays, the New Acropolis Museum, located in the historic Makrigiánni district at the foot of the Acropolis, is complete. This €130-million, multi-storey showpiece has been designed by Bernad Tschumi to house the stunning treasures found on the Acropolis hill. It is constructed over excavations of an early Christian settlement and a glass walkway hovers over the ruins.

The collection has been installed in chronological order and begins with finds from the slopes of the Acropolis, including statues and reliefs from the Shrine of Asklepios.

The **Archaic Collection** is set out in a magnificent double-height gallery and contains fragments of pedimental statues such as the statue of *Moschophoros*, or the Calf-Bearer (c.570 BC).

The sky-lit **Parthenon Gallery** on the top floor is undoubtedly the highlight. Here, looking out onto the Acropolis hill itself, the remaining parts of the Parthenon frieze are displayed in their original order.

View of the Parthenon from the southwest at sunrise

Benáki Museum ❼
Μουσείο Μπενάκη

Corner of Koumpári & Vasilíssis Sofías, Kolonáki. **Tel** 210 367 1000.
3, 7, 8, 13. ⏰ 9am–5pm Mon & Wed–Fri, 9am–midnight Thu, 9am–3pm Sun. ● main public hols. (free Thu). 🅿 ♿ limited.

This outstanding museum contains a diverse collection of Greek art and crafts, jewelry, regional costumes and political memorabilia from the 3rd century BC to the 20th century. It was founded by Antónios Benákis (1873–1954), the son of Emmanouíl Benákis, a wealthy Greek who made his fortune in Egypt. Antónios Benákis was interested in Greek, Persian, Egyptian and Ottoman art from an early age and started collecting while living in Alexandria. When he moved to Athens in 1926, he donated his collection to the Greek State, using the family house as a museum which was opened to the public in 1931. The elegant Neo-Classical mansion was built towards the end of the 19th century by Anastásios Metaxás, who was also the architect of the Kallimármaro stadium.

The Benáki collection is made up of gold jewellery, some dating from as far back as 3000 BC, as well as icons, pieces of liturgical silverware, Egyptian artifacts, Greek embroideries and the work of the late artist Chatzikyriákos-Gkíkas.

Museum of Cycladic Art
Μουσείο Κυκλαδικής
και Αρχαίας Ελληνικής
Τέχνης

Neof'ytou Doúka 4 (new wing at Irodótou 1), Kolonáki. **Tel** 210 722 8321. 3, 7, 8, 13. ⏰ 10am–4pm Mon, Wed & Fri; 10am–8pm Thu; 10am–3pm Sat. ● main public holidays. 🅿 🅾 ♿

Opened in 1986, this modern museum offers the visitor the world's finest collection of Cycladic art. Assembled by Nikólaos and Dolly Goulandrís and helped by the donations

of other Greek collectors, it has brought together a fine selection of ancient Greek art, spanning 5,000 years of history.

Spread over five floors, the displays start on the first floor, which is home to the Cycladic collection. Dating back to the 3rd millennium BC, the Cycladic figurines were found mostly in graves, although their exact usage remains a mystery. One of the finest examples is the *Harp Player*. Ancient Greek art is exhibited on the second floor and the Charles Polítis collection of Classical and Prehistoric art on the fourth floor, highlights of which include some terracotta figurines of women from Tanágra, central Greece. The third floor of the museum is used for temporary, visiting exhibitions.

A new wing was opened in the adjoining Stathátos Mansion in 1992, named after its original inhabitants, Otto and Athiná Stathátos. It houses the Greek Art Collection of the Athens Academy. Temporary exhibitions are also on display on the first floor of the Stathátos Mansion.

Seated Cycladic figure

National Gallery of Art ❾
Εθνική Πινακοθήκη

Vasiléos Konstantínou 50, Ilísia. **Tel** 210 723 5937. 3, 13. ⏰ 9am–3pm Mon & Wed–Sat; 10am–2pm Sun. ● main public hols. 🅿 🅾 ♿

This modern, low-rise building holds a permanent collection of European and Greek art. The first floor is devoted mainly to European art and includes works by Van Dyck, Cézanne, Dürer and Rembrandt, as well as Picasso's *Woman in a White Dress* (1939) and Caravaggio's *Singer* (1620). Most of the collection is made up of Greek art from the 18th to 20th centuries. The 1800s feature paintings of the War of Independence *(see pp42–3)*. There are also some excellent portraits including *The Loser of the Bet* (1878) by Nikólaos Gýzis (1842–1901), *Waiting* (1900) by Nikifóros Lýtras (1832–1904) and *The Straw Hat* (1925) by Nikólaos Lýtras 1883– 1927).

Temporary exhibitions are on the ground floor.

Icon of the *Adoration of the Magi* from the Benáki Museum

Getting Around Athens

Trolleybus stop sign

The sights of Athens' city centre are closely packed, and almost everything of interest can be reached on foot. This is the best way of sightseeing, especially in view of the appalling traffic congestion, which can make both public and private transport slow and inefficient. The expansion of the metro system, though not yet complete, already provides a good alternative to the roads for some journeys. However, the bus and trolleybus network still provides the majority of public transport in the capital for Athenians and visitors alike. Taxis are a useful alternative and, with the lowest tariffs of any EU capital, are worth considering even for longer journeys.

Orange and white regional bus for the Attica area

One of the fleet of yellow, blue and white buses

BUS SERVICES IN ATHENS

Athens is served by an extensive bus network. Bus journeys are inexpensive, but can be slow and uncomfortably crowded, particularly in the city centre and during rush hours; the worst times are from 7am to 8:30am, 2pm to 3:30pm and 7:30pm to 9pm.

Tickets can be bought individually or in a book of ten, but either way, they must be purchased in advance from a *períptero* (street kiosk), a transport booth, or certain other designated places. The brown, red and white logo, with the words *eisitíria edó*, indicates where you can buy bus tickets. The same ticket can be used on any bus or trolleybus, and must be stamped in a ticket machine to validate it when you board. There is a penalty fine for not stamping your ticket and tourists who are unfamiliar with this system are often caught out when inspectors board buses to carry out random checks. Tickets are valid for one ride only, regardless of the distance and, within the central area, are not transferable from one vehicle to another.

Athens ticket b

USEFUL ROUTES IN ATHENS

There are currently three metro lines; expansions are ongoing. Kerameikos station opened in 2008 and line three now extends to Egaleo.

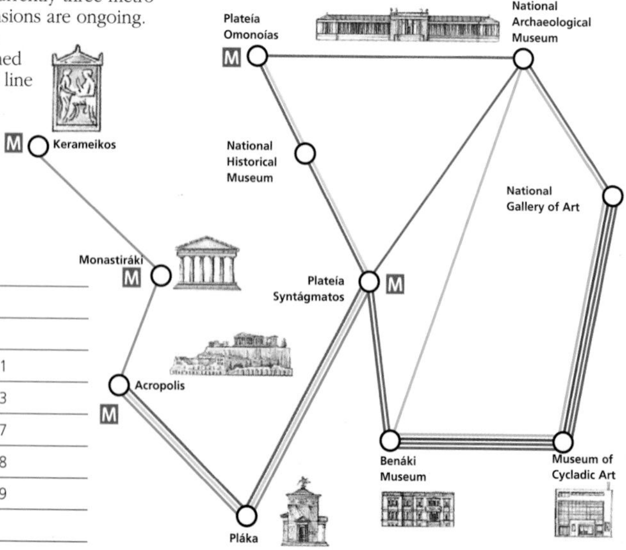

KEY
— Bus A5
— Bus 230
— Trolleybus 1
— Trolleybus 3
— Trolleybus 7
— Trolleybus 8
— Trolleybus 9
Ⓜ Metro

Kerameikos · Monastiráki · Acropolis · Pláka · Plateía Omonoías · National Historical Museum · Plateía Syntágmatos · Benáki Museum · National Archaeological Museum · National Gallery of Art · Museum of Cycladic Art

Monastiráki metro sign

ATHENS BUS NETWORKS

There are three principal bus networks serving greater Athens and the Attica region. They are colour coded blue, yellow and white; and green. Blue, yellow and white buses cover an extensive network of over 300 routes in greater Athens, connecting districts to each other and to central Athens. In order to reduce Athens' smog, some of these are being replaced with green and white "ecological" buses running on natural gas.

Orange and white buses serve the Attica area (*see pp146–7*). On these you pay the conductor and, as distances are greater, fares are also more expensive. The two terminals for orange and white buses are both situated on Mavrommatáin, by Pedío tou Áreos (Areos Park). Though you can board at any designated orange stop, usually you cannot get off until you are outside the city area. These buses are less frequent than the blue, yellow and white service, and on some routes stop running in the early evening.

Green express buses, the third category, travel between central Athens and Piraeus. Numbers 040 and 049 are very frequent – about every 6 minutes – running from Athinas, by Plateía Omonoías, to various stops in Piraeus, including Plateía Karaïskáki, at the main harbour.

TROLLEYBUSES IN ATHENS

Athens has a good network of trolleybuses, which are purple and yellow in colour. There are over 20 routes that criss-cross the city. They provide a good way of getting around the central sights. All routes pass the Pláka area. Route 3 is useful for visiting the National Archaeological Museum from Plateía Syntágmatos, and route 1 links Lárissis railway station with Plateía Omonoías and Plateía Syntágmatos.

![An Athens trolleybus]

An Athens trolleybus

ATHENS' METRO

The metro, which has three lines, is a fast and reliable means of transport in Athens.

Line 1 runs from Kifissiá in the north to Piraeus in the south, with central stations at Thiseío, Monastiráki, Omónoia and Victoria. The majority of the line is overland and only runs underground between Attikí and Monastiráki stations through the city centre. The line is used mainly by commuters, but offers visitors a useful alternative means of reaching Piraeus.

Lines 2 and 3 form part of a huge expansion of the system, most of which was completed in time for the 2004 Olympic Games. The new lines have been built 20 m (66 ft) underground in order to avoid material of archaeological interest. Sýntagma and Acropolis stations have displays of archaeological finds.

Line 2 runs from Agios Antónios in northwest Athens to Agios Dimitrios in the southeast. Line 3 runs from Egaleo to Ethniki Amyna in the northeast with some trains continuing to Eleftheríos Venizélos airport. Two extensions – westbound to Haidari and northbound to Agia Paraskeví – are scheduled for completion by 2009.

One ticket allows travel on any of the three lines and is valid for 90 minutes in one direction. You cannot exit a station, then go back to continue your journey with the same ticket. A cheaper ticket is sold for single journeys on Line 1. Tickets can be bought at any metro station and must be validated before entering the train – use the machines at the entrances to all platforms. Trains run every five minutes from 5am to 12:30am on Line 1, and from 5:30am to 12:30am on Lines 2 and 3.

Archaeological remains on display at Sýntagma metro station

DRIVING IN ATHENS

Driving in Athens can be a nerve-racking experience, especially if you are not accustomed to Greek road habits. Many streets in the centre are pedestrianized and there are also plenty of one-way streets, so you need to plan routes carefully. Finding a parking space can also be very difficult. Despite appearances to the contrary, parking in front of a no-parking sign or on a single yellow line is illegal. There are small car parks at street level for legal parking, as well as underground car parks, though these usually fill up quickly.

In an attempt to reduce dangerously high air pollution levels, there is an "odd-even" driving system in force. Cars that have an odd number at the end of their licence plates can enter the central grid, also called the *daktýlios,* only on dates with an odd number, and cars with an even number at the end of their plates are only allowed on dates with an even number. To avoid this, some people have two cars – with an odd and even plate. The rule does not apply to foreign cars but, if possible, it is better to avoid taking your car into the city centre.

No parking on odd-numbered days of the month

No parking on even-numbered days of the month

Yellow Athens taxi

ATHENIAN TAXIS

Swarms of yellow taxis can be seen cruising around Athens at most times of the day or night. However, trying to persuade one to stop for you can be difficult, especially between 2pm and 3pm when taxi drivers usually change shifts. Then, they will only pick you up if you happen to be going in a direction that is convenient for them.

To hail a taxi, stand on the edge of the pavement and shout out your destination to any cab that slows down. If a cab's "TAXI" sign is lit up, then it is definitely for hire, (but often a taxi is also for hire when the sign is not lit). It is also common practice for drivers to pick up extra passengers along the way, so do not ignore the occupied cabs. If you are not the first passenger, take note of the meter reading immediately: there is no fare-sharing, so you should be charged for your portion of the journey only, (or the minimum fare of 2.7 euros, whichever is greater).

Athenian taxis are extremely cheap by European standards – depending on traffic, you should not have to pay more than 2.7 euros to go anywhere in the downtown area, and between 5 and 8 euros from the centre to Piraeus. Double tariffs come into effect between midnight and 5am, and for journeys that exceed certain distances from the city centre. Fares to the airport, which is out of town at Spáta, are about 25 euros. There are also small surcharges for extra pieces of luggage weighing over 10 kg (22 lbs), and for journeys from the ferry or railway terminals. Taxi fares are increased during holiday periods, such as Christmas and Easter.

For an extra charge, (1.5 euros), you can make a phone call to a radio taxi company and arrange for a cab to pick you up at an appointed place and time. Radio taxis are plentiful in the Athens area. Listed below are the telephone numbers of a few companies:

Express
Tel 01 993 4812.

Kosmos
Tel 18300.

Hermes
Tel 210 411 5200.

WALKING

The centre of Athens is very compact, and almost all major sights and museums are to be found within a 20- or 25-minute walk of Plateía Syntágmatos, which is generally regarded as the city's centre. This is worth bearing in mind, particularly when traffic is congested, all buses are full and no taxi will stop. Athens is still one of the safest European cities in which to walk around, though, as in any sizeable metropolis, it pays to be vigilant, especially at night.

Sign for pedestrian area

Visitors to Athens, walking up Areopagos Hill

ATHENS TRANSPORT LINKS

The hub of Athens' city transport is the area around Plateía Syntágmatos and Plateía Omonoías. From this central area trolleybuses or buses can be taken to the airport, the sea port at Piraeus, Athens' two train stations, and its domestic and international coach terminals. In addition, three new tram lines connect the city centre with the Attic coast.

Bus E95 runs between the airport and Syntágma and bus E96 between the airport and Piraeus. Buses 040 and 049 link Piraeus harbour with Syntágma and Omonoías in the city centre. The metro also extends to Piraeus harbour and the journey from the city centre to the harbour takes about half an hour.

Trolleybus route 1 goes past Lárissis metro station, as well as Lárissis train station, with the Peloponnísou station a short walk away from them. Bus 024 goes to coach terminal B, on Liosíon, and bus 051 to coach terminal A, on Kifisoú.

Tram line 1 (T1) runs from Syntágma to Néo Fáliro on the coast; T2 runs from Néo Fáliro to the Athens suburb of Glyfáda; T3 runs from Glyfáda to Syntágma.

Though more expensive than public transport, the most convenient way of getting to and from any of these destinations is by taxi. The journey times vary greatly but, if traffic is free-flowing, from the city centre to the airport takes about 40 minutes; the journey from the city centre to the port of Piraeus takes around 40 minutes; and the journey from Piraeus to the airport takes about 60 minutes. Taxis are abundant in Athens and are relatively inexpensive compared with most other European cities.

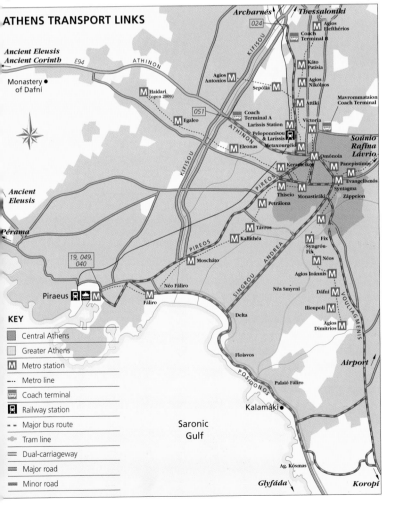

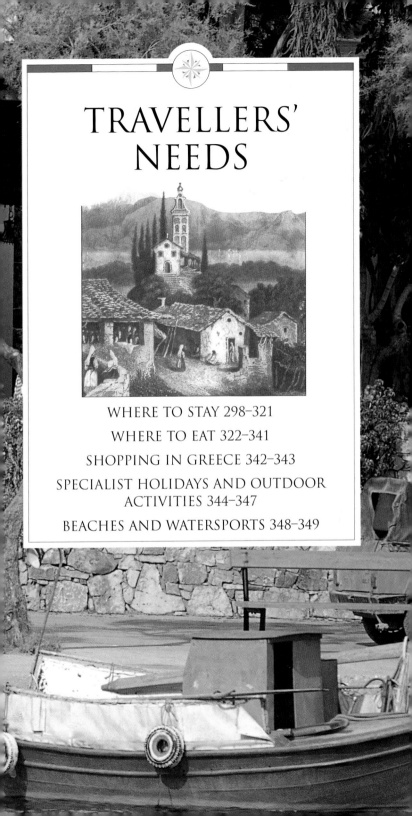

TRAVELLERS' NEEDS

WHERE TO STAY

Accommodation in the Greek islands has improved enormously in recent years. Prices, however, have increased steeply and while luxury hotels are still among the cheapest in Europe, at the lower end of the market, Greece no longer enjoys a price advantage over other Mediterranean destinations. Despite the effects of mass tourism, hospitality off the beaten track can still be warm and heartfelt. Various types of accommodation are described over the next four pages. Information is also given for camping and hostelling. The listings section on pages 302–321 includes over 230 places to stay, ranging from informal *domátia* (rooms) and mountain refuges, to luxurious spa resorts, boutique hotels and accommodation in restored buildings.

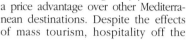

**Holiday apartment sign,
Ionian Islands**

HOTELS

In the most popular resorts on islands such as Corfu, Crete, Kos, Rhodes and Zákynthos, large, purpose-built hotels, erected to serve the needs of package holiday companies, prevail. These are usually located on or near the best beaches and are block-booked by tour operators. In high season such hotels do not offer rooms to individual visitors. However, they may have rooms available – often at bargain prices – during the off-peak spring and autumn weeks. These resort hotels have a wide range of facilities including outdoor pools (usually unheated), buffet restaurant and bar, and rooms with en-suite shower and bathroom, but their architecture is usually bland and unimaginative.

Large hotels at the top end of the market – in destinations such as Rhodes, Eloúnta on Crete, Kos and Skiáthos – offer luxurious facilities such as health spas, beauty centres, floodlit tennis courts, indoor and outdoor pools, a choice of gourmet restaurants and bars. The service in such hotels tends to be more attentive than that offered by the package holiday hotels.

BOUTIQUE AND STYLE HOTELS

Boutique hotels in restored traditional buildings are a recent phenomenon and have added a much-needed dash of character to Greece's accommodation portfolio. These hotels put a premium on style, design and location and are usually found in the heart of historic towns and villages (rather than on the beach). They are most numerous in the Cyclades, Crete and Rhodes, but at least one can be found on most islands. Few have more than a dozen rooms, and booking ahead is essential. More than a dozen stylish hotels in the Greek islands are members of the Small Luxury Hotels consortium which has its own website, www.slh.com, where hotels can be viewed and booked online.

CHAIN HOTELS

Major international hotel brands have made few inroads into Greek island territory. However, several major Greek and Cypriot hotel chains have luxury properties on the most popular islands. These include **Louis Hotels**, with properties on Corfu, Kefalloniá, Zákynthos, Kos, Rhodes and Crete; **Aldemar Hotels**, which has luxury hotels on Crete and Rhodes; and **Astir Hotels**; and **Capsis Hotels** with hotels on Crete and Rhodes.

Hotel Alykí *(see p312)*, **with boats at the waterfront, on Sými**

◁ **Steni Valá Taverna on the island of Alónnisos**

Skiáthos Palace Hotel *(see p307)*

RESTORED SETTLEMENTS AND BUILDINGS

A programme initiated by the EOT (the Greek National Tourist Office) during the 1970s encouraged the restoration of many derelict examples of traditional architecture for use as guesthouses. More recently, grants and tax breaks have encouraged the owners of many fine old buildings to convert more of these into accommodation. Standards, room sizes and facilities vary widely, depending on the constraints of preserving the original building. Some offer en-suite accommodation, others have only shared bathrooms. Such developments are found on Ydra, Crete, Sými, Mýkonos, Sýros, Lésvos, Folégandros, Kálymnos and Rhodes.

VILLAGE ROOMS

Rooms to rent by the night or week, with or without a reservation, are easy to find on all the islands – owners with vacant rooms meet every arriving ferry, even in the small hours, and most island tourist offices can also help find rooms. Look out for signs saying "rent rooms", *domátia* (bedrooms) or, where there is a regular German clientele, *zimmer frei*.

In the early days of tourism, islanders often rented their spare bedrooms, or even offered cots on the flat roof of their homes or in their gardens. Nowadays they are almost always in small, purpose-built blocks with solar-heated hot water, en-suite shower and WC, marble floors and pine

furniture, but do not usually have air-conditioning or heating. Most also have a balcony or veranda. Those in the centre of harbour towns can be noisy at night.

GRADING

Hotels, village rooms and apartments are graded by the EOT. Hotels are graded from A-class to E-class, though in practice there are very few D-class hotels and no E-class hotels in the islands. There is also an L category for the most luxurious hotels. Rooms and apartments are graded A to C. The grading system places more emphasis on services, fixtures and fittings than on style and quality. A room with carpet, phone and a small TV will score more highly than a room with marble or pine floors and no phone or TV. This means gradings for many smaller hotels, boutique hotels and restored guesthouses can be misleading.

C-class hotels must have en-suite bathrooms and at least one bar and basic restaurant. B-class hotels must have a full service restaurant, pool and other sports facilities. A-class and deluxe hotels offer the full range of in-room services

Dodecanese window

and facilities and an array of luxury extras such as gourmet dining, satellite TV, several pools, tennis courts and water sports, conference and business centres.

PRICES

EOT sets guideline prices each year for all classes of accommodation except deluxe hotels. However, market forces mean rates in practice vary widely. A stay of more than three nights in a village room usually entitles you to a 10 per cent discount and rates in all types of accommodation are up to 50 per cent lower in the "shoulder" seasons (April–early June, except Easter, and mid-September to late October).

OPENING SEASONS

Almost all hotels, guesthouses and village rooms close from the end of October until the Orthodox Easter (which may fall any time from early to late April). Some larger hotels in Mýkonos, Rhodes and some other islands do stay open throughout the winter, but if you plan a visit to any of the smaller islands between the months of November and March you should book accommodation in advance. The listings in this guide indicate when hotels are closed.

Volissos Traditional Apartments *(see p307)* at Volissós, Chíos

BOOKING

Most visitors to Greece choose to travel with a package holiday company, reserving their flights and accommodation in advance. The easiest way for independent travellers to reserve all kinds of accommodation is via the internet. Almost all accommodation providers, except for the smallest, family-run village rooms, can be contacted via email and there are numerous booking websites. Small Luxury Hotels of the World (www.slh.com) is one such site specialising in independently-owned hotels in Greece and elsewhere. A credit card deposit is normally required when booking and is forfeited if you fail to arrive.

SELF CATERING APARTMENTS

Self-catering apartments are the best-value accommodation in Greece. Most are in small complexes of 20–30, built by local owners to meet the demands of the big package holiday companies. You can expect a balcony, a kitchenette, small fridge, en-suite shower and WC.

Many apartment complexes have shared pools. Studios sleep two, usually in twin beds. Apartments sleep up to four, with a separate double bedroom as well as a twin-bedded or sofa-bed room. Most islands are over-supplied with apartments, and individual travellers can usually negotiate bargain rates.

Monastery of Agios Geórgios on Skýros *(see pp116–17)*

Travel á la Carte *(see p347)* and **Greek Options** both offer good quality self-catering apartments.

Beware signs advertising *garsonieres* – these are old-fashioned, cramped and often dirty studios, usually above a noisy restaurant or shop.

VILLAS

Luxury villas are often booked several years in advance, by specialist holiday companies. The widest range of villa accommodation is to be found on Corfu, Paxós, Kefaloniá, Skiáthos, Skópelos and Rhodes. The largest and most expensive villas sleep up to ten, offering facilities such as a private pool, hire car, satellite TV, DVD player, fully-equipped kitchen and maid service. Some even come complete with cook and household staff. Simpler and smaller villas usually have similar fittings and facilities, but without luxury

Self-catering apartment on Santorini

extras such as a pool. Specialist villa companies include Abercrombie & Kent (www.abercrombiekent.co.uk) and the Greek Islands Club (www.greekislandsclub.com).

YOUTH HOSTELS

Only two youth hostels, on Corfu and Santoríni, are recognised by the **International Youth Hostel Federation** and can be booked on its website (www.hihostels.com). Unofficial "youth hostels" offering dormitory rooms can be found on some islands but are generally poorly equipped and often dirty and overcrowded.

MOUNTAIN REFUGES

There are three mountain refuges on Mount Psiloreítis and in the White Mountains on Crete. Only one, at Kallergi, is open year round. Keys can be obtained from **EOS** (Greek Mountaineering Club) branches in Irákleio and Chaniá.

RURAL TOURISM

Conceived during the 1980s to give women in the Greek provinces a measure of financial independence, rural tourism allows foreigners to stay on a bed-and-breakfast basis in a village house, and also provides the opportunity to participate, if desired, in the daily life of a farming community. Information on agrotourism can be obtained from the Greek National Tourist Organisation website *(see p353)* or from visiting www.agrotour.gr

MONASTERIES

Some monasteries in the remoter parts of Crete, Rhodes and other large islands operate *xenones* (hostels) intended primarily for Orthodox pilgrims. Accommodation is rarely available at short notice and is in spartan dormitories with a frugal evening meal and morning coffee. It is customary when staying at a monastery to leave a donation.

CAMPING

The Panhellenic Camping Union lists officially recognised campsites on Amorgós, Andros, Astypálaia, Evvoia, Corfu, Crete, Folégandros, Ios, Kéa, Kefalloniá, Kos, Lefkáda, Léros, Mílos, Mýkonos, Náxos, Páros, Santoríni, Sérifos, Sífnos, Sýros, Tínos, Pátmos and Rhodes on ts website, www.panhellenic-camping-union.gr. Some of hese are very basic, offering ittle more than cold-water showers and toilets and space o pitch a tent. Others are quite sophisticated, with swimming pools, restaurants, ennis and volleyball courts, aundry and mini-market. Most have space for motor caravans as well as tents.

Green Hotel *(see p320)* in Spili, Crete

Lakka Paxi Camping, on the Ionian Islands

DISABLED TRAVELLERS

The organisation **Tourism For All** produces an information sheet on travel around Greece for disabled people. Other organizations providing useful information about access to hotels and places of interest in Greece include **SATH** (Society for the Advancement of Travel for the Handicapped) and **Door to Door**, an online travel guide. Only the largest and most modern hotels in Greece have even the most basic facilities (such as lifts and wheelchair accessible toilets) for people with disabilities. In the hotel listings of this guide we have indicated which establishments have facilities, such as lifts and wheelchair ramps.

FURTHER INFORMATION

The most useful source of information on accommodation in the Greek islands is the *Greek Travel Pages* which lists details of all kinds of accommodation with links to individual properties and booking sites. The **Hellenic Chamber of Hotels** publishes a yearly *Guide to Hotels* listing all officially registered hotels.

A list of registered guesthouses and hotels is also available from the Greek National Tourist Organisation.

DIRECTORY

CHAIN HOTELS

Aldemar Hotels
262 Kifissiás Avenue,
Kifissiá, 14562 Athens.
Tel 210 623 6150.
www.aldermarhotels.com

Astir Hotels
www.astir-palace.com

Capsis Hotels
10 Parnonos St, Marousi,
15125 Athens.
Tel 210 614 2083.
www.capsis.gr

Louis Hotels
Louis House, 20
Amphipoleos St, 2025
Strovolos, Nicosia, Cyprus.
Tel 357 225 88211.
www.louishotels.com

Mamidakis Hotels of Greece
Panepistimíou 56,
10678 Athens.

Tel 210 381 9781-6.

SELF-CATERING

Greek Options
Tel 0844 8004787.
www.greekoptions.co.uk

YOUTH HOSTELS

IYHF (UK)
1st Floor, Fountain House,
Parkway, Welwyn Garden
City, Herts AL8 6JH.
Tel 01707 324170.

IYHF (Greece)
Víktoros Ougó 16,
10438 Athens.
Tel 210 523 2049.
@ y-hostels@otenet.gr

MOUNTAIN REFUGES

EOS (Ellinikós Orivatikós Sýndesmos)

(Greek Alpine Club)
Filadelfías 126, 13671
Acharnés, Attica.
Tel 210 246 1528.

CAMPING

Greek Camping Association
Sólonos 102,
10680 Athens.
Tel 210 362 1560.
www.panhellenic-camping-union.gr

DISABLED TRAVELLERS

Door to Door
www.dptac.gov.uk/door-to-door/

SATH
347 Fifth Avenue,
Suite 610, New York,
NY 10016.
Tel 212 447 7284.
www.sath.org

Tourism For All
c/o Vitalise, Sharp Road
Industrial Estate, Kendal,
Cumbria LA9 6NZ.
Tel 08451 249971.
www.tourismforall.
org.uk

FURTHER INFORMATION

Greek Travel Pages
Psýlla 6, 10557 Athens.
Tel 210 324 7511.
www.gtp.gr

Hellenic Chamber of Hotels
Stadiou 24,
10564 Athens.
Tel 210 331 2535.
@ grhotels@otenet.gr

Choosing a Hotel

The hotels in this guide have been selected across a wide price range for their facilities, good value, and location. The entries are listed by region, starting with the Ionian Islands, then alphabetically by island name or area. Wheelchair access is minimal in all but the more deluxe hotels in the Greek islands – phone ahead for details.

PRICE CATEGORIES
Price categories are for a standard double room for one night in peak season, including tax, service charges and breakfast:

€ under 50 euros
€€ 50–80 euros
€€€ 80–120 euros
€€€€ 120–180 euros
€€€€€ over 180 euros

THE IONIAN ISLANDS

CORFU Pink Palace
€

Agios Gordios, Sinaradhes, 49084 **Tel** *26610 53103* **Fax** *26610 53025* **Rooms** *20*

A legend among fun-seeking budget travellers for more than 20 years, the Pink Palace is strictly for the young and uninhibited. On one of Corfu's best sandy beaches, it has comfortable single, double, multi-occupancy and dormitor rooms and rates include breakfast and three course dinner. Closed Oct–Apr. **www.thepinkpalace.com**

CORFU Fundana Villas
€€

Palaiokastrítsa, 49083 **Tel** *26630 22532* **Fax** *26630 22453* **Rooms** *12*

The old farm buildings of this complex have been converted into cottages and villas sleeping 2–7 people. The complex boasts a peaceful setting with fine views over woods and vineyards. Two pools, one large and one for toddlers. Access to Palaiokastrítsa beach, restaurants and shops is by car. **www.fundanavillas.com**

CORFU Belle Helene
€€€

Agios Giorgios, 49081 **Tel** *26630 96201-3* **Fax** *26610 96209* **Rooms** *54*

This family-run hotel is located in the picturesque bay of Agios Giorgios, on the north west coast and 35km (22 miles) from Corfu town. This is not as crowded as some of the resorts, and there is a lovely, long beach 50m from the hotel. The hotel has parking, an outdoor pool and disabled access. **www.bellehelenehotel.com**

CORFU Corfu Secret
€€€

Agios Markos, 49083 **Tel** *26610 97921* **Fax** *26610 97931* **Rooms** *23*

This boutique hotel is located in a quiet hillside location above the popular tourist resort of Ipsos, on the east coast, 15km (9 miles) from Corfu town. There are fabulous panoramic views of the sea, an outdoor pool, and all modern comforts in the rooms, which are designed in the traditional Venetian style. **www.corfusecret.gr**

CORFU Fiori
€€€

Tzavros, 49100 **Tel** *26610 80110/ 91921* **Fax** *26610 80111* **Rooms** *29*

Just 2km (1 mile) from the popular tourist resort of Dassia, the Fiori offers the best of both worlds - the peace and quiet of its secluded location and private garden with pool, and the beach activities of nearby Dassia. Car recommended - there is ample parking in the private carpark. **www.fiorihotel.gr**

CORFU Louis Corcyra Beach
€€€

Gouviá, 49100 **Tel** *26610 90196* **Fax** *26610 91591* **Rooms** *370*

Large, luxury resort complex near the lively resort of Gouviá. Accommodation includes rooms, studios, suites and bungalows in vast wooded grounds 8 km (5 miles) from Corfu town. The resort boasts five semi-private beaches, three pools and live entertainment. All-inclusive packages available. Closed Oct–Apr. **www.louishotels.com**

CORFU Akrotiri Beach
€€€€

T Desylla 155, Palaiokastrítsa, 49100 **Tel** *26630 41237* **Fax** *26630 41277* **Rooms** *127*

Largest and best of the hotels around popular Palaiokastrítsa. The hotel offers a plethora of facilities including 2 pools, restaurant, 3 bars, live entertainment, and watersports including scuba diving nearby. Comfortable, well-equipped rooms with balconies and some of Corfu's best views. Closed Oct–Apr. **www.akrotiri-beach.com**

CORFU Bella Venezia
€€€€

N. Zambeli 4, Corfu Town, 49100 **Tel** *26610 46500* **Fax** *26610 20707* **Rooms** *32*

Close to the centre of Corfu town, this dignified Neo-Classical mansion has been restored and converted into a pleasant boutique hotel. High-ceilinged rooms are comfortable and tasteful without being extravagant. Hospitable, friendly staff. Public areas and some rooms are wheelchair-accessible. **www.bellaveneziahotel.com**

CORFU Cavalieri
€€€€

Kapodistriou 4, 49100 **Tel** *26610 39336* **Fax** *26610 39283* **Rooms** *50*

Striking 17th-century Venetian townhouse overlooking Corfu town's main public green space, the Spianádha. Public rooms are decorated in Venetian style and furnished with antique and reproduction furniture. Guest rooms have high ceilings, some with balconies, but are a little on the small side. **www.cavalieri-hotel.com**

Key to Symbols *see back cover flap*

CORFU Pelecas Country Club
☐☐☐☐☐ €€€€

Pelekas, 49100 **Tel** *26610 52239* **Fax** *26610 52919* **Rooms** *10*

Formal gardens surround this fabulous 19th-century manor, with luxurious suites and studios, all furnished with Venetian and Corfiot antiques. The club is favoured by some of the wealthiest people in Greece for its seclusion and attentive personal service. Excellent pool. Closed Nov–Apr. **www.country-club.gr**

CORFU Corfu Palace
☐☐☐☐☐☐☐ €€€€€

Dimoktratias 2, Corfu Town, 49100 **Tel** *26610 39485* **Fax** *26610 31749* **Rooms** *115*

The "grande dame" of Corfu hotels, built in 1954 and still the best place to stay in Corfu town. The marble bathrooms, cheerfully decorated rooms and suites, and grand public areas lend a touch of class. The rooms at the front have fine sea views. Large pool with shaded terrace. **www.corfupalace.com**

CORFU Divani Palace
☐☐☐☐☐☐☐ €€€€€

Nafsiká 20, Corfu Town, 49100 **Tel** *26610 38996* **Fax** *26610 35929* **Rooms** *162*

Probably the most expensive hotel in Corfu, but with a range of facilities and services that justify the high price, including a choice of restaurants, health and beauty facilities, and attentive staff. Standing on a wooded hillside overlooking the lagoon of Kanóni, 3km (2 miles) outside Corfu Town. Closed Oct–Apr. **www.divanis.gr**

CORFU Grecotel Corfu Imperial
☐☐☐☐☐☐☐ €€€€€

Kommeno, 49100 **Tel** *26610 88400* **Fax** *26610 91881* **Rooms** *320*

One of the most luxurious resort hotel complexes on Corfu, offering bungalows, villas and suites as well as standard double rooms. The rooms here are more imaginatively furnished than most big resorts. Good choice of watersports and other activities and several bars and restaurants. Closed Oct–Apr. **www.grecotel.gr**

ITHAKI Mentor
☐☐☐☐☐☐ €€€

Paralia, Vathý, 28300 **Tel** *26740 33033* **Fax** *26740 32293* **Rooms** *36*

Small, family run hotel with fine harbour views from some rooms. Services include a café-bar with a terrace overlooking the harbour, and an internet corner. Rooms are clean and comfortable and the service good. No restaurant, but there are plenty to choose from in nearby Vathý. There is a sandy beach 800 metres (2,600 ft) away. **www.hotelmentor.gr**

ITHAKI Perantzada 1811
☐☐☐☐☐ €€€€

Odissea Androutsou, Vathý, 28300 **Tel** *26740 23914* **Fax** *26740 33493* **Rooms** *19*

Neo-Classical with a 21st century twist, this stylish small hotel is located on the harbour front. The 19th-century façade disguises a pop-art interior, with rooms painted white and brightened by vivid striped fabrics in bright greens, blues, pink and orange. Renowned for its designer chic. Great food and service. **www.arthotel.gr**

KEFALLONIA Linardos
☐☐ €€

Asos, 28085 **Tel** *26740 51563* **Fax** *26740 51563* **Rooms** *12*

Located in the centre of the pretty village of Asos, Linardos offers rooms with fine views across the bay to the ruins of the Venetian castle. A choice of doubles, triples, twins and family rooms, each with a well-equipped mini-kitchen. Hire a motor boat at the harbour to explore the nearby coast. Closed Oct–May. **www.linardosapartments.gr**

KEFALLONIA Olga
☐☐☐☐ €€€

Antoni Tritsi 82, Argostóli, 28100 **Tel** *26710 24981* **Fax** *26710 24985* **Rooms** *43*

A small, modern hotel in the centre of Argostóli with a mix of double, twin and family rooms, each with its own balcony. All have fridges, phones and en-suite shower and WC. Well priced for those who want to stay in the centre of Argostóli. The beach, however, is 3 km (2 miles) from the hotel. **www.olgahotel.gr**

KEFALLONIA Apostolata Island Resort and Spa
☐☐☐☐☐☐☐ €€€€€

Skála, 28082 **Tel** *26710 83581* **Fax** *26710 83583* **Rooms** *137*

Set on a hillside 3 km (2 miles) from Skála village, this is the first luxury all-inclusive resort hotel on Kefallonia. Rooms are stylish and comfortable with panoramic sea or mountain views. Buffet meals, snacks, local drinks and entertainment are all included in the rate. Excellent facilities. Closed Oct–Apr. **www.louishotels.com**

KEFALLONIA Emelisse Art Hotel
☐☐☐☐☐☐☐ €€€€€

Eblisi, Fiskárdo, 28084 **Tel** *26740 41200* **Fax** *26740 41026* **Rooms** *65*

Fourteen traditional-style stone houses with stylish rooms boasting four-poster beds and private terraces. Facilities include a poolside bar and restaurant, breakfast terrace, tennis court, gym and billiards. Mountain bikes are also available for guests to use. Good views over bay area. Closed Oct–Apr. **www.arthotel.gr**

LEFKADA Agios Nikitas
☐☐☐☐☐ €€€

Agios Nikítas, 31100 **Tel** *26450 97460-1* **Fax** *26450 97462* **Rooms** *36*

This attractive small hotel comprises 28 twin and double rooms and eight suites, all with large balconies. The rooms occupy three traditional-style buildings that overlook the beach and harbour at Agios Nikitas, just 150 metres (490 ft) away. Pleasant gardens and sea views from most rooms. Closed Nov–Apr. **www.hotelagiosnikitas.com**

LEFKADA Odyssey
☐☐☐☐☐ €€€

Agios Nikítas, 31080 **Tel** *26450 97351-2* **Fax** *26450 97421* **Rooms** *40*

This hotel is located in the attractive traditional village of Agios Nikitas, on the west coast of the island, and 12km (7 miles) from Lefkada town. There are plenty of fish tavernas and bars nearby. Guests can stay in the main building, or in one of three self-catering units. **www.odyssey-hotel.gr**

LEFKADA Porto Fico

Vasiliki, 31100 **Tel** *26450 31402* **Fax** *26450 31467* **Rooms** *20*

This is the best hotel in the pretty, peaceful village of Vasilikí, 38 km (24 miles) from Lefkáda town. All rooms have balconies with sea or mountain views. Pondi beach, a long sweep of sand and white pebbles, is only 60 metres (1,968 ft) from the hotel. Two pools including separate children's pool. Closed Oct–Apr. **www.portoficohotel.com**

MEGANISI Porto Vathi Studios

Vathý, Meganísi, 31083 **Tel** *26450 51622* **Rooms** *13*

Located on the tiny island of Meganísi, a few minutes across the water from Lefkáda, Vathý Studios offers basic self-catering accommodation decorated in typical island style – white walls, blue woodwork, pine ceilings. Pretty views, plenty of peace and quiet. Closed Oct–May. **panosmeg@otenet.gr**

PAXOS Paxos Beach Hotel

Gáïos, 49082 **Tel** *26620 32211* **Fax** *26620 32695* **Rooms** *42*

Popular, family-run hotel offering well-appointed, simply decorated, comfortable rooms with sea and mountain views. Facilities include a private beach, tennis court and wind-surfing equipment. There is also a jetty where you can rent a motor boat to explore the island. Closed Nov–Apr. **www.paxosbeachhotel.gr**

PAXOS Paxos Club

Gáïos, 49082 **Tel** *26620 32451* **Fax** *26620 32097* **Rooms** *26*

This comfortable, family-run hotel comprises 26 purpose-built studios and two-room apartments sleeping up to five people. Each has its own veranda or balcony overlooking the gardens or large pool. Rooms are clean with well-equipped mini-kitchens. Closed Oct–Apr. **www.paxosclub.gr**

ZAKYNTHOS Tsivouli Park

Lithákia, 29100 **Tel** *26950 55018* **Rooms** *7*

Surrounded by unspoilt farmland, Tsivouli Park offers whitewashed rooms with iron beds and balconies in traditional-style stone cottages. The owners keep their own livestock here and grow fruit and vegetables. Full board is reasonably priced and is by prior arrangement. Pretty beach within easy walking distance.

ZAKYNTHOS Leedas Village

Lithákia, Agios Sóstis, 29092 **Tel** *26950 51305* **Fax** *26950 29934* **Rooms** *20*

Self-catering apartments for 2–9 people in 5 attractive stone villas, all with terraces. Apartments are plain and cool, with terracotta paved floors and fully equipped kitchens. The villas are set in pretty, flower-filled gardens and are a short walk from the beach. Children's playground. Closed Nov–Apr. **www.leedas-village.com**

ZAKYNTHOS Paliokaliva Village

Tragaki, 29100 **Tel** *26950 63770* **Fax** *26950 65144* **Rooms** *18*

Rooms in ten stone cottages set around a pool and among olive trees. Each villa has its own terrace and is furnished with wrought-iron beds and simple wooden furniture. All have basic self-catering facilities, but there is also a good bar-restaurant that serves lunch and dinner. Sandy beach less than 2 km (1 mile) away. **www.paliokaliva.gr**

ZAKYNTHOS Villa Tzogia

Kambi, 29100 **Tel** *26950 94060* **Rooms** *4*

Located on the outskirts of a small village, on the unspoilt west side of the island, Villa Tzogia comprises a villa and apartments sleeping 2–4 people. Set in lush gardens with views over rolling hills and vineyards. Shared pool and a couple of tavernas nearby. Closed Oct–Apr. **www.tzogia.gr**

THE ARGO-SARONIC ISLANDS

AIGINA Pension Rena

Agia Irini, Aigina Town, 18910 **Tel** *22970 24760* **Fax** *22970 24244* **Rooms** *8*

Quirky, friendly and affordable, Pension Rena has a loyal following. The rooms are plain but attractive, light and airy – book early for those on the second floor which have balconies and views over the Gulf. There are also several restaurants and a beach nearby. **www.pension-rena.gr**

AIGINA Nafsiká

N Kazantzaki 55, Aigina Town, 18910 **Tel** *22970 22333* **Fax** *22970 22477* **Rooms** *36*

The Nafsiká is a village-style resort with comfortable rooms, some with views of the Gulf and the ruins of the Temple of Apollo. The hotel boasts lushly-filled courtyards and a large pool. It is only 50m (164 ft) from the island's best beach 'Kolóma'. Closed Nov–Mar. **www.hotelnafsika.com**

AIGINA Aeginitikou Archontiko

Ag. Nikólaou and Thomaidou 1, 18910 **Tel** *22970 24968* **Fax** *22970 26716* **Rooms** *11*

Built in the early 19th century, this delightful island mansion has played host to Orthodox saints, sea-captains, poets and musicians. Superbly restored, it boasts marvelous painted ceilings, stained-glass windows, flagstoned floors and two tranquil interior courtyards. Closed Nov–Mar.

KYTHIRA Kamares Apartments

Aroniadika, 80200 **Tel** *27360 33420* **Fax** *27360 34124* **Rooms** *6*

This sturdy stone building with its blue-painted shutters and arched ceilings looks out over a plateau of vineyards and terraced fields. Apartments have basic kitchenettes with fridge and two-ring cooker. Breakfast is served in the rooms. Beaches are a 15–25-minute walk away. Closed Nov–Mar. **www.kamareshotel.gr**

KYTHIRA Margarita

Chóra, 80100 **Tel** *27360 31711* **Fax** *27360 31325* **Rooms** *11*

Located on the outskirts of Kýthira's beautiful main village with dazzling views of the sea, Margarita has white-washed walls and blue wooden shutters. Bedrooms are bright and modern with plain furniture, and there is a sunny flagstoned terrace. Beaches are 1.5 km (1 mile) below the village. **www.hotel-margarita.com**

KYTHIRA Nostos Inn

Chóra, 80100 **Tel** *27360 31056* **Fax** *27360 31834* **Rooms** *6*

In the heart of one of the Aegean's most beautiful villages, Nostos is an inn in the proper sense, with a cheery bar-café on the ground floor and spotless rooms on the upper floor. Some rooms look out over the village to the sea, all have high ceilings, polished wood floors and modern amenities. Closed Oct–Mar. **www.nostos-kythera.gr**

KYTHIRA Vasilis Bungalows

Chóra, 80100 **Tel** *27360 31125* **Fax** *27360 31553* **Rooms** *12*

Twelve bungalows, some with self-catering facilities, set among olive trees just outside the harbour village of Kapsáli. White walls, painted woodwork, stylish bedrooms and spacious breakfast room. Several bars and restaurants in nearby Kapsáli. All rooms have Wifi and balconies. Closed Nov–Mar. **www.kythirabungalowsvasili.gr**

POROS Sto Roloi

Póros, 18020 **Tel** *22980 25808* **Rooms** *7*

Typical island house divided into three well-equipped self-catering apartments sleeping 2–7 people. Accommodation can be rented separately, in combination, or as a whole house. The separate Anemone House, has two houses, which share a swimming pool. Plenty of bars and restaurants nearby. **www.storoloi-poros.gr**

SPETSES Armàta Hotel

Enouà Agiou Antomiou, 180-50 Spetses **Tel** *22980 72683* **Fax** *22980 75403* **Rooms** *20*

This elegant boutique hotel is modern is design and decor. It is located in a quiet area, close to the shops and restaurants of Spetses island. It is within walking distance to both the port and the beach. The hotel offers guests a range of amenties in the rooms, with TV and Wifi, 16 of the rooms have a private balcony. **www.armatahotel.gr**

SPETSES Zoe's Club

Spétses Town, 18050 **Tel** *22980 74447* **Fax** *22980 72841* **Rooms** *22*

Delightful, modern complex of 11 studios, 8 suites and 4 maisonettes surrounding a large pool on the outskirts of Spétses Town. Prettily designed, well managed, and in a quiet location but within easy distance of the harbour area with its tavernas, bars and cafes. Excellent base for a long family holiday. Closed Oct–Apr. **www.zoesclub.gr**

SPETSES Economou Mansion

Kounoupitsa, Spétses Town, 18050 **Tel** *22980 73400* **Fax** *22980 74074* **Rooms** *8*

Wonderful, family run luxury guesthouse within a dynastic mansion. Large rooms with beamed ceilings, marble floors and soft-coloured rugs. The two suites also have kitchenettes and sea-view balconies. Breakfast is served on the terrace. About 10 minutes' walk from Spétses Town. Closed Nov–Mar **www.septsestravel.gr**.

SPETSES Orloff

Palió Limáni, Spetses, 18050 **Tel** *22980 75444* **Fax** *22980 74470* **Rooms** *20*

Named for an 18th-century Russian adventurer and would-be liberator of Greece, this is a welcome addition to the Spétses hotel scene. Super pool within high stone walls, a mix of twin and double rooms, self-catering studios, suites and a separate villa that sleeps up to 10 people. Stylishly decorated rooms. Closed Dec–Mar. **www.orloffresort.com**

YDRA Delfini

Harbour, Ydra Town, 18040 **Tel** *22980 52082* **Fax** *22980 53828* **Rooms** *11*

The cheapest spot for a short stay in Ydra, and only a few steps from the ferry quay. Rooms are clean and basic, some have en-suite WC and shower. A couple of rooms look out over the harbour, but are exposed to a lot of early morning noise as boats and hydrofoils arrive outside. Closed Oct–Apr. **www.delfini-hotel.gr**

YDRA Nefeli

Tsamadoú 8-14, Ydra Town, 18040 **Tel** *22980 53297* **Fax** *22980 53297* **Rooms** *9*

This tiny hotel is perched above Ydra's pretty harbour. All rooms have good views and are set above three little courtyard terraces where breakfast is served. As the climb up from the harbour is steep, it is not suitable for small children or people with disabilities, but it is a pleasant place for a short stay. Closed Nov–Feb. **www.hotelnefeli.eu**

YDRA Leto

Chóra, Ydra, 18040 **Tel** *22980 53385* **Fax** *22980 53806* **Rooms** *21*

Argo-Saronic grandeur is the keynote of this boutique hotel housed in an old Ydriot mansion. Antiques grace public areas and bedrooms, the best of which look out onto the paved courtyard. One of very few island hotels to offer dedicated accommodation for wheelchair users and their partners. Closed Nov–Mar. **www.letohydra.gr**

YDRA Angelica   €€€€

Ydra town, 18040 **Tel** *22980 53264* **Fax** *22980 53542* **Rooms** *14*

Traditional Ydriot house which has been sensitively transformed into a stylish small hotel. The rooms have stone walls, beamed ceilings, wood or marble floors and some have balconies that look out over Ydra Town and the harbour. Well managed, quiet and with plenty of places to eat and drink nearby. **www.angelica.gr**

YDRA Mirànda Hotel   €€€€

Ydra town, 18-40 **Tel** *22980 552230* **Fax** *22980 53510* **Rooms** *14*

Converted into a hotel in 1992, this hotel still holds some traditional features from its days as a mansion. In the special rooms, which are available to book, there are hand-painted ceilings, executed by Florentine and Ventian artists. Located close to the beach and port, a traditional breakfast is served on the patio. **www.mirandahotel.gr**

YDRA Orloff   €€€€

Ydra Town, 18040 **Tel** *22980 52564* **Fax** *22980 53532* **Rooms** *9*

One of the very first comfortable hotels on Ydra, the Orloff is named after an 18th-century Russian adventurer and combines style with authenticity – rooms and public areas have antique furniture, lamps and mirrors as well as modern amenities. Located on a quiet square, with its own pretty courtyard. Closed Nov–Mar. **www.orloff.gr**

YDRA Bratsera   €€€€€

Chóra, Ydra, 18040 **Tel** *22980 53971* **Fax** *22980 53626* **Rooms** *28*

Compared with boutique hotels elsewhere, the Bratsera is amazingly affordable and boasts a big pool, breezy public areas and excellent restaurant. The individually decorated rooms have polished wood floors, tall windows and antique furnishings. It is worth visiting Ydra just to stay here. Closed Nov–Mar. **www.bratserahotel.com**

YDRA Xelidonia   €€€€€€

Chóra, Ydra, 18040 **Tel** *00 32 234 75814* **Fax** *00 32 234 36880* **Rooms** *3*

Superb self-catering apartment-mansion sleeping up to six people in three bedrooms, high above Ydra's scenic harbour. Rooms have iron four-poster beds, white linen and wood or tiled floors. The terrace boasts unbeatable views over Ydra Town and harbour. Closed Nov–Feb. **www.xelidonia.com**

THE SPORADES AND EVVOIA

ALONISSOS Konstantintas Studios   €

2 km from Patitíri, 37005 **Tel** *24240 66165* **Fax** *24240 66165* **Rooms** *9*

Traditionally-designed studios and rooms in the old village of Alónissos, but with modern facilities including en-suite shower and WC, fridge and basic kitchen amenities. Good value for money and an excellent base for exploring the island or for a short island-hopping stopover. Closed Nov–Mar. **www.konstantinasstudios.gr**

ALONISSOS Haravgi   €€

Patitiri, 37005 **Tel** *24240 65090* **Fax** *24240 65189* **Rooms** *19 (9 Studios, 10 rooms)*

Alónnisos has a limited range of accommodation and the Haravgi is a no-frills hotel offering simple but clean rooms, all with balconies. Acceptable for a short stay for those planning on island hopping around the Sporades, but not the best choice for a stay of more than a few days. Closed Nov–Mar.

EVVOIA Apollon   €€€

Kárystos Bay, 34001 **Tel** *22240 22045* **Fax** *22240 22049* **Rooms** *36*

Situated on the edge of a lush bay, this modern hotel offers suites sleeping up to five people. Rooms have en-suite shower and WC and overlook the sea. There is also a terrace with views and a lovely garden. Facilities include a restaurant and watersports. Close to beach. Closed Nov–Mar.**www.apollonsuiteshotel.com**

EVVOIA Béis   €€€

Kými Beach, 34003 **Tel** *22220 22604* **Fax** *22220 29113* **Rooms** *38*

This comfortable hotel has rooms with good views of the port. All rooms have en-suite shower and WC and there is an excellent restaurant serving freshly-caught fish. The hotel is located close to the beach and offers watersports facilities. Wheelchair accessible.**www.hotel-beis.gr**

SKIATHOS Atrium   €€€€

Paralia Platanias, Agía Paraskevi, 73002 **Tel** *24270 49345* **Fax** *24270 49444* **Rooms** *75*

Large, well-appointed modern hotel on a pine-covered hillside, 150 metres (492 ft) above the sandy beach at Paralías. Built with a package-tour clientele in mind, it has a large pool and excellent choice of watersports. Some distance from the restaurants and nightlife of Skiáthos Town. Closed Nov–Apr. **www.atriumhotel.gr**

EVVOIA Candíli   €€€€€

1 km (0.5 miles) from seafront, Prokópi, 34004 **Tel** *69740 62100* **Rooms** *12*

Set in a splendid estate, Candíli is a hotel and seminar centre offering creative courses in mosaics and painting. The estate comprises a manor house (suitable for families), two cottages and charming bedrooms in converted granaries and stables. Well-equipped art studio, dining and games room. A landrover is available for use. **www.candili.gr**

Key to Price Guide *see p302* **Key to Symbols** *see back cover flap*

SKIATHOS Aegean Suites
⊞▤🍸⓪⌧ €€€€€

Fteliá, 73002 **Tel** *24270 24068* **Fax** *24270 24070* **Rooms** *20*

A stylish, purpose-built hotel enjoying views out to sea and over a large pool. The price reflects the attentive service and state-of-the-art facilities. This hotel is more geared towards romantic couples than those on family holidays. Closed Nov–Apr. **www.aegeansuites.com**

SKIATHOS Esperides
⊞⊞🏃▤🍸⓪⌧ €€€€€

Achládia, 73002 **Tel** *24270 22535* **Fax** *24270 21580* **Rooms** *181*

Huge resort hotel overlooking a sandy beach, with excellent facilities including tennis courts. The rooms are spacious and those in front have fine views. The Esperides is a full-service hotel with a wide choice of restaurants and bars. Closed Nov–Apr. **www.esperidesbeach.gr**

SKIATHOS Palace
⊞🏃▤🍸⓪⌧ €€€€€

Koukounariés Beach, 37002 **Tel** *24270 49700* **Fax** *24270 49666* **Rooms** *258*

Situated on a white sandy beach backed by pine trees, this resort-style hotel enjoys fantastic sea views. All rooms have ensuite bathroom and WC, satellite TV, direct dial phone, a fridge and balcony or terrace. Site facilities include a shop, mini-market, indoor and outdoor bars, tennis courts and sauna. Closed Nov–Mar. **www.skiathos-palace.gr**

SKOPELOS Thea Home
▤🍸⓪ €€

Skópelos Town, 37003 **Tel** *24240 22859* **Fax** *24240 23556* **Rooms** *12*

Medium-sized complex of self-catering studios and apartments designed in mock-traditional style and laid out around a large pool. The Zanetta offers good facilities for families and enjoys an attractive location. Reasonably priced. Closed Oct–Apr. **www.skopelosl.gr/theahome**

SKOPELOS Aegeon
▧⊞🏃▤🍸⓪⌧ €€€

Skópelos Town, 37003 **Tel** *24240 22619* **Fax** *24240 22194* **Rooms** *15*

Small, cheerful and affordable hotel with bright rooms, all with balconies overlooking Skópelos Town and its bay. The rooms, all in a three-storey building in mock-traditional style, are simple but bright and clean, with en-suite shower and WC. Below the hotel is a terraced garden and a children's play area. Closed Oct–Apr.

SKOPELOS Pleoussa Studios
▤⌧ €€€

Ambeliki, 37003 **Tel** *24240 23141* **Fax** *24240 23844* **Rooms** *10*

A small complex of ten pretty, self-catering studios, painted in ochre and cream and built around an arcaded courtyard which is paved in typical Skopelos fashion. Upper-floor rooms have attractive wooden balconies with views of the sea. 50 metres (164 ft) from a pebbly beach. Closed Oct–Apr. **www.pleoussa-skopelos.gr**

SKYROS Nefeli
⊞🏃▤🍸⓪⌧ €€€

Chòra, Skýros Town, 34007 **Tel** *22220 91964* **Fax** *22220 92061* **Rooms** *22*

A well designed and serviced hotel, with double or twin-bedded rooms, two suites and four self-catering studio apartments laid out in low-rise buildings around an attractive pool. Not far from the island's unspoilt main town, which has plenty of restaurants, and within walking distance of the beach. **www.skyros-nefeli.gr**

THE NORTHEAST AEGEAN ISLANDS

CHIOS Filoxènia Hotel
▨▤ €€

Chios, Roidou 2 and Vaupàlou 8 **Tel** *22710 22813* **Fax** *22710 28447* **Rooms** *20*

This hotel is situated in a scenic area of the Aegean islands. It is close to the port and the towns main shops and restaurants. Previousy a mansion, built in the 1900s, it was renovated in 2004, with much of the original features still intact, for example the 80-year-old tiles. Car parking facilities are available on site.

CHIOS Aeriko
▧⊞▤⓪⌧ €€

Karfás, 82100 **Tel** *22710 32336* **Fax** *22710 32335* **Rooms** *6*

Within walking distance of a pebbly beach at Karfas, this small pension stands among olive groves. Traditional buildings around a pleasant pool where breakfast is served at iron tables under shady trees. Family run and friendly, Aeriko is a good base from which to explore the island or just laze. Closed Nov–Mar. **www.benovias.gr**

CHIOS Volissos Traditional Apartments
▨▤⓪⌧ €€

Volissós, 82100 **Tel** *22740 21421* **Fax** *22740 21521* **Rooms** *16*

Designer and sculptor Stella Tsakiri has lovingly restored these old-fashioned cottages. All have self-catering facilities, larger cottages have two bedrooms, some with views of the sea or surrounding countryside. Inside are exposed beams and stonework, whitewashed walls, platform beds and traditional fabrics. Closed Nov–Mar. **www.volissostravel.gr**

CHIOS Plaka Studios
▨▤⓪🍴⌧ €€€

Karfás, 82100 **Tel** *22710 32955* **Fax** *22710 32966* **Rooms** *10*

Small, well-appointed studios with self-catering facilities and balconies with sea views. All rooms have TV, kitchenette with fridge and small cooker and there is a small breakfast bar. Next to Karfas beach, the island's best beach. Well-priced, especially by Chiot standards.

CHIOS Perleas Mansion   €€€

Odos Vítiadou, Kámbos, 82100 **Tel** *22710 32217* **Fax** *22710 32364* **Rooms** *7*

Large rooms with iron beds, polished wood floors and spotless linen in the stone buildings of a former farm, hidden away in the Chiot countryside among olive trees. There are lovely, pine-shaded terraces and a lily pond which was formerly the farmhouse reservoir. **www.perleas.gr**

CHIOS Argentikon   €€€€€

Odos Argenti, Kámbos, 82100 **Tel** *22710 33111* **Fax** *22710 31645* **Rooms** *8*

One of the most lavishly luxurious places to stay in the Greek islands. No expense has been spared in this former mansion of a dynasty of Chiot-Genoese aristocrats. Fine dining, a pool in a stone-walled courtyard, and opulent surroundings make a stay here an experience to savour. **www.argentikon.gr**

CHIOS Ta Petrina   €€€€€

Volissós, 82103 **Tel** *22740 21128* **Fax** *22740 21013* **Rooms** *18*

This collection of stone houses stands above the hillside village of Volissós in the midst of island farmland, olive groves and vineyards. Family run, with rooms in a mini-village of six cottages available separately, in combination or as a whole. A charming mix of old and new, with modern kitchen facilities. **www.tapetrina.gr**

IKARIA Cavos Bay   €€

Armenistis, 83301 **Tel** *22750 71381* **Fax** *22750 71380* **Rooms** *63*

A good value hotel with modern facilities, a large pool and world class views. Rooms have satellite TV and balconies with views; only the studios have air conditioning. Located on the outskirts of the village and about 10 minutes' walk from an excellent sandy beach. Closed Nov–Apr. **www.cavosbay.com.gr**

IKARIA Erofili Beach   €€€

Armenistis, 83301 **Tel** *22750 71058* **Fax** *22750 71483* **Rooms** *31*

Unpretentious small hotel with a great location above a huge sweep of white sandy beach. All rooms have balconies with good views. The hotel sits on the outskirts of the village where there are plenty of cafés, bars and restaurants. Can be a little noisy at night in high season. Closed Oct–Apr.

IKARIA Messakti Village   €€€€

Gialiskari, 83301 **Tel** *22750 71331* **Fax** *22750 71330* **Rooms** *55*

This is a modern hotel next to a fabulous sandy beach. The interior boasts exposed stonework, flagstoned floors, wooden beams and sleeping galleries. A mix of large studio rooms and apartments, with two suites built into an old stone tower. Shared WC and bathrooms. Closed Nov–Apr. **www.mesakti-village.com**

LESVOS Nassos Guest House   €

Mólyvos, 81108 **Tel** *22530 71432* **Rooms** *7*

This guesthouse is located in the centre of Mólyvos, one of the nicest villages on Lésvos. The simple pretty rooms are decorated in pastel colours. The best rooms are those with views over the village roofs to the sea. One large double has en-suite WC and shower. **www.nassosguesthouse.com**

LESVOS Olive Press   €€€

Mólyvos Beach, 81108 **Tel** *22530 71205* **Fax** *22530 71647* **Rooms** *80*

Charming hotel converted from an old olive press with rooms and self-catering studios around a grassy, shady courtyard. Located next to a clean, pebbly beach and close to the many restaurants and bars of Mólyvos. Some rooms have balconies overhanging the beach. **www.olivepress-hotel.com**

LESVOS Vaterá Beach   €€

Vaterá Beach, 81300 **Tel** *22520 61212* **Fax** *22520 61164* **Rooms** *24*

A modern, medium sized family-run establishment located on an excellent beach. The comfortable rooms are light and breezy with tall windows and pine furniture. There are some family rooms and some with self-catering facilities. Good restaurant. Internet access. Closed Nov–Apr. **www.vaterabeach.gr**

LESVOS Clara   €€€€

Avláki, Pétra, 81109 **Tel** *22530 41532* **Fax** *22530 41535* **Rooms** *42*

Not far from Pétra's long sweep of beach and lively village, the Clara is a medium-sized resort with a wide range of facilities including tennis courts and pool. Rooms are plain but well-equipped and all have balconies with views of the sea and of Mólyvos. A comfortable place for a longer stay. Closed Oct–Mar. **www.clarahotel.gr**

LESVOS Loriet   €€€€€

Vareiá Beach, 81100 **Tel** *22510 43111* **Fax** *22510 41629* **Rooms** *35*

A luxurious, stylish hotel located 2 kms (1 mile) from Mytilini Town. The hotel is housed in a restored 19th-century mansion where there are seven luxury suites. Additional rooms and studio apartments occupy recently added wings. 24-hour room service, two restaurants and a poolside cocktail bar. **www.loriet-hotel.com**

LIMNOS Porto Myrina Palace   €€€€

81400 Myrina, Limnos **Tel** *22540 24805* **Fax** *22540 24858* **Rooms** *150*

A quiet hotel situated close to the beach, in the tranquil bay of Avlon. All the main sites are close by, including the archaeological museum and Myrina village. All rooms in the hotel have private balconies and facilities include; pool bar, tennis court, basketball court and gym. **www.ellada.net/portomyr/**

Key to Price Guide *see p302* **Key to Symbols** *see back cover flap*

LIMNOS Villa Afroditi

Platí Beach, Platí, 81400 **Tel** *22540 23141* **Fax** *22540 25031* **Rooms** *24*

A small, well-run hotel offering clean, simple rooms at affordable prices. There are three suites and three apartments as well as twin rooms available but booking well ahead is essential. The attached restaurant is popular and facilities are good. Close to beach. Closed Nov–Apr. **www.afroditi-villa.gr**

SAMOS Kalidon

Kokkári, 83100 **Tel** *22730 92800* **Fax** *22730 92608* **Rooms** *28*

An attractive, small hotel offering well-designed rooms and friendly service. The owners also run the slightly larger and pricier Kalidon Palace nearby and the facilities there are open to guests of the Kalidon. Good value and a good base for exploring the rest of Sámos. Closed Nov–Apr. **www.kalidon.gr**

SAMOS Kerkis Bay

Ormos Marathokampos, 83102 **Tel** *22730 37202* **Fax** *22730 37372* **Rooms** *29*

Located in the heart of a pleasantly relaxed harbour village and housed in an attractive, traditional-style building. The well-appointed rooms offer all modern comforts and there is an attractive leafy courtyard. Sun loungers and umbrellas on the pebbly beach a few steps away and excellent watersports nearby. Closed Nov–Mar.

SAMOS Olympia Beach

Kokkári Beach, 83100 **Tel** *22730 92420* **Fax** *22730 92457* **Rooms** *12*

This charming, small hotel is built in traditional style and offers light and airy rooms, some with sea views. Located right on Kokkári's clean, pebbly beach. An excellent base for a range of activities including climbing, mountain biking, sea kayaking, diving and windsurfing. Closed Oct–Apr. **www.olympia-hotels.gr**

SAMOS Arion

Kokkári, 83100 **Tel** *22730 92020* **Fax** *22730 92006* **Rooms** *108*

Large but well-designed, low-rise resort hotel located on a tree-covered hillside on the outskirts of one of the island's prettiest resort villages. Buffet restaurant, bars and good facilities for families. There are white pebble beaches nearby and plenty of restaurants and nightlife to choose from. Free shuttle service. Closed Oct–Apr. **www.arion-hotel.gr**

SAMOTHRAKI Kastro

Palaiópolis, 68002 **Tel** *25510 89400* **Fax** *25510 41000* **Rooms** *50*

This modern, well-managed hotel is designed to fit in with the local architecture. It has a large pool and good range of facilities and is one of the few international-standard hotels on Samothraki, which suffers from a shortage of full-service accommodation. Closed Oct–Apr. **www.kastrohotel.gr**

THASOS Arsinoe Cottages

Limenaria, 64002 **Tel** *25930 52796* **Fax** *25930 52295* **Rooms** *7*

Beautiful stone cottages surrounded by lush greenery on a slope above a sandy beach. Accommodation is in five cottages – the smaller ones sleep two, the larger houses up to four. Inside are whitewashed walls, stone fireplaces, cast-iron beds, tiled floors and simple pine furniture. Shared kitchen. Closed Oct–Apr.

THASOS Miramare

Skála Potamiás, 64004 **Tel** *25930 61040* **Fax** *25930 61043* **Rooms** *30*

A modern, well-equipped hotel with a great location among pine woods at the southern end of Chryssí Ammoudiá, the most spectacular sandy beach on the island. Comfortable, reasonably-priced rooms, and a good place for a family holiday. Closed Oct–Apr. **www.hotelmiramare.gr**

THASOS Alexandra Beach

Potós beach, 64004 **Tel** *25930 52391* **Fax** *25930 51185* **Rooms** *124*

Large resort hotel on the beach at Potós, near Limenária, with accommodation in bungalows and twin and double rooms. Among the facilities are a restaurant, three bars and an array of sports, health and fitness activities including tennis, volleyball, windsurfing and waterskiing. Closed Nov–Apr.

THE DODECANESE

ASTYPALAIA Australia Studios

Péra Gialós, 85900 **Tel** *22430 61275* **Fax** *22430 61067* **Rooms** *15*

Modern, open-plan hotel with a choice of rooms sleeping 2–3, and self-catering studios sleeping up to 4. Balconies have sea views and there is a tree-shaded garden. Below the hotel is a restaurant offering fresh fish and meat grills. The beach is 50 metres (164 ft) away. Closed Nov–Apr.

CHALKI Argyrenia

Nimporió, 85101 **Tel** *22460 45205* **Fax** **Rooms** *9*

Located midway between Nimporió and Potamós beach is this small, unassuming, pension-style guesthouse with simply furnished chalet-style rooms. Most rooms have terraces and there is a lovely, leafy garden. Numerous bars and restaurants within easy walking distance. Good value. Closed Oct–Apr.

CHALKI Captain's House
Nimporió, 85110 **Tel** *22460 45201* **Fax** *22460 45201* **Rooms** *3*

This tiny guesthouse has just one double and two twin-bedded rooms and is much in demand. Housed in an old-fashioned villa with red-tiled roof and large, simple rooms surrounding a shaded terrace-garden. There is a small sunbathing terrace. Rooms have fridges. Closed Oct–Apr.

CHALKI Villa Praxithea
Nimporió, 85110 **Tel** *22410 70172* **Fax** *22410 70175* **Rooms** *6*

These two self-catering apartments sleep up to eight people and occupy an attractive seaside villa on the outskirts of Chálki's main village. Floors are traditional tile or wood, rooms are painted in pale pastel colours and there is a small terrace for sunbathing. The upper rooms have harbour views. Closed Oct–Apr. **www.villapraxithea.com**

KALYMNOS Galini
Póthia, 85200 **Tel** *22430 31241* **Fax** *22430 31100* **Rooms** *14*

This family-run pension has better than average facilities and overlooks Vathý Bay and the boatyard. Rooms are simple but comfortable and are good value for money. There is a pleasant terrace where breakfast is served and a restaurant for snacks and light meals. Closed Oct–Apr.

KALYMNOS Panorama
Póthia, 85200 **Tel** *22430 23138* **Fax** *22430 23138* **Rooms** *13*

Set back from the bustling seafront at Póthia, the Panorama lives up to its name with great views of the bay. Bedrooms are comfortable, modern and simply furnished and most have balconies. The hotel has a small bar and breakfast restaurant and there are many tavernas and cafés nearby. Closed Oct–Apr. **www.panorama-kalymnos.gr**

KARPATHOS Amoopi Bay
Amoopí, 85700 **Tel** *22450 81184* **Fax** *22450 81105* **Rooms** *65*

A well-appointed but unassuming small hotel offering reasonably-priced accommodation. All rooms have a balcony or terrace, phone and satellite TV, and the hotel has its own garden restaurant (rates including dinner are negotiable) 300 metres (984 ft) from sandy Amoopí beach. Closed Nov–Apr.

KASTELLORIZO Karnayo
Kastellórizo Town, 85111 **Tel** *22460 70626* **Fax** *22460 49266* **Rooms** *8*

The most stylish and attractive place to stay on Kastellórizo, this lovingly-restored traditional building has four rooms, some with wooden balconies, and four self-catering apartments. Set on a cobbled square with palm trees and bougainvillea. A bathing ladder gives access to the harbour.

KASTELLORIZO Mediterraneo
25 Martiou, Megísti, 85111 **Tel** *22460 49007* **Fax** *22460 49007* **Rooms** *9*

One of the prettiest hotels in the Dodecanese, with colourful bedrooms in shades of blue and terracotta and a superb, huge suite on the ground floor. A bathing ladder gives access to clear blue water literally on the doorstep. Rooftop terrace serving breakfast and, on request, dinner. Fine views across the beautiful bay. **www.medterraneopension.org**

KOS Afendoulis
Evripidou 1, Kos Town, 85300 **Tel** *22420 25321* **Fax** *22420 25797* **Rooms** *23*

A small, family hotel in the centre of Kos town with friendly, English-speaking management and comfortable, simple rooms. Fairly quiet (by Kos Town standards) and a short walk from the good town beach. Breakfast and drinks are served in a lovely garden filled with jasmine and bougainvillea. Closed Nov–Apr.

KOS Karavia Beach
Aginaropi, Marmári, 85300 **Tel** *22420 41291* **Fax** *22420 41215* **Rooms** *295*

This vast, all-inclusive resort is favoured by Italian, German, Dutch and Belgian clientele. The resort offers a wide range of activities including boat trips and picnics on nearby islands such as Psérimos. Rate includes buffet breakfast, lunch and dinner, alcoholic and soft drinks, and watersports. Closed Nov–Apr.

KOS Porto Bello Beach
Kardámaina, 85302 **Tel** *22420 91217* **Fax** *22420 91168* **Rooms** *293*

Billed as an 'ultra all-inclusive' resort, the luxury Porto Bello is located on a 5 km (3 miles) long beach, 2 km (1 mile) from Kardámaina. Rate includes buffet breakfast, lunch and dinner, snacks, drinks, non-motorised water sports, a wide range of land sports and live entertainment. Closed Nov–Apr.

KOS Grecotel Kos Imperial
Psalídi, 85300 **Tel** *22420 58000* **Fax** *22420 25192* **Rooms** *384*

This hotel, 4 km (2.4 miles) from Kos town, is lavishly landscaped with free-form pools linked by flowing streams. Accommodation includes luxury villas, family suites, and double or twin rooms. All have a balcony or terrace and a choice of garden or sea views. There is a Thalassotherapy spa. Closed Nov–Apr. **www.grecotel.com**

KOS Grecotel Kos Royal Park
Agios Geórgios, Marmári, 85300 **Tel** *22420 41488* **Fax** *22420 41373* **Rooms** *268*

This well-appointed and well-located luxury resort on sandy Marmári beach, offers all-inclusive packages as well as standard rates. Facilities and activities include pools for adults and children, windsurfing, pedaloes, canoes, beach sports, and live entertainment. Bicycle rental and horse riding are available nearby. Closed Nov–Apr. **www.grecotel.com**

Key to Price Guide *see p302* **Key to Symbols** *see back cover flap*

OS Louis Helios Beach
ardámaina, 85300 **Tel** 22420 91602 **Fax** 22420 91390 **Rooms** 108

pgraded to 'A' (five star) category in 2001, this luxury resort on a semi-private beach is pleasingly aloof from the ubbub of Kardámaina. Excellent facilities and a wide range of activities for families. All-inclusive packages are vailable and the hotel offers a choice of sea, pool or mountain view rooms. Closed Nov–Mar. **www.louishotels.com**

EROS The Nest
gía Marína, 85400 **Tel** 00 39 051 234 974 **Fax** 00 39 051 239 086 **Rooms** 1

robably the most delightful place to stay on Léros, this luxury cottage, on a hillside between the harbour and the astle, has a double bedroom with platform bed, and a sofa bed. Beautifully decorated with silk and soft linen curtains, nd fine views from the terrace over the bay. Minimum 7-night stay. Closed Nov–Mar.

NISYROS Porfyris
Mandráki, 85303 **Tel** 22420 31376 **Fax** 22420 31176 **Rooms** 38

he most sophisticated option on Nísyros – a pleasant, reasonably-priced hotel set among citrus groves at Mandráki, ose to the harbour. There are good views from the hotel terrace across to the coast of Kos and the tiny island of iali. Rooms are comfortable and simply furnished. Closed Oct–Apr.

ATMOS Artemis
rikos, 85500 **Tel** 22470 31555 **Fax** 22470 34016 **Rooms** 24

his small, resort-style hotel on the outskirts of the island's second largest seaside village offers rooms in a village-tyle array of whitewashed bungalows. Each room has its own balcony or terrace with views of the sea. Facilities re adequate and include satellite TV, direct dial phones and fridges. Closed Oct–Apr.

ATMOS Asteri
Merichas, Skála, 85500 **Tel** 22470 32465 **Fax** 22470 31347 **Rooms** 37

his pleasant, small family-run hotel overlooks Mérichas bay on the outskirts of Skála. Its rooms are comfortable and odern, 11 of them have been recently added to the original building. The hotel is in a quiet location within easy valking distance of the waterfront bars and restaurants, and ferry harbour. Closed Nov–Mar.

ATMOS Blue Bay
kála, 85500 **Tel** 22470 31165 **Fax** 22470 32303 **Rooms** 25

. five-minute walk from the Skála ferry dock, the aptly named Blue Bay has a quiet location and is run by a friendly reek-Australian family. Rooms are modern and comfortable, with twin beds, neutral furnishings and wooden nutters. Breakfast is served on the waterfront terrace. Internet access. Closed Oct–Apr. **bluebayhotel@yahoo.com**

ATMOS 9 Muses
apsila, Grikos, 85500 **Tel** 22470 34079 **Fax** 22470 33151 **Rooms** 12

ungalow apartments with tiled floors and private balconies or terraces, plus fabulous views, snack bar, restaurant, aby sitting and wheelchair-accessible rooms. Gríkos, the nearest village with tavernas and a beach, is around 2 km mile) away. Scooter rental and car parking available. Booking essential. **www.9muses-gr.com**

RHODES Apollo
Omirou 82, 85100 **Tel** 22410 32003 **Rooms** 5

his restored old house has bright, breezy rooms with four-poster beds – spend a bit extra for the more expensive op-floor rooms at the front which have wonderful Old Town views. Set in a quieter part of the Old Town with a retty inner courtyard. Short walk to sights and shopping. Closed Nov–Apr. **www.apollo-touristhouse.com**

RHODES Domus Rodos
Plateía Plátonos, Rhodes Old Town, 85100 **Tel** 2410 25965 **Fax** 2410 24766 **Rooms** 18

iood, quiet location on a square close to the centre of the Old Town. The building has antique features such as vooden ceilings and staircases, and rooms are simply furnished. No restaurant but several places to eat and drink ust a few steps from the front door. Closed Nov–Apr. **www.domusrodoshotel.gr**

RHODES Mango
Plateía Dorieos 3 (Old Town), 85100 **Tel** 22410 24877 **Fax** 22410 24877 **Rooms** 14

ituated on one of the Old Town's quieter squares, Mango is clean, cheap and cheerful with basic en-suite rooms vith fridges. There is a roof terrace with views of the Old Town, broadband internet access and cheap drinks. A avourite with scuba divers and yachties. Five-minute walk from ferry quay. Closed Nov–Mar. **www.mango.gr**

RHODES Annapolis Inn
28 Oktobriou and Ionos Dragoumi, Rhodes New Town, 85100 **Tel** 22410 24538 **Fax** 22410 31910 **Rooms** 44

his comfortable hotel in Rhodes New Town comprises 44 studios, suites and apartments with well-equipped nini-kitchens. All rooms have balconies and most have luxury bathrooms. Suites with wheelchair access are lso available. Mini-market. 24-hour reception. **www.annapolisinn.gr**

RHODES Camelot
Themistokléous 18, 85100 **Tel** 22410 26649 **Fax** 22410 26549 **Rooms** 3

mall pension in a medieval building a short walk from the Old Town's busy main squares. Choice of simple double or triple rooms with en-suite WC and shower or twin rooms with separate toilet and shower. Pretty pebbled inner ourtyard, lavish breakfasts and attractive surroundings. Closed Nov–Apr. **www.camelothotel.gr**

RHODES Paris   €€€

Agíou Fanouríou 88 (Old Town), 85100 **Tel** *22410 26356* **Fax** *22410 21095* **Rooms** *20*

Basic, affordable rooms surrounding a small courtyard in the heart of the Old Town. Single, twin and three-bed rooms, some en-suite, some with shared shower and WC. Within walking distance of the ferry harbour, and (after the bar closes) quiet at night. Plenty of places to eat and drink nearby. Closed Nov–Mar

RHODES San Nikolis Hotel   €€€

Ipodamou, Rhodes Old Town, 85100 **Tel** *22410 34561* **Fax** *22410 32034* **Rooms** *18*

Nestling under the walls of the Old Town is this charming, small hotel housed in an old Rhodian townhouse with stone walls covered with ivy and bougainvillea. Rooms are very comfortable, with antique furnishings, and some with balconies. Excellent views from the rooftop terrace. Closed Nov–Apr. **www.s-nikolis.gr**

RHODES Aldemar Paradise Royal Mare and Paradise Village   €€€€

Koskinoú, 85100 **Tel** *22410 67040* **Rooms** *1270*

This huge, all-inclusive resort, 6 km (3.5 miles) from Rhodes Town, combines two hotels, the five-storey Royal and the low-rise Paradise Village. There are nine pools, including two toddler pools, and a semi-private beach. Wheelchair access to all public areas plus five wheelchair-friendly poolside bungalows. Closed Nov–Mar. **www.aldemarhotels.com**

RHODES Archontiko Angelou   €€€€

Waterfront, Alinda, 85400 **Tel** *22470 22749* **Fax** *22470 24403* **Rooms** *8*

Built in 1895, this lovely pink and white mansion has been prettily restored and is now a well-kept, family-run guest house with traditionally-furnished rooms. The guesthouse is set in an attractive garden full of jasmine and geranium. No restaurant but picnic lunches can be arranged. Short walk to beach. Closed Nov–Mar. **www.hotel-angelou-leros.com**

RHODES Marco Polo Mansion   €€€€

Agíou Fanouríou 40-42, Rhodes Old Town, 85100 **Tel** *22410 25562* **Fax** *22410 29115* **Rooms** *7*

Gorgeously evocative of medieval Rhodes, with double and twin rooms in a superbly restored 16th-century Ottoman townhouse. Wooden floors, painted ceilings, gorgeous fabrics and a leafy inner courtyard where à la carte meals can be ordered. Closed Nov–Apr. **www.marcopolomansion.web.gr**

RHODES Fashion Hotel Nikos Takis   €€€€€

Panetíou 26, Rhodes Old Town, 85100 **Tel** *22410 70773* **Fax** *22410 24643* **Rooms** *7*

Colourful and stylish new hotel in the Old Town, owned by two of Greece's best known fashion designers. Great location next to the Palace of the Grand Masters. Suites have four-poster beds, carved wooden furniture, embroidered silk soft furnishings and luxury bathrooms. Views over the Old Town. **www.nikostakishotel.com**

RHODES Melenos Lindos   €€€€€

Lindos, 85107 **Tel** *22440 32222* **Fax** *22440 31720* **Rooms** *12*

Located in the village centre, Melenos is a haven of style and luxury. Its white, village-style rooms, with wooden furniture and comfortable platform beds, are in true Lindian style. There are flower-filled terraces and a fine rooftop restaurant. Closed Nov–Apr. **www.melenoslindos.com**

RHODES Rodos Palace   €€€€€

Leofóros Triádon, Ixia, 85100 **Tel** *22410 25222* **Fax** *22410 25350* **Rooms** *785*

Built in the 1960s and completely renovated in 2001, the Rodos Palace is the island's flagship luxury hotel. Rooms and suites surround a central 20-storey tower with world-class views from the upper floor rooms. Choice of bars and restaurants, indoor and outdoor pools, excellent leisure and business facilities. **www.rodos-palace.gr**

RHODES Rodos Park Suites   €€€€€

Riga Fereou 12, 85100 **Tel** *22410 89100* **Fax** *22410 24613* **Rooms** *60*

The most luxurious hotel in Rhodes New Town, the Rodos Park Suites was opened in 1962 and was completely rebuilt in 2002. Excellent facilities including a fine pool and courtyard, a good restaurant, stylish bar, and rooms with modern facilities including satellite TV and internet access. 24-hour room service. **www.rodospark.gr**

SYMI Les Catherinettes & Marina Studios   €€

Harani, Sými town, 85600 **Tel** *22460 71671* **Fax** *22460 72698* **Rooms** *11*

A lovely old mansion overlooking the harbour (with a traditional taverna at street level). All twin rooms have fridges, high coffered and painted ceilings, tall windows and small balconies with views. Across the narrow alley, the studios and apartments sleep 2–4 and have fully-equipped kitchens and verandas. Closed Oct–Apr. **marina-epe@rho.fortnet.gr**

SYMI Alyki   €€€€

Waterfront, Sými town, 85600 **Tel** *22460 71665* **Fax** *22460 71655* **Rooms** *15*

A delightful converted ship-owners' mansion right on the waterfront. The bedrooms do not live up to the grandeur of the lobby, but the best have balconies and lovely harbour views. Café-terrace on the water's edge, and a bathing ladder gives access to the clean water of the harbour. Quiet location. Closed Oct–Apr **www.hotelaliki.gr**.

TELENDOS On the Rocks   €€

Chochlakas, Télendos, 85200 **Tel** *22430 48260* **Fax** *22430 48261* **Rooms** *3*

Three comfortable rooms with balconies overlooking the neighbouring island of Kálymnos. On the Rocks is situated right on the beach and has an excellent fish restaurant and small friendly bar. Only for those seeking real island peace and quiet. Closed Oct–Apr.

Key to Price Guide *see p302* **Key to Symbols** *see back cover flap*

THE CYCLADES

AMORGOS Lakkí Village

Lakkí Beach, Aigiáli, 84008 **Tel** *22850 73506* **Fax** *22850 73244* **Rooms** *50*

€€€

Delightful village of double rooms, suites, bungalows and family apartments in blue and white split-level buildings surrounding a garden of flowers and organic vegetables (served in the hotel's own taverna). Children's playground and shallow, clean, sandy beach nearby. Shuttle bus to harbour. Closed Nov–Apr. **www.lakkivillage.gr**

AMORGOS Aegialis

On hillside above Aigiáli Village, 84008 **Tel** *22850 73393* **Fax** *22850 73395* **Rooms** *50*

€€€€

This is the smartest hotel in Aigiáli and boasts fine views of the sea and nearby islands from its hillside location. Ideal setting for a peaceful holiday with sandy beaches and village tavernas within walking distance. A good base for exploring Amorgós. **www.amorgos-aegialis.com**

ANDROS Eleni Mansion

Empirikou 9, Chóra, 84500 **Tel** *22820 22270* **Fax** *22820 22294* **Rooms** *8*

€€€€

A fine example of an Andros sea-captain's mansion, converted into a cosy hotel with single, twin and triple rooms. Rooms have tall ceilings and windows, wood floors, dark furniture and pink soft furnishings. Only the upstairs rooms have views, one room has its own roof terrace, and all have a small fridge. Closed Oct–Apr. **www.elenimansion.gr**

ANDROS Paradise Andros

Chóra, Andros, 84500 **Tel** *22820 22187* **Fax** *22820 22340* **Rooms** *44*

€€€€

This elegantly appointed hotel is set in a gracious Neo-Classical building with an ornate interior of antiques, chandeliers and mirrors. The hotel is favoured by well-off Athenians and is situated on the outskirts of Andros town, 700 metres (2,300 ft) from the beach. Good restaurant and attentive service. Closed Nov–Mar. **www.paradiseandros.gr**

ANTIPAROS Lilly's

Chóra, Antíparos, 84007 **Tel** *22840 61411* **Fax** *22840 28328* **Rooms** *15*

€€€

Tiny Antíparos is where you go to escape the summer bustle of neighbouring Páros and Lilly's is the perfect retreat. Twin and double rooms, self-catering apartments and two cottages sleeping up to four people are set around a garden with palm trees. The Cycladic-style rooms are bright and breezy. Closed Oct–May. **www.lillyisland.com**

FOLEGANDROS Kifines tou Aegaiou

Chóra, 84011 **Tel** *22860 41274* **Fax** *22860 41274* **Rooms** *4*

€€€

Three studios sleeping up to four people and one split-level apartment sleeping up to five, amid fields just outside Chóra. All have basic kitchenettes and are prettily decorated in blues and yellows. Breakfast is served on the terrace and Chóra, 500 metres (1,640 ft) away, has plenty of restaurants, bars and cafés. Closed Nov–Apr. **www.kifines.gr**

FOLEGANDROS Anemomilos

Chóra, 84111 **Tel** *22860 41309* **Fax** *22860 41407* **Rooms** *17*

€€€€€

This immaculate, clifftop complex of cottage apartments has views to match any in the Aegean. Each studio has a traditional stone platform bed and separate sitting area. The views from each balcony are stupendous. There is a small bar beside a circular pool. Closed Nov–Apr. **www.anemomilosapartments.com**

IOS Liostasi Ios

Germanoli, Chóra, 84001 **Tel** *22860 92140* **Fax** *22860 92680* **Rooms** *27*

€€€€

This stylish but affordable little resort is a good choice on budget-conscious Ios. There is a great pool with deck and bar, and all the rooms have balconies or verandas with harbour views. The nightlife of Chóra is just 1 km (0.5 miles) away. No extra charge for children under four sharing with parents. Closed Nov–Apr. **www.liostasi.gr**

KEA Brillante Zoi

Korissia, 84002 **Tel** *22880 22685* **Fax** *22880 22687* **Rooms** *22*

€€€

An attractive small hotel on the way to one of Kéa's better beaches and favoured by a clientele of Athenian weekenders. Rooms are large with colourful tiled bathrooms and a choice of sea and mountain views. Ioulis Village, 5 km (3 miles) away, has bars and tavernas. **www.hotelbrillante.gr**

KEA Keos Katikies

Korissia, 84002 **Tel** *22880 21661* **Fax** *22880 21659* **Rooms** *15*

€€€

Beautiful sunset views from the west-facing rooms of this tiny, friendly hotel. Rooms with verandas or balconies overlook the bay and natural harbour below. The small café-bar, where breakfast and snacks are served, also has a fine panoramic outlook. Closed Oct–Apr. **www.keos.gr**

KYTHNOS Porto Klaras

Loutrá Beach, 84006 **Tel** *22810 31276* **Fax** *22810 31600* **Rooms** *20*

€

Well-appointed, small apartment complex near the beach and Kýthnos's natural hot springs. The springs attract a steady flow of Greek sufferers from arthritis, rheumatism and other ailments, so the rooms have better than usual wheelchair access. Family suites, twins and doubles with sea views. Closed Nov–Apr. **www.porto-klaras.gr**

KYTHNOS Kalypso

Loutrá, 84006 **Tel** *22810 31418* **Fax** *22810 31163* **Rooms** *12*

Set above the village of Loutrá, with its beach and hot springs, the Kalypso offers twin and double-bedded rooms, some with basic self-catering facilities including two-ring cooker and fridge. The building is in traditional Cycladic style, with stone, white plaster and blue woodwork. Tavernas and cafés nearby in Loutrá. Closed Oct–Apr.

MILOS Aeolis Hotel

84801 **Tel** *22870 23985* **Rooms** *12*

Newly-built collection of apartments in Cycladic village-style cottages with wooden balconies, whitewashed terraces and views of the village and the sea. Single, double and family apartments available, each with a fully-equipped kitchenette. Lagada beach is 100 metres (328 ft) away. Closed Oct–Apr.

MILOS Alba

Adámas, 84801 **Tel** *22870 23239* **Fax** *22870 23239* **Rooms** *5*

Tiny, affordable and friendly, with stylish studios. Rooms have tiled floors, queen-sized beds, good bathrooms and a fully equipped mini-kitchen. Each has its own terrace with white sun umbrellas and wide views of the bay. Very good value for money. Closed Oct–Apr.

MILOS Popi's Windmill

Off main square, Trypiti, 84801 **Tel** *22870 22286* **Fax** *22870 22396* **Rooms** *8*

Popi's Windmill is a luxuriously converted windmill sleeping up to five people on three levels, with veranda and views towards Adamas port. There is a separate living room, dining room and kitchen, shared WC and shower room (but no bath) and twice-daily maid service. Closed Sep–Jan.

MYKONOS Pension Matina

Chóra, 84600 **Tel** *22890 23049* **Fax** *22890 26423* **Rooms** *40*

Situated on the outskirts of Mýkonos town, this small hotel complex has all modern conveniences including a good-sized swimming pool and rooms with air conditioning and balconies. Built in typical mock-Cycladic style, it is within walking distance of the restaurants and shops of Chóra. Closed Oct–Apr. **www.pension-matina.gr**

MYKONOS Rohari

Rohari, Chóra, 84600 **Tel** *22890 23107* **Fax** *22890 24307* **Rooms** *60*

This mid-sized hotel on the outskirts of Mýkonos Town has light and breezy rooms, and views overlooking the white roofs of the village and the harbour. Relatively quiet, though some noise from the main road, and just under 1 km (0.6 miles) from the ferry port. Closed Nov–Mar.

MYKONOS Villa Konstantin

Agios Vassílios, Chóra, 84600 **Tel** *22890 26204* **Fax** *22890 26205* **Rooms** *14*

Fine hillside location overlooking Mýkonos Town, with flagstoned paths and a courtyard filled with geraniums and jasmine. A mix of apartments, double and triple rooms, all self-catering with fridge and small cooker. Each room has its own balcony or miniature terrace. Closed Oct–Mar. **www.villakonstantin.com**

MYKONOS Zorzis

Kalogeras, Chóra, 84600 **Tel** *22890 22167* **Fax** *22890 24169* **Rooms** *10*

A charming small hotel with lots of individual style, located in the centre of Mýkonos town but on a quiet pedestrian street. The rooms have luxury en-suite shower and WC with power shower, Louis XV antique beds, beamed ceilings and roof fans. Small terrace on the street and another, more secluded, at rear. **www.zorzishotel.com**

MYKONOS Belvedere

Skoli Kalon Technon, Rohari, Chóra, 84801 **Tel** *22890 25122* **Fax** *22890 25126* **Rooms** *50*

A pleasant boutique hotel with friendly, professional service. Rooms are in whitewashed buildings with green-painted shutters, those at the front look out over the roofs of Mýkonos Town to the sea. Good restaurant and bar, modern facilities including satellite TV. Quiet at night, but some traffic noise from main road at rear. **www.belvedere.com**

MYKONOS Cavo Tagoo

Thesi Tagkou, Mýkonos, 84600 **Tel** *22890 23692* **Fax** *22890 24923* **Rooms** *80*

Located on a hillside about 2 km (1 mile) from the centre of Mýkonos Town, Cavo Tagoo stands in splendid isolation away from the bustle of the island capital, but within a short drive of its shops, restaurants and nightlife. Built in island style, with white walls and blue woodwork. Shady eucalyptus trees and fine views. Closed Nov–Mar. **www.cavotagoo.gr**

MYKONOS Mykonos Theoxenia

Chóra, 84600 **Tel** *22890 22230* **Fax** *22890 23800* **Rooms** *52*

Stylish and colourful hotel operated by the Louis group. Designer rooms with luxury bathrooms and balconies over-looking the bay or the huge pool. Located beside the town's famous row of windmills, five minutes walk from the centre. Excellent service, good food and pleasant atmosphere. Closed Nov–Mar. **www.mykonostheoxenia.com**

MYKONOS Princess of Mýkonos

Agios Stéfanos, 84600 **Tel** *22890 24713* **Fax** *22890 23031* **Rooms** *38*

Situated on one of the island's most exclusive beaches, 4 km (2.5 miles) from Mýkonos town, this hotel is chic and stylish. Its small size allows for personal, attentive service and, like most Mýkonos hotels, it is designed in mock-Cycladic style. The pool is on the small side, but the beach is nearby. Closed Nov–Mar. **www.princessofmykonos.gr**

MYKONOS Semeli

Rohari, Chóra, 84600 **Tel** *22890 27466* **Fax** *22890 27467* **Rooms** *62*

A very attractive hotel with the atmosphere of a private manor house. Rooms and public areas are beautifully light and breezy, decorated in off-white and pale green shades. Four-poster beds in some rooms. Large pool in a walled courtyard full of shrubs and potted flowers. 500 metres (1,640 ft) from the beach. Closed Oct–Apr.

MYKONOS Mýkonos Grand

Agios Ioánnis, 84600 **Tel** *22890 25555* **Fax** *22890 25111* **Rooms** *107*

One of the very best hotels on Mýkonos, the Mýkonos Grand overlooks Agios Ioánnis Bay. Two-storey, village-style building with whitewashed walls, arches and terracotta pots full of flowering shrubs. Rooms are light and airy and very well-furnished. Excellent service and equally good food. Closed Nov–Mar. **www.mykonosgrand.gr**

NAXOS Castro

Kástro, Chóra, 84300 **Tel** *22850 25201* **Fax** *22850 25200* **Rooms** *2*

Book well in advance to stay in one of the two apartments in this old whitewashed building inside the ramparts of the old Venetian town. Each has basic self-catering facilities and is furnished with antiques. Shared sun terrace and views over the roofs of the old town. Closed Nov–Apr.

NAXOS Chateau Zevgoli

Bourgos, Chóra, 84300 **Tel** *22850 25201* **Fax** *22850 25200* **Rooms** *9*

Wonderfully atmospheric with whitewashed rooms in an old island home that overlooks a tiny flower-filled courtyard. The rooms are small but cosy and furnished with island antiques – stripy rugs and old wooden beds. Rooftop terrace with super views across the harbour. Closed Nov–Mar.

NAXOS Grotta

Náxos town, 84300 **Tel** *22850 22101* **Fax** *22850 22000* **Rooms** *40*

Beautifully situated on a headland north of the Venetian Kástro, this hotel is built in the same style as the medieval castle and enjoys good views of the whitewashed town and the sea. Rooms are neutrally decorated, with traditional striped rugs and bedspreads, and have satellite TV and fridge. **www.hotelgrotta.gr**

NAXOS Kavos

Agios Prokopios, 84300 **Tel** *22850 23355* **Fax** *22850 26031* **Rooms** *19*

Located on one of the best beaches in the Aegean, Kavos offers a collection of excellent value studios, apartments and suites with self catering facilities. Rooms are in simple white cottages, surrounded by greenery, and each has painted furniture and galleried bed spaces. Closed Nov–Mar. **www.kavos-naxos.com**

PAROS Dina

Main shopping street, Paroikia, 84400 **Tel** *22840 21325* **Fax** *22840 23525* **Rooms** *8*

A small and friendly establishment, centrally placed on a bustling street in the heart of Paroikiá. Rooms have wrought iron balconies and fridges and some look out over the village rooftops or across to the old church of Agia Triada. Affordable comfort, plain clean rooms. Closed Nov–Apr. **www.hoteldina.com**

PAROS Anthippi

Paroikiá, 84400 **Tel** *22840 21601* **Fax** *22840 21601* **Rooms** *9*

This pretty guesthouse is a bargain, with rooms decorated in beachcomber style – shells, model boats, driftwood and other finds from the sea – with mural paintings and beamed ceilings. Lots of greenery on the garden terrace and a small bar by the pool. Closed Oct–Apr. **www.anthippi.com**

PAROS Heaven Naoussa

Náousa, 84401 **Tel** *22840 51549* **Fax** *22840 51575* **Rooms** *9 (5 Suites, 4 rooms)*

Not far from the centre of Náousa, this complex of rooms in traditional-style stone cottages surrounds a walled pool. Rooms are light, airy, prettily decorated and furnished. Two self-catering apartments are also available by the week. Harbour-side shops and restaurants are just a couple of minutes' walk away. Closed Sep–May. **www.heaven-naoussa.gr**

PAROS Petres

Agios Andréas, Náousa, 84401 **Tel** *22840 52467* **Fax** *22840 52759* **Rooms** *16*

The rooms here have high, beamed ceilings, whitewashed walls, big brass beds and antique wardrobes. Those at the front look out over the bay, others over lush gardens or a large pool surrounded by greenery. Snack bar, tennis court and gym. Closed Oct–Apr. **www.petres.gr**

PAROS Astir of Páros

Kolympíthres, 84400 **Tel** *22840 51976* **Fax** *22840 51985* **Rooms** *57*

Luxury is the hallmark of this village-style resort next to one of Paros's best beaches. The resort is set in lush tropical grounds full of bougainvillea and palm trees. There is a children's pool, sauna, putting green, tennis courts, choice of bars and restaurants, art gallery, shuttle bus to Paroikia and even a heli-pad. Closed Nov–Apr. **www.astirofparos.gr**

PAROS Lefkes Village

Léfkes, 84400 **Tel** *22840 41827* **Fax** *22840 42398* **Rooms** *20*

Great location in the hills of Páros – some way from the beach but with a gorgeous pool and magnificent views to compensate. Neo-Classical style villas with spacious rooms, each with balcony, and four-poster beds. Set in lush gardens and with its own quirky museum of island life. Closed Oct–Apr. **www.lefkesvillage.gr**

SANTORINI Keti

€€€

Thíra, 84700 **Tel** *22860 22324* **Fax** *22860 22380* **Rooms** *7*

Just off the steep steps that lead from Thira Town to the harbour below, Hotel Keti shares the same stunning crater view as the much more costly boutique hotels elsewhere on the island. The whitewashed rooms have arched ceilings, marble floors and are simply furnished and decorated. Closed Oct–Apr. **www.hotelketi.gr**

SANTORINI Chelidonia

€€€€

Oía, 84702 **Tel** *22860 71287* **Fax** *22860 71649* **Rooms** *8*

Collection of eight traditional "cave-houses" restored by the owners since the early 1980s and now comprising a choice of eight houses sleeping up to four. Each has a private balcony and all are decorated in dazzling white, blue woodwork and splashes of yellow. Views to match any in Oía. **www.chelidonia.com**

SANTORINI Katikies

€€€€

Oía, Santoríni, 84700 **Tel** *22860 71401* **Fax** *22860 71129* **Rooms** *27*

Stunning views from the horizon pool make this the best hotel on Santoríni. The elegantly-appointed, traditional cave-rooms are whitewashed, each has a private terrace, some terraces also have whirlpools. Fine restaurant and attentive service. Closed Nov–Apr. **www.katikies.com**

SANTORINI Chromata

 €€€€€

Imerovígli, 84700 **Tel** *22860 24850* **Fax** *22860 23728* **Rooms** *26*

Dazzlingly colourful, with rooms decorated in hot pink or cool turquoise, and mock-leopard and zebra prints. A clear plexiglass platform spans the pool, and a table can be set on it for a floodlit dinner. Rooms are set in tiers, with small terraces. Under the same management as Katikies in Oia. Closed Nov–Mar. **www.chromata-santorini.com**

SANTORINI Artemis Villas

 €€€€€

Imerovígli, 84700 **Tel** *22860 22712* **Fax** *22860 23638* **Rooms** *10*

Friendly, affordable complex of Santorinian *skaftes* (cave houses), on the lip of the famous caldera and with the requisite soaring views over the crater and its islands. Rooms are whitewashed and have private balconies, furnishing are a mix of antique and modern. The sunset views are as good as any on Santorini. **www.artemisvillas.gr**

SANTORINI Esperas

 €€€€€

Oía, 84702 **Tel** *22860 71088* **Fax** *22860 71613* **Rooms** *20*

Superb sunset views and bedrooms tunneled into the cliffside are features of this wonderful collection of studios, suites and villas. All en-suite, but shower rooms are small; each unit has a fridge and kitchenette. Apartments have gallery beds, villas have two bedrooms. An excellent pool with shady grottoes. Closed Nov–Mar. **www.esperas.gr**

SANTORINI Notos Therme & Spa

 €€€€€

Vlycháda, 84703 **Tel** *22860 81115* **Fax** *22860 81266* **Rooms** *28*

Marvellous hillside location with views over the vineyards of Santoríni and its spectacular crater. Rooms are luxuriously but simply appointed in a mix of double or twin, superior double, junior and senior suites. Lovely poolside bar and outstanding programme of health and beauty therapeutic treatments. Closed Oct–May. **www.snotos.com**

SANTORINI Perivolas

 €€€€€

Oía, 84702 **Tel** *22860 71308* **Fax** *22860 71309* **Rooms** *20*

This is one of the boutique hotels which put Oía on the style map and it is immaculately designed and decorated with service and facilities to match. Some rooms have private whirlpool or plunge pool and there is a superb infinity pool overlooking the caldera. Gourmet restaurant. Closed Nov–Mar. **www.perivolas.gr**

SANTORINI Sun Rocks

 €€€€€

Firostefani, 84700 **Tel** *22860 23241* **Fax** *22860 23991* **Rooms** *17*

Strictly for romantic couples, Sun Rocks is one of the classiest operations on Santoríni, with four-poster beds in white-vaulted rooms, a pretty pool with breathtaking views, a pleasant bar from which to relish the scenery, and two excellent restaurants. Service to match, but it is a steep 150-step climb to the car park. Closed Nov–Mar. **www.sunrocks.gr**

SANTORINI Zannos Melathron

€€€€€

Pýrgos, 84700 **Tel** *22860 28220* **Fax** *22860 28229* **Rooms** *19*

Superbly opulent hotel housed in a 19th-century mansion and retaining original features such as fine murals and painted ceilings. Located in the quiet village of Pýrgos, the hotel enjoys views of the surrounding vineyards from its terraces. Large rooms and suites. Caviar on the menu and fine cigars and vintages in the bar. Closed Nov–Apr. **www.zannos.gr**

SIFNOS Noble Apartments

€

Kástro, Sífnos, 84003 **Tel** *62589 6953* **Fax** *13105 455140* **Rooms** *1*

An authentic Sifniot village home, sleeping up to 3 people. Whitewashed stone walls, olive wood furniture, striped island fabrics and small but well-equipped kitchen. Available only by the week, but a unique island experience and good value for money. Must be booked well in advance. **www.nobleapartments.com**

SIFNOS Aperanto

€€

Fáros, Sífnos, 84003 **Tel** *22840 71473* **Fax** *22840 71473* **Rooms** *9*

Fabulous and affordable guesthouse with a great location in a charming village. Rooms are attractively furnished with iron or platform beds and decorated with island ceramics and some antique furniture. Peaceful village with a handful of tavernas. The guesthouse is a short walk away from Apokoftos beach. Closed Oct–Apr.

Key to Price Guide *see p302* **Key to Symbols** *see back cover flap*

SIFNOS Petali Village
⊞▦▮▯▨ €€€

Ano Petali, Apollonía, 84003 **Tel** *22840 33024* **Fax** *22840 33391* **Rooms** *23*

Surpisingly modern within, belying its village-style architecture. Rooms are in plain, whitewashed cottages with luxury bathrooms and modern, neutral furnishings. Facilities include satellite TV, and a pleasant terrace restaurant. Located on the outksirts of Apollonía, some way from the beach. **www.hotelpetali.gr**

SYROS Apollonos
▦▮▨ €€€€

Apóllonos 8, Ermoupoli, 84100 **Tel** *22810 81387* **Fax** *22810 83082* **Rooms** *3*

Extremely plush accommodation in a lovely Sýros mansion with wrought-iron balconies looking straight out over the harbour. Prettily painted ceilings, gleaming chandeliers, polished wood floors and antique dressers recreate the 19th-century heyday of Ermoúpoli – once the wealthiest city in the islands. Book well ahead. **stathopoulosg@syr.forthnet.gr**

TINOS Voreades
▦▮▯▨ €€

Foskolou 7, Chóra, 84200 **Tel** *22830 23845* **Rooms** *12*

With its whitewashed walls, arched doorways, exposed patterned stonework and flagstone floors, this little guesthouse is typically Tinos. Choice of single and double rooms and one two-bedroom suite, all with fridge and private balcony. Small café-bar and views of the sea and the nearby islands. Closed Nov–Feb. **www.voreades.gr**

TINOS Carlo
⊞▦▮▯▨ €€€

Agios Ioánnis, 84200 **Tel** *22830 24159* **Fax** *22830 24169* **Rooms** *24*

This attractive small bungalow complex is in the familiar Cycladic style and sits on a hillside with fine views. Rooms have modern facilities including fridge, each has a balcony with views. Facilities include broadband internet access. Shuttle bus service to the island port. Closed Nov–Mar. **www.carlobungalows.com**

CRETE

AGIA ROUMELI Tara-Calypso
▦▮▯▨ €

Agía Rouméli, 73011 **Tel** *28250 91231* **Fax** *28250 91431* **Rooms** *30*

Most people pass straight through Agía Rouméli after walking the Samariá Gorge. For those who want to linger a little longer, this small hotel is something of a bargain, with clean, simple rooms close to the beach, some of them with views out over the Libyan Sea. No restaurant, but plenty of tavernas in the village. Closed Nov–Mar.

AGIOS NIKOLAOS Minos Beach Art 'Otel
⊞▮▦▮▯▨ €€€€€

Ammoudí, 72100 **Tel** *28410 22345* **Fax** *28410 22548* **Rooms** *180*

A medium-sized complex of rooms and bungalows just outside Agios Nikólaos. The complex boasts a large pool, semi-private beaches, lush grounds with a collection of specially commissioned sculptures, and super views of the lovely Gulf of Mirabello. Seafront suites have private pools. Good restaurant. Closed Nov–Mar. **www.bluegr.com**

AGIOS NIKOLAOS St Nicolas Bay
⊞▮▦▮▯▨ €€€€€

Agios Nikólaos, 72100 **Tel** *28410 25041* **Fax** *28410 24556* **Rooms** *108*

Magnificent complex of bungalows and luxury suites – some with private pool – on an enviable seaside site just outside Agios Nikólaos. Landscaped gardens full of citrus and olive trees surround the buildings and the hotel has its own virtually private beach. Watersports and choice of 8 restaurants and bars. Closed Nov–Mar. **www.stnicolasbay.gr**

ARCHANES Villa Arhanes
▨⊞▮▯▨ €€€

Ano Archánes, 70100 **Tel** *28103 90770* **Fax** *28103 90778* **Rooms** *9 in 6 apartments*

This 19th-century farmhouse offers half-board or bed-and-breakfast accommodation. Rooms are cosy with traditional furnishings and antiques. There is a shared pool and views over the rolling vineyards. This professionally-run villa is a short drive from Knossos and Irakleio.

CHANIA Nostos
▮▨ €€€

Zamnbeliou 46, 73113 **Tel** *28210 94743* **Fax** *28210 94740* **Rooms** *12*

This charming, brightly painted little hotel sits on a traffic free lane in the heart of Chaniá's old quarter. From its shaded roof terrace there are fine views out to sea and to the peaks of the White Mountains. Studio rooms have balconies and gallery beds. One block back from the harbour. **www.nostos-hotel.com**

CHANIA Palazzo di Pietro
€€€

Agion Déka 13, 73100 **Tel** *28210 20410* **Fax** *28210 58338* **Rooms** *7*

Housed in an 800-year-old townhouse in Chaniá's most atmospheric, traffic-free street. Lovely studios and apartments with mini-kitchen, stone fireplaces, four-poster beds and designer bathrooms. No views, but on the plus side, none of the noise that you get when you stay on the waterfront. **www.palazzodipietro.com**

CHANIA Pandora Suites
▦▮▯▨ €€€

Lithinon 27–29, 73132 **Tel** *28210 43588* **Fax** *28210 57864* **Rooms** *12*

Eight double/twin rooms, plus four apartments, perched high above Chaniá harbour and with panoramic views of the White Mountains. Rooftop terrace, pretty interior courtyard with tropical plants, helpful and attentive staff, and the cafés and restaurants of the harbourfront just a short walk away. Closed Nov–Mar. **www.pandorahotel.gr**

CHANIA Amfora
目 ⓘ ☒ €€€€

Parodos Theotokopoulou 20, 73131 **Tel** *28210 93224* **Fax** *28210 93226* **Rooms** *21*

This 13th-century Venetian mansion has tastefully appointed rooms and a charming roof terrace overlooking Chaniá's picturesque harbour. Situated in the heart of Chaniá's old quarter, on a mostly traffic free street just a few steps from the waterfront with its numerous restaurants, shops and bars. **www.amphora.gr**

CHANIA Villa Andromeda
目 ☲ ⓘ ☒ €€€€

Venizelou 150, 73133 **Tel** *28210 28300* **Fax** *28210 28303* **Rooms** *8*

Oozing period dignity, this 19th-century building was once the German consulate. There are eight plainly-furnished suites. The grand sitting rooms, by contrast, glow with yellow stucco. Large pool and terrace outside. Located 2 km (1 mile) from the harbour front. Closed Nov–Mar. **www.villandromeda.gr**

CHANIA Casa Delfino
目 Ⓨ ⓘ ☒ €€€€€

Theofanou 9, 73100 **Tel** *28210 87400* **Fax** *28210 96500* **Rooms** *22*

One of the most stylish and luxurious boutique hotels in Crete, with beautifully designed and furnished rooms and suites surrounding a fountain courtyard. Originally an aristocrat's mansion, Casa Delfino has gleaming marble floors and serves truly splendid buffet breakfasts. The penthouse suite has harbour views. **www.casadelfino.com**

CHANIA Metohi Kindelis
☲ ☲ ⓘ €€€€€

Perivólia, Chaniá, 73100 **Tel** *28210 41321* **Fax** *28210 43930* **Rooms** *3 apartments*

Each of these superb villa apartments within a large Venetian farmhouse has its own pool, set in huge, lush gardens. Inside are cool marble floors, wood-burning fireplaces for cooler evenings, fully equipped kitchens and luxurious bathrooms. Extras include satellite TV and DVD players. **www.metohi-kindelis.gr**

CHANIA La Perle Resort Hotel and Health Spa Marine
☲ ☆ 目 Ⓨ ⓘ ☒ €€€€€

Stavrós, Akoritíri, 73100 **Tel** *28210 39400* **Fax** *28210 39650* **Rooms** *126*

Opened in 2002, the La Perle offers a range of health and beauty programmes, good facilities for children, comfortable rooms and suites, bars, a good restaurant, and an indoor heated pool as well as an outdoor pool. The nearest beach is at Stavros, 4 km (2.5 miles) from the hotel. Closed mid-Oct–Mar. **www.perle-spa.com**

CHERSONISOS Creta Maris
☲ ☆ 目 Ⓨ ⓘ ☒ €€€€

Chersónisos, 70014 **Tel** *28970 27000* **Fax** *28970 22130* **Rooms** *180*

This is one of the area's longest established luxury resort complexes, with comfortably appointed bungalows in well-maintained grounds. Facilities and service are excellent, with a choice of bars, restaurants and activities. The resort even has its own outdoor theatre and open-air cinema. Good facilities for children. Closed Nov–Mar. **www.maris.g**

CHERSONISOS Galaxy Villas
☲ ☆ 目 Ⓨ ⓘ ☒ €€€€

Agiou Konstantinou, Koutouloufari, 70014 **Tel** *28970 22910* **Fax** *28102 11211* **Rooms** *53*

A low-rise resort of apartments, built in traditional style using natural materials. The resort is surrounded by lawns and palm trees and is only 1 km (0.6 miles) from Chersónisos. Each villa has a private veranda or patio with sea or mountain views. Facilities include children's playground, billiards and TV room. Closed Nov–Mar. **www.galaxy-villas.com.gr**

ELOUNTA Eloúnda Island Villas
ⓘ ☒ €€

Kolokytha, 72053 **Tel** *28410 41274* **Fax** *28410 41276* **Rooms** *30 in 10 apartments*

Modern split-level apartments with basic self-catering facilities on their own island, connected with mainland Eloúnta by a bridge. Attractive terraces with fantastic views out to sea and a tiny, virtually private beach, as well as a tennis court. The bars and tavernas of Elounta are a 5–10 minute walk away. Closed Nov–Mar. **www.eloundaisland.gr**

ELOUNTA Eloúnda Beach Hotel
☲ ☲ ☆ 目 Ⓨ ⓘ ☒ €€€€€

Eloúnta, 72053 **Tel** *28410 63000* **Fax** *28410 41373* **Rooms** *258*

The "grande dame" of Greek resorts, the Eloúnda Beach offers luxurious villas – some with private pools – scattered around a picturesque headland and a private beach. Excellent water sports, fine dining, very attentive service. Closed Nov–Mar. **www.eloundamare.gr**

ELOUNTA Eloúnda Mare Hotel
☲ ☆ 目 Ⓨ ⓘ ☒ €€€€€

Eloúnta, 72053 **Tel** *28410 41102* **Fax** *28410 41307* **Rooms** *215 suites and villas*

One of Greece's most luxurious resort hotels, with a mix of suites and villas set in lush grounds. Facilities include a choice of bars and restaurants, watersports and in-room extras such as satellite TV and DVD players. There is also a golf course and other activities nearby. Excellent service. Closed Nov–Mar. **www.eloundamare.gr**

IERAPETRA Eden Rock
☲ ☲ ☆ 目 Ⓨ ⓘ ☒ €€€

Agia Fotiá, 72200 **Tel** *28420 61723* **Fax** *28420 61734* **Rooms** *100*

Very comfortable, mid-range hotel on a quiet beach, close to Galíni village and 15km (9 miles) west of Ierápetra Town. Accommodation is a mix of rooms with balconies, self-catering studios and apartments, and one grand villa. Rooms have satellite TV and the hotel is just a short walk from the beach. **www.edenrock.gr**

IRAKLEIO Lato Hotel
目 Ⓨ ⓘ ☒ €€€€

Epimenidou 15, 71202 **Tel** *28102 28103* **Fax** *28103 34955* **Rooms** *58*

Formerly comfortable but undistinguished, the Lato – in the centre of the old town – has been reborn as Irákleio's first boutique hotel. Its rooms are stylish with balconies, terraces or glassed-in mini-conservatories, satellite TV and internet access. There is also a roof garden, mini-gym and sauna. **www.lato.gr**

Key to Price Guide *see p302* **Key to Symbols** *see back cover flap*

IRAKLEIO Out of the Blue Capsis Elite Resort 🔲🏖️🏃🍴🍸🅿️🛏️ €€€€€
Agía Pelagía, 71000 **Tel** *28108 11112* **Fax** *28018 11314* **Rooms** *465*

This huge, self-contained luxury complex stands on its own promontory and combines two hotels and a conference centre. A full range of watersports and other facilities and activities are on offer, including its own zoo. Rooms range from standard doubles to large suites with private pool, butler and maid service. Private beach. Closed Nov–Mar. **www.capsis.gr**

IRAKLEIO Villa Helidona 🏖️🛏️ €€€€€
Episkopí, 70008 **Tel** *69726 23671* **Rooms** *4*

One of the few villas with pools that can be rented independently by the week. Villa Helidona sits on the outskirts of a small market town among fields and vineyards, not far from Crete's capital. There is a large pool and a fully-equipped kitchen. Shops and restaurants are nearby. Closed Nov–Mar. **www.villahelidona.com**

KASTELLI KISSAMOU Mirtilos 🏖️🏃🍸🛏️🅿️ €€
Kastélli, 73400 **Tel** *28220 23079* **Fax** *28220 23079* **Rooms** *35*

Comfortable if a little bland, this hotel offers well-appointed rooms in self-catering suites or apartments. Rooms have satellite TV and balcony or veranda with a choice of sea, mountain or garden view. There is a large pool surrounded by lawns and palm trees. Internet facilities available. Closed Nov–Mar. **www.mirtilos.com**

LASITHI Zeus's House 🏖️🛏️🅿️ €€€€
Agios Konstantínos, 72052 **Tel** *28102 22218* **Fax** *28102 288240* **Rooms** *2*

In the middle of the Lasíthi plateau, Zeus's House offers apartment rooms in a beautiful restored traditional house with flagstone floors and stone arches. The rooms are decorated with local antiques and sleep up to four people. Good pool in a large, verdant garden. Closed Nov–Mar. **www.cretanvillas.gr**

LOUTRO Hotel Porto Loutró 🍴🍸🛏️🅿️ €
Loutró, 73011 **Tel** *28250 91433* **Fax** *28250 91091* **Rooms** *45*

Hotel Porto Loutró is in the heart of a tiny village that is only accessible by boat. The rooms are elegantly simple but very comfortable. All have a terrace or balcony; the best have views over the bay. This is a perfect base for exploring the White Mountains. Closed Nov–Mar. **www.hotelportoloutro.com**

LOUTRO The Blue House 🍴🍸🛏️🅿️ €
Loutró, 73011 **Tel** *28250 91337* **Fax** *28250 91127* **Rooms** *15*

For an affordable stay in Loutró, this prettily decorated, modest guesthouse is one of the best choices. It has its own taverna and bar, and most of the rooms have balconies looking out over the bay and the White Mountains. Particularly handy as an overnight stop. Closed Nov–Mar.

MAKRYGIALOS Aspros Potamos 🍴🛏️🅿️ €
Aspros Potamós, 72055 **Tel** *28430 51694* **Fax** *28430 52292* **Rooms** *17*

Delightful, simple stone cottages with solar-powered electricity and stone fireplaces in the picturesque gorge of the Aspros Potamós, just inland from Makrýgialos Beach. Basic self-catering facilities, but plenty of restaurants and tavernas nearby. This is the perfect place for a peaceful holiday. **www.asprospotamos.com**

MAKRYGIALOS White River Cottages 🏖️🛏️🅿️ €€€
Aspros Potamós, 72055 **Tel** *28430 51120* **Fax** *28430 51120* **Rooms** *17*

A 15-minute walk from the beach, bars and restaurants of Makrýgialos, this village of little stone houses surrounds a small swimming pool and is in turn surrounded by olive groves, pines and rugged hillsides. Simple, stylish and peaceful, with self-catering facilities. Closed Nov–Mar. **wriver@otenet.gr**

MALIA Malia Studios 🏖️🍴🍸🛏️🅿️ €€
Stalida Coastal Road, 70007 **Tel** *28970 31655* **Fax** *28102 13378* **Rooms** *9*

Small, comfortable self-catering hotel with a mix of studios, one-bedroom apartments and larger split-level apartments, all with kitchenette. The hotel also has a snack bar and there is a mini market nearby. Close to the nightlife and watersports of Mália and 150 metres (492 ft) away from the beach. Closed Nov–Mar. **www.malistudioshotel.com**

PALAIOCHORA Hotel Rea 🍴🍸🛏️ €€
Antoníou Peraki, 73001 **Tel** *28230 41307* **Fax** *28230 41605* **Rooms** *14*

Small, friendly, family-run hotel in the centre of peaceful Palaiochóra, about a five-minute walk from the village's long, sandy beach. Breakfast and cold drinks are served on a shady, flower-decked terrace. This is an unpretentious little place offering good value for money. Closed Nov–Mar.

RETHYMNO Mythos Suites Hotel 🏖️🍸🛏️ €€€
Plateía Karáoli 12, 74100 **Tel** *28310 53917* **Fax** *28310 51036* **Rooms** *15*

The most tranquil little haven in Réthymno, hidden away in a tiny back alley. The friendly and helpful owners have converted several old buildings into a delightful hotel, with cool, stylishly-furnished bedrooms set around a pretty courtyard and small pool – just big enough for an afternoon dip. Closed mid-Nov–mid-Mar. **www.mythos-crete.gr**

RETHYMNO Palazzino di Corina 🏖️🍴🍸🛏️🅿️ €€€€
Dambergi 7–9, 74100 **Tel** *28310 21205* **Fax** *28310 21204* **Rooms** *21*

An excellent recent addition to Réthymno's hotel portfolio. Inside the converted Venetian building are stylish, air-conditioned suites – some of which are split-level – and most of which have their own balconies. Public areas are graced by stone arches, columns and antique furniture. There is a courtyard with a pool. **www.corina.gr**

RETHYMNO Palazzo Vecchio
🖥 🍸 🔟 €€€€

Iroon Polytechníou/Melissinou, 74100 **Tel** *28310 35351* **Fax** *28310 25479* **Rooms** *23*

New meets old in this 15th-century townhouse which has been converted into a stylishly grand modern hotel with in-room facilities including satellite TV, mini-kitchen and full-sized bathrooms (a rarity in most historic hotels in Réthymno) Pleasant courtyard bar with fountain, attentive service, central location. Closed Nov–Mar. **www.palazzovecchio.gr**

RETHYMNO Avli Suites Hotel
🖥 🍸 🔟 €€€€€

Xanthoudidou 22, 74100 **Tel** *28310 58250* **Fax** *28310 58255* **Rooms** *7*

Seven colour-themed suites in an old Venetian mansion, above a good restaurant. The hotel opened in 2005 and the rooms have modern facilities including satellite TV and internet access. There are great views from the roof terrace, which also has a whirlpool big enough for ten people. Closed Nov–Mar. **www.avli.gr**

SPILI Green Hotel
🗐 €€

Spíli, 74100 **Tel** *28320 22225* **Fax** *28320 22225* **Rooms** *11*

This small, simple hotel, in the centre of Spili, offers basic twin-bedded rooms with en-suite shower and WC, and balconies that overlook the main square of the village. In summer, the hotel is made colourful by huge tubs of geraniums. Services include sauna, massage and aromatherapy. Closed Nov–Mar. **www.greenhotel.gr**

SPILI Villa Panorama
🗐 🔟 🖼 €€€

Agia Paraskevi, Kerames, 74053 **Tel** *21080 49602* **Fax** *21080 49602* **Rooms** *5 in 2 apartments*

This sturdy stone house has sweeping views over the Libyan Sea from its perch high above the south coast of Crete. Each apartment has a large living room, fully-equipped kitchen (with washing machine), and simply decorated bedrooms. There are shops and tavernas nearby at Spíli, Agia Fotiá or Kerames. **www.villapanorama-crete.gr**

STALIDA Villa Anna
🖥 🍸 🔟 🖼 €

PO Box 29, Stalida, 70014 **Tel** *28970 31506* **Fax** *28970 31985* **Rooms** *16*

This small apartment complex offers excellent value for money, with a large pool and in-room facilities including mini-bar, safe and refrigerator. There is a snack bar, pool bar and breakfast room, and some rooms have sea or mountain views while others overlook the pool and garden. Closed mid-Oct–Apr. **www.villamary-anna.gr**

VLATOS Milia Traditional Settlement
🗐 🍸 🔟 🖼 €€

Vlátos, 73012 **Tel** *28210 46774* **Fax** *28220 51569* **Rooms** *14*

Perched high in the mountains of Crete's wild west, this village of traditional stone cottages is a comfortable place to stay, with simple rooms featuring stone floors and old wood furniture. The restaurant serves great traditional food and local wine. The surrounding scenery is stunning. **www.milia.gr**

ATHENS

AIRPORT Sofitel Athens Airport
📶 🏊 🖥 🍸 €€€€€

Eleftherios Venizélos International Airport, 19019 **Tel** *21035 44000* **Fax** *20135* **Rooms** *345*

The most convenient overnight stop for those flying in late or leaving early, and used mainly by business travellers. Facilities include indoor pool, gym, restaurants, a bar, and business centre. All rooms have internet connection. Good value for money for leisure travellers too. Three non-smoking floors. **www.sofitel.com**

EKALI Life Gallery
📶 🏊 🖥 🍸 🔟 🖼 €€€€€

Thisseos 103, 14565 **Tel** *21062 60400* **Fax** *21062 29353* **Rooms** *30*

This ultra-stylish boutique hotel in Ekali, northwest of the centre, is a member of the Small Luxury Hotels consortium. Its rooms and suites have been individually designed and all have state of the art facilities. The hotel also has two pools and boasts a very good restaurant. Book well in advance. **www.bluegr.com**

KIFISSIA Kefalari Suites
📶 🖥 🍸 🔟 €€€€€

Pendelis 1, 14562 **Tel** *21062 33333* **Fax** *21062 33330* **Rooms** *13*

Quirky, colourful and luxurious, with themed suites in an eccentric 19th-century mansion. Suites have mini-kitchens, luxury bathrooms, and modern facilities including wireless internet connection. Most have their own balcony or veranda. The shared rooftop terrace features a whirlpool tub. **www.kefalarisuites.gr**

KIFISSIA Pentelikon
📶 🏊 🖥 🍸 🔟 🖼 €€€€€

Deligiannis 66, 14562 **Tel** *21062 30650* **Fax** *21080 10314* **Rooms** *101*

An opulent hotel in a grand Neo-Classical palace in fashionable Kefalari, on the outskirts of Kifissiá. Room facilities include PC and fax modem, internet access, satellite TV and mini-bar. The hotel is set in luxurious gardens and boasts one of Greece's few Michelin-starred restaurants, Vardis. **www.pentelikon.gr**

KOLONAKI Saint George Lycabettus
📶 🏊 🖥 🍸 🔟 🖼 €€€€€

Kleomenous 2, 10675 **Tel** *21072 90711* **Fax** *21072 90439* **Rooms** *162*

This excellent hotel is a well-kept secret and can be good value for money in off-peak season. Located beneath Lykavittós Hill at the top end of Kolonáki, its rooftop pool terrace has super views, as does Le Grand Balcon restaurant. The service here is old-fashioned and attentive. The main sights are within walking distance.

AGONISSI Grand Resort Lagonissi

Kilometre 40, Athens-Soúnio Highway, 19010 **Tel** *22910 76000* **Fax** *22910 24514* **Rooms** *289*

The grounds of this luxurious resort cover more than 120 acres of beach and gardens and offer a choice of rooms, suites with private beach decks and villas with heated pools. Facilities include whirlpool spas, DVD players and nine restaurants and bars. Limousine, yacht and executive jet charter service available. Closed Nov–Mar. **www.grandresort.gr**

OMONIA The Alassia (City Boutique Hotel)

Sokrátous 50, Omònia Square **Tel** *21052 7400* **Fax** *21052 74029* **Rooms** *82*

Modelled on the Italian style, with marble walls and columns, this hotel is ideal for exploring the city on foot. Ideal location to the tube station and all amenties. Acropolis, Sýntagma and Pláka are all a short walking distance away. **www.thealassia.com.gr**

PLAKA Kimon

Apóllonos 27, 105-57 **Tel** *21033 14658* **Fax** *21032 14203* **Rooms** *14*

Much the cheapest hotel in this part of Pláka, the Kimon - if you overlook a total lack of communal amenities - is a good option. The rooms are all en suite, painted cheerfully and come complete with mock-antique furniture (such as iron bedsteads), as well as TV and telephones.

PLAKA Acropolis View

Vebster 10, 11742 **Tel** *21092 17303* **Fax** *21092 30705* **Rooms** *32*

Comfortable hotel conveniently close to the Acropolis and Pláka, but far enough away to escape most of the late night noise. Some rooms have a view of the Acroplis, most have balconies and there is a roof garden. Affordable and competently managed. **www.acropolisview.gr**

PLAKA Art Gallery

Erechthiou 5, 11742 **Tel** *21092 38376* **Fax** *21092 33025* **Rooms** *21*

Affordable, friendly, family-run pension on a quiet street within easy walking distance of the Acropolis and Pláka. Some rooms have balconies from which you can see the Parthenon and public areas are decorated with original paintings. Single, twin and triple rooms available. Helpful, Greek-Canadian management.

PLAKA Central Hotel

Apóllonos 21, 10557 **Tel** *21032 34537* **Fax** *21032 25244* **Rooms** *84*

One of the capital's more affordable yet stylish places to stay with glass and marble interior and designer touches. There is a roof terrace with whirlpool and bar and great views of the Acropolis. The rooms at the back of the hotel also have Acropolis views. Upper-floor rooms are quieter. **www.centralhotel.gr**

PLAKA Magna Grecia

Mitropoleos 54, 10563 **Tel** *21032 40314* **Fax** *21032 40317* **Rooms** *10*

Remarkably good value for a four-star boutique hotel, and in an excellent location for a few days exploring the sights of Athens. Simply furnished and attractively decorated in neutral colours. Wooden floors add a homely touch, as do the original paintings that adorn each room. 24-hour reception and room service. **www.magnagreciahotel.com**

PLAKA Parthenon

Makri 6, 11742 **Tel** *21092 34594* **Fax** *21092 35797* **Rooms** *79*

Close to the Acropolis metro station, this is one of the best-value and most convenient hotels in the heart of Athens. Rooms have satellite TV, minibar, trouser press, internet connection and tea and coffee-making facilities. There is a bar and restaurant, but is also close to many cafés, bars and tavernas in the area.

PSYRRI Arion

Agíou Dimitríou 18, 10554 **Tel** *21032 22707* **Fax** *21032 22412* **Rooms** *57*

Stylish, friendly and affordable hotel in fashionable Psyrrí, with cleanly-designed rooms, en-suite bathrooms and roof terrace. The front-facing rooms on the upper floors have Acropolis views and are the most sought-after. There are also good views of the Acropolis from the rooftop bar. **www.arionhotel.gr**

SYNTAGMA Athens Hilton

Vas. Sofias 46, 11528 **Tel** *21072 81000* **Fax** *21072 81111* **Rooms** *524*

Renovated in time for the 2004 Olympics, the Hilton has two large swimming pools, several restaurants and a location opposite the National Art Gallery. Full 24-hour service and stunning views of the Acropolis or Lykavittós from the upper floor rooms and from the rooftop Galaxy bar. **www.hiltonathens.gr**

SYNTAGMA Grande Bretagne

Plateia Sýntagma, 10563 **Tel** *21033 30000* **Fax** *21033 28034* **Rooms** *321*

This grand, old hotel is located opposite the Parliament building on Syntágma and boasts a Neo-Classical façade, public areas and rooms furnished with fine antiques, and lots of marble, gilt and polished wood. There are indoor and outdoor pools, a spa and fine dining. Rooftop garden with magnificent views. **www.grandebretagne.gr**

VOULIAGMENI The Margi

Litous 11, 16671 **Tel** *21089 29000* **Fax** *21089* **Rooms** *88*

Perfectly poised between the sights of central Athens and the stylish Vouliagméni seaside, this hotel exudes style and character. The bedrooms are light and airy, with opulent marble bathrooms, the best have lovely views of the Saronic Gulf. All have internet access, satellite TV, and mini-bar. Excellent restaurant. **www.themargi.gr**

WHERE TO EAT

To eat out in Greece is to experience the democratic tradition at work. Rich and poor, young and old, all enjoy their favourite local restaurant, taverna or café. Greeks consider the best places to be where the food is fresh, plentiful and well cooked, not necessarily where the setting or cuisine is the fanciest. Visitors too have come to appreciate the simplicity and health of the traditional Greek kitchen – olive oil, yoghurt, vegetables, a little meat or

Tsikoudiá, a strong spirit from Crete

fish and some wine, always shared with friends. The traditional three-hour lunch and siesta is still the daily rhythm of the islands, and only in the main tourist areas will you find the Western European routine of a substantial breakfast, a larger and briefer lunch (1pm–2:30pm) and an earlier dinner (7:30–11pm). Greeks prefer a quick breakfast coffee, heavy lunch, and an evening *mezédes* selection, before a long, late dinner that can stretch well into the night.

TYPES OF RESTAURANT

Often difficult to find in more developed tourist resorts, the *estiatórion*, or traditional Greek restaurant, is one of Europe's most enjoyable places to eat. Friendly, noisy, and sometimes in lovely surroundings, *estiatória* are reliable purveyors of local recipes and wines, particularly if they have been owned by the same family for decades. Foreigners unfamiliar with Greek dishes may be invited into the kitchen to choose their fare. In Greece, the entire family dines together and takes plenty of time over the meal, especially at the weekends.

Many traditional restaurants specialize in either a regional cuisine, a method of cooking, or a certain type of food. In some Northeast Agean islands such as Lésvos, where a small minority of Greeks from Asia Minor have settled, food may be spicier than the Greek norm, with lots of red peppers and such dishes as *giogurtlú* (kebabs drenched in yoghurt and served on pitta bread).

The menu in a traditional restaurant tends to be short, comprising at most a dozen *mezédes* (appetizers or snacks), eight main dishes, four or five vegetable dishes and salads, plus a dessert of fresh or cooked fruit, and a selection of local and national wines.

Restaurants vary from very expensive in the main island towns to the magnificently inexpensive. The cheapest of the traditional restaurants is known as a *mageirió*, though they are becoming increasingly rare. Here there is little choice in either wines or dishes, all of which will be *mageireftá* (ready-cooked), but the food is home-made and tasty and the barrel wine is at the very least drinkable and is often good if it comes from the owner's village.

Many hotels have restaurants

Traditional restaurant on Pátmos

open to non-residents. Large island hotels generally offer more expensive, international cuisine. Some will also offer a Greek menu, which tends to be a more elaborate presentation of traditional dishes. Smaller country hotels, however, occasionally have excellent kitchens, and serve good local wines; it is worth checking on any close by.

In the last few years a new breed of young Greek chefs has emerged in *"kultúra"* restaurants, developing a style of cooking that encompasses the country's magnificent raw materials, flavours and colours. These dishes are served with exciting new Greek wines such as Erodios (a rosé), Mackedon (a sauvignon/roditis blend) or a Chardonnay.

TAVERNAS

One of the great pleasures for the traveller in Greece is the tradition of the taverna, a place to eat and drink, even if you simply snack on some *mezédes*. Traditional tavernas open mid-evening and stay open late; occasionally they are also open for lunch.

Windmill restaurant *(see p332)*, Skiáthos town

Outside diners at a taverna in Plakiás, Crete

Menus are short and seasonal – perhaps six or eight *mezédes* and four main courses comprising casseroles and dishes cooked *tis óras* (to order), along with the usual accompaniments of vegetables, salads, fruit and wine.

Some tavernas specialize in the foods and wines of the owner's home region, some in a particular cooking style and others in certain foods.

A *psarotavérna* is the place to find good fish dishes. In small fishing villages you may find the rickety tables of a *psarotavérna* literally on the beach. Close to the lapping waves the owner may serve fish, such as red mullet, bass and octopus, that he himself caught that morning. The large fish restaurants in the tourist areas may serve frozen or imported fish, although the law stipulates that menus must state whether fish is fresh or frozen. For delicious grills try a *psistariá*, a taverna that specializes in spit-roasts and chargrilling *(sta kárvouna)*. In the countryside, you may find lamb, kid, pork, chicken, game, offal, lambs' heads and even testicles char-grilled, and whole lamb is roasted on the spit. At the harbourside, fish and shellfish are grilled (broiled) and served with fresh lemon juice and olive oil. Family-run country tavernas and cafés provide simple meals, such as omelettes and

ordian player in
averna on Sými

salads at any time of day, but many close quite early in the evening. After your meal in the taverna, follow the Greeks and enjoy a visit to the local *zacharoplasteío (see p324)* for a range of desserts.

CAFES AND BARS

Cafes, known as *kafeneía*, are the pulse of Greek life and even the tiniest hamlet has a place to drink coffee and wine. Equally important is its function as the centre of communication – mail is collected here, telephone calls made, and newspapers read, dissected and discussed.

All *kafeneía* serve Greek coffee, sometimes *frappé* (instant coffee served cold, in a tall glass), soft drinks, beer, ouzo and local wine. Most also serve some kind of snack to order. All open early in the morning and remain open until late at night. As the social hub of their communities, country *kafeneía*, as well as many in island towns, open seven days a week.

A *galaktopoleío*, or "milk shop", has a seating area where you can enjoy fine yoghurt and honey. A *kapileió* (wine shop with a café-bar attached) is the place to try local wines from the cask, and you may find a few bottled wines as well. The owner is invariably from a wine village or family, and will often cook some simple regional specialities to accompany the wine.

In a *mezedopoleío*, or *mezés* shop, the owner will not only serve the local wine and the *mezédes* that go with it, but also ouzo and the infamous spirit raki, both distilled from the remnants of the grape harvest. Their accompanying *mezédes* are less salty than those served with wine.

No holiday in Greece is complete without a visit to an *ouzerí*. You can order a dozen or more little plates of savoury meats, fish and vegetables and try the many varieties of ouzo, served in small jugs, with a glass of water to wash the ouzo down. It is a noisy and fun place to eat and drink.

Artemónas restaurant *(see p338)* on the island of Sífnos

A waterside restaurant at Skála Sykaminiás, Lésvos

FAST FOOD AND SNACKS

Visitors can be forgiven for thinking Greeks never stop eating, for there seem to be snack bars on every street and vendors selling sweets, nuts, rolls, seasonal corn and chestnuts at every turn.

Although American-style fast-food outlets dominate tourist centres, it is easy to avoid them by trying the traditional Greek eateries. Try the food of the extremely cheap *souvlatzídiko*, which offers a mostly take-away service of *souvláki* – chunks of meat, fish or vegetables, grilled (broiled) or roasted on a skewer – with fresh bread. The *ovelistírio* serves *gýros* – meat from a revolving spit in a pitta bread pocket. The food is sold "*sto chéri*" (in the hand, or to take away).

Many bakeries sell savoury pies and an array of flavourful bread rolls, and in busy areas you will always be able to find a café serving substantial snacks and salads.

If you have a sweet tooth you will love the *zacharoplasteío* (literally, "shop of the sugar sculptor"). The baker prepares traditional sweet breads, tiny sweet pastries and a whole variety of fragrant honey cakes.

Baklavás, a sweet cake of wheat, honey and nuts

BREAKFAST

For Greeks, this is the least important meal of the day. In traditional homes and *kafeneía* a small cup of Greek coffee accompanies *paximádia* (slices of bread rusks) or *koulourákia* (firm, sesame-covered, or slightly sweet, rolls in rings or s-shapes) or pound cakes, filled with traditional home-made jam.

Elsewhere, and in many tourist cafés, this has been replaced by a large cup of brewed coffee and French-style croissants or delicious brioche-style rolls. During summer, some *kafeneía* will still serve fresh figs, thick yoghurt, pungent honey and slightly sweet currant bread, as well as a variety of English and continental breakfasts to cater for visitors' tastes.

RESERVATIONS

Although island restaurants generally have a casual atmosphere, they can, of course, be very popular; if it is possible to make a reservation, it is probably best to do so. Also, it is local practice to visit the restaurant or taverna earlier in the day to check on the dishes to be served. The proprietor will then take your order and reserve any special dish that you request.

WINE

The grape varieties that abound in Greece today produce wines that are quite distinct from those of Western Europe. However, restaurateurs are only now learning to look after bottled wines. If the wine list contains the better wines, such as Ktima Merkoúri, Seméli or Strofiliá, the proprietor probably knows how to look after them and it will be safe to order a more expensive bottle. For a little less, good-value bottles include the nationally known Cambás and Boutári wines.

Traditional restaurants and tavernas may only stock carafe wine, which is served straight from the barrel and is always inexpensive. Carafe wines are often of the region, and the Greek rosé in particular is noted for having an unusual but pleasing flavour.

Wine from Límnos

HOW TO PAY

Greece is very much a cash society. If you need to pay by credit card, check first that the restaurant takes your credit card – many proprietors do not accept the whole range. *Kafeneía* almost never take credit cards, and café-bars very rarely do, although many will be happy to take travellers' cheques. Country restaurants, tavernas, *kafeneía* and bars will only accept cash.

The restaurant listings in this guide on pages 330–41 indicate whether or not credit cards are accepted at each establishment.

Kástro's bar *(see p337)* in the town of Mýkonos

Views of the Acropolis from the marble roof terrace of Pil Poul *(see p341)*

SERVICE AND TIPPING

Greeks take plenty of time when they eat out and expect a high level of attention. This means a great deal of running around on the part of the waiter, but in return they receive good tips – 15 per cent if the service has been especially attentive, though more often a tip is about 10 per cent. Prices in traditional establishments do include service, but the waiters still expect a tip so always have coins ready to hand. Tap water is offered free with the meal.

Western-style restaurants and tourist tavernas sometimes add a service charge to the bill; their prices can be considerably higher because of additional trimmings, such as air-conditioning and phones.

DRESS CODE

The Greeks dress quite formally when dining out. Visitors should wear whatever is comfortable, but skimpy tops and shorts, and active sportswear are usually only acceptable near the beach – though most tourist establishments rarely turn away custom. Some hotel restaurants have policies requesting formal dress; in the listings we indicate which restaurants fall into this category.

In summer, if you dine outside, take a jacket or sweater for later in the evening.

CHILDREN

Children become restaurant and taverna habitués at a very early age in Greece – it is an essential part of their education. Consequently, children are welcome everywhere in Greece except the drinking bars. In formal restaurants children are expected to be well behaved, but in summer, when the Greeks enjoy long hours eating outside, it is perfectly acceptable for the children to play and enjoy themselves too. Special facilities, such as high chairs, are unknown in all but the most considerate hotel dining rooms, but generally, casual restaurants and tavernas are perfect for dining with children of any age.

Basket of local bread from Rhodes

SMOKING

Smoking is commonplace in Greece and until recently establishments maintaining a no-smoking policy have been difficult to find. However, new EU regulations make it obligatory for all restaurants to have no-smoking areas. In practice, of course, change is slow but for at least half the year you can always dine outdoors.

WHEELCHAIR ACCESS

In country areas, where room is plentiful, there are few problems for wheelchair users. But in crowded tourist restaurants access is often restricted. The streets themselves can have uneven pavements (sidewalks) on the islands, and many restaurants have narrow doorways and steps. There are several organizations for assisting disabled vacationers, and those listed on pages 301 and 353 provide specific information for visitors travelling to the Greek islands.

VEGETARIAN FOOD

Greek cuisine provides plenty of choice for vegetarians. Greeks enjoy a variety of dishes for each course, so it is easy to order just vegetable dishes in any traditional restaurants, tavernas or *kafeneía*. Greek vegetable dishes are substantial, inexpensive and very satisfying. Usually they are prepared in imaginative ways to complement or enhance their flavour.

Vegans may have a little more difficulty but, as Greek cooking relies very little on dairy products, it is possible to follow a vegan diet on any of the Greek islands.

PICNICS

The best time to picnic in Greece is in spring, when the countryside is at its most beautiful and temperatures are not too hot. Traditional seasonal foods, such as Lenten olive oil breads, sweet Easter breads, pies filled with wild greens, fresh cheese and young retsina wine, make perfect picnic fare. In summer, peaches and figs, yoghurt, hard cheese, tomatoes, bread and olives are the ideal beach snacks.

People drinking coffee at the Liston in Corfu town

The Flavours of Greece

The ancient Greeks regarded cooking as both a science and an art – even a topic for philosophy. In out-of-the-way places on the mainland and on the more far-flung islands, you will still find dishes, ingredients and culinary styles untouched by time. Elsewhere, Greek cookery has been much influenced by the Ottoman Empire, with its spiced meat dishes, and filled pastries and vegetables. In the recent past, Greek cuisine was often thought of as peasant food. Today, it is that very simplicity, and its reliance on seasonal, local produce, that makes Greek food so popular with visitors.

Oregano and thyme

Island fisherman returning to harbour with the day's catch

ATHENS AND THE PELOPONNESE

The capital is essentially a city of immigrants from the countryside, the islands and the shores of the eastern Mediterranean. That diversity is reflected in its markets and its cuisine. Street food is a quintessential part of Athens life. In the Peloponnese ingredients are as varied as the terrain: fish from the sea and, from the mountains, sheep, goat and game. From the hills come several varieties of cheese, olives and honey.

CENTRAL AND NORTHERN GREECE

Mainland Greece, with its long and chequered history, is a place where regional food boundaries are blurred and a variety of cooking traditions coexist. The meat and fruit dishes of Thessaloníki show a Jewish influence; the spices, sausages and oven cooking of Ioánnina stem from Ottoman times; while a love of sheep's cheese, pies and offal came to Métsovo and the Epirus mountains with the Vlach shepherds. The spicy food of the North is the legacy of the 1922 Greek immigrants from Asia Minor, while the Balkan influence is obvious in the use of pickles, walnuts and yogurt.

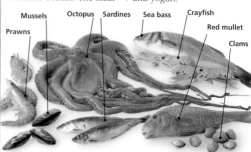

Mussels   Octopus   Sardines   Sea bass   Crayfish
Prawns   Red mullet
                                           Clams

Selection of seafood from the clear waters of Greece

REGIONAL DISHES AND SPECIALITIES

Sweets such as nougat, *pastéli* (honey-sesame candy), *loukoúmia* (yeast doughnuts in syrup) and *chalvás* (halva, or sweetmeats) have been a part of Greek street life since the days of Aristotle. They are sold in small shops or stalls. *Píttes*, or pies, are a speciality of the western Epirus region. Fillings range from game or offal to cheese and vegetables, often combined with rice or pasta. Reflecting Middle-Eastern influences, *Soutzoukákia*, a speciality of northern Thrace and Macedonia, are meat patties flavoured with coriander, pepper and cumin. *Choirinó kritikó*, the classic dish of inland Crete villages, is thick pork cutlets baked until tender, while *Sýka me tyrí* is a summer *mezés*, dessert or snack, of fresh figs with *miztkýra* cheese, made from whey.

Olives

Fakés *is a sour Peloponnese soup of green lentils, lemon juice or wine vinegar, tomatoes, herbs and olive oil.*

Produce on sale in a typical Greek market

THE ISLANDS

Each group of islands has a distinct culinary identity reflecting its geographical location and history. Many Ionian dishes are pasta based, a legacy of the era of Venetian occupation. Those of the Cyclades are intensely flavoured. The cooks of the Dodecanese and Northeast Aegean benefit from the rich harvest of the surrounding sea. Crete is unique in its long Turkish occupation and taste for highly spiced dishes, and Cretan cooking has a number of recipes unique to the island. The use of pork, a legacy of antiquity, is more popular here than anywhere else in Greece. Some lovely kitchen utensils and unusual ingredients from Minoan times have been excavated by archaeologists on Crete.

FISH AND SEAFOOD

The warm and sheltered waters of the Aegean are the migratory path for tuna and swordfish, and a feeding ground for tasty anchovies and sardines. Coves and caves around the hundreds of rocky islands shelter

Bread being baked in an outdoor communal oven

highly prized red mullet, dentex and parrot fish, while the long shoreline is home to shellfish and crustaceans. Fish are usually served with their heads on: to Greeks this is the tastiest part, and it helps to identify the variety.

OTHER PRODUCE

Greece is home to the largest variety of olives in the world. They are cured by methods used for thousands of years. The best quality olive oil, extra-virgin, is made by pressing just-ripe olives only. Greece produces sheep's, cow's and goat's cheeses, usually named by taste and texture, not place of origin.

WHAT TO DRINK

Wine has been part of Greek cultural life from the earliest times. Major wine-producing areas include Attica, Macedonia and the Peloponnese. Mavro-daphne is a fortified dessert wine from Pátra. Greek specialities include *tsípouro*, distilled from the residue of crushed grapes; retsina, a wine flavoured with pine resin *(see p147)*; and the strong, aniseed-flavoured spirit, ouzo *(see p140)*. Coffee in Greece is traditionally made from very finely ground beans boiled up with water in a long handled *mpríki* (coffee pot) and drunk from a tiny cup. It is served in cafés rather than tavernas.

Spetzofáï, *from central Greece, is sautéed slices of spicy country sausage with herbs and vegetables.*

Barboúnia, *or red mullet, has been the most esteemed fish in Greece since antiquity. It is usually simply fried.*

Loukoumádes *are a snack of small deep-fried doughnuts soaked in honey-syrup and sprinkled with cinnamon.*

The Classic Greek Menu

The traditional first course is a selection of *mezédes*, or snacks; these can also be eaten in *ouzerís*, or bars, throughout the day. Meat or fish dishes follow next, usually served with a salad. The wine list tends to be simple, and coffee and cakes are generally consumed after the meal in a nearby pastry shop. In rural areas traditional dishes can be chosen straight from the kitchen. Bread is considered by Greeks to be the staff of life and is served at every meal. Village bakers vary the bread each day with flavourings of currants, herbs, wild greens or cheese. The many Orthodox festivals are celebrated with special breads.

Greek pitta breads

Souvlákia *are small chunks of pork, flavoured with lemon, herbs and olive oil, grilled on skewers.*

Choriátiki saláta, *Greek salad, combines tomatoes, cucumber, onions, herbs, capers and feta cheese.*

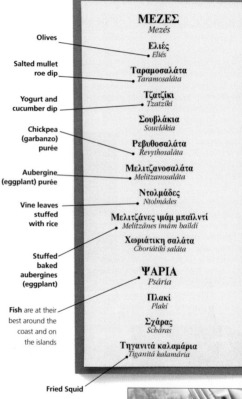

ΜΕΖΕΣ
Mezés

Olives — **Ελιές**
Eliés

Salted mullet roe dip — **Ταραμοσαλάτα**
Taramosaláta

Yogurt and cucumber dip — **Τζατζίκι**
Tzatzíki

Σουβλάκια
Souvlákia

Chickpea (garbanzo) purée — **Ρεβυθοσαλάτα**
Revythosaláta

Aubergine (eggplant) purée — **Μελιτζανοσαλάτα**
Melitzanosaláta

Vine leaves stuffed with rice — **Ντολμάδες**
Ntolmádes

Μελιτζάνες ιμάμ μπαϊλντί
Melitzánes imám baïldí

Stuffed baked aubergines (eggplant) — **Χωριάτικη σαλάτα**
Choriátiki saláta

ΨΑΡΙΑ
Psária

Fish are at their best around the coast and on the islands — **Πλακί**
Plakí

Σχάρας
Scháras

Fried Squid — **Τηγανιτά καλαμάρια**
Tiganitá kalamária

Psária plakí *is a whole fish baked in an open dish with vegetables in a tomato and olive oil sauce.*

Scháras *means "from the grill". It can be applied to meat or fish, or even vegetables. Here, grilled swordfish has been marinated in lemon juice, olive oil and herbs before being swiftly char-grilled.*

MEZEDES

Mezédes are eaten as a first course or as a snack with wine or other drinks. *Taramosaláta* is a purée of salted mullet roe and bread crumbs or potato. Traditionally a dish for Lent, it is now on every taverna menu. *Melitzanosaláta* and *revithosaláta* are both purées. *Melitzanosaláta* is grilled aubergines (eggplant) and herbs; *revithosaláta* is chickpeas (garbanzos), coriander and garlic. *Melitzánes imám baïldí* are aubergines filled with a purée of onions, tomatoes and herbs. *Ntolmádes* are vine leaves stuffed with currants, pine nuts and rice.

Revithosaláta

Ntolmádes

Taramosaláta

Melitzánes imám baïldí

Typical selection of mezédes

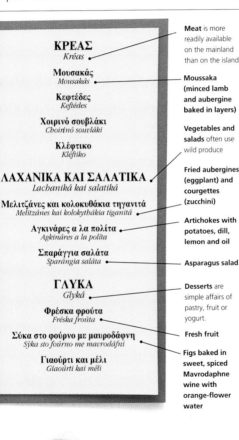

ΚΡΕΑΣ
Kréas

Μουσακάς
Mousakás

Κεφτέδες
Keftédes

Χοιρινό σουβλάκι
Choirinó souvláki

Κλέφτικο
Kléftiko

ΛΑΧΑΝΙΚΑ ΚΑΙ ΣΑΛΑΤΙΚΑ
Lachaniká kai salatiká

Μελιτζάνες και κολοκυθάκια τηγανιτά
Melitzánes kai kolokythákia tiganitá

Αγκινάρες α λα πολίτα
Agkináres a la políta

Σπαράγγια σαλάτα
Sparángia saláta

ΓΛΥΚΑ
Glyká

Φρέσκα φρούτα
Fréska froúta

Σύκα στο φούρνο με μαυροδάφνη
Sýka sto foúrno me mavrodáfni

Γιαούρτι και μέλι
Giaoúrti kai méli

Meat is more readily available on the mainland than on the islands

Moussaka (minced lamb and aubergine baked in layers)

Vegetables and salads often use wild produce

Fried aubergines (eggplant) and **courgettes** (zucchini)

Artichokes with potatoes, dill, lemon and oil

Asparagus salad

Desserts are simple affairs of pastry, fruit or yogurt.

Fresh fruit

Figs baked in sweet, spiced Mavrodaphne wine with orange-flower water

Keftédes *are meatballs of pork with egg and breadcrumbs, flavoured with herbs and cumin and fried in olive oil.*

Kléftiko *is usually goat meat wrapped in parchment paper and cooked so that the juices and flavours are sealed in.*

Sweet pastries *filled with nuts and honey, syrup-drenched cakes, pies, doughnuts and glyká (candied fruits) are mainly eaten in cafés. The most famous of all are baklavas, with layers of filo pastry and nuts, and kataïfi, known to tourists as "shredded wheat".*

Giaoúrti kai méli *(yogurt with honey) is served in speciality "milk shops", to be eaten there or taken home.*

Choosing a Restaurant

The restaurants in this guide have been selected across a wide price range for their good value, excellent food and interesting location. The entries below are listed by region, starting with the Ionian Islands, then alphabetically by island name or area. For *Flavours of Greece* and *The Classsic Greek Menu* see pages 326–9.

PRICE CATEGORIES
Price categories are for a three-course meal for one, including a half-bottle of house wine, tax and service.

€ Under 12 euros
€€ 12–18 euros
€€€ 18–24 euros
€€€€ 24–32 euros
€€€€€ over 32 euros

THE IONIAN ISLANDS

CORFU Rouvas   🍴   €€
St. Desillas 13, Corfu Town **Tel** *26610 31182*

One of the most affordable traditional eating places in Corfu Town, with a well-established reputation for some of the best Greek and Corfiot dishes. Located in the centre of town, near the main produce market, so the ingredients are usually very fresh. Closed dinner; Sun.

CORFU Rex   🍴 🍴   €€
Kapodistriou 66, Corfu Town **Tel** *26610 39649*

Housed in a mid-19th-century building, one block behind the arcades and cafés of the Spianádha, the Rex is a Corfu institution – a truly traditional Greek restaurant serving genuine Corfiot food. Specialities include swordfish *bourthéto*, a spicy fish stew with tomatoes, peppers and potatoes. Good, attentive service.

CORFU Toula's Taverna   🍴   €€€
Agni, just south of Kalámi **Tel** *26630 91350*

Toula's Taverna, converted from a 19th-century olive press, is located on the beach and has outside tables. The restaurant specializes in seafood and traditional Corfu dishes and recently won the Greek Gourmet award for seafood restaurants. There is even a cookery book, which is for sale at the restaurant. Closed Nov–Apr.

CORFU Eu-Lounge Café   🍴 🍴 🍴   €€€€
Kapodistriou 32, Corfu Town **Tel** *26610 80670*

Stylish, modern café-restaurant with tables indoors and out. The menu is international-modern fusion, with a good choice of pasta dishes, appetisers and salads, and the wine list is extensive. An ideal spot for lunch or snacks on a visit to Corfu Town, or for an evening out in smart surroundings.

ITHACA Kantoúni   🍴   €€
Limáni **Tel** *26740 32549*

A simple taverna serving a reasonable selection of oven-cooked standard meals, along with a small selection of grilled meat dishes and the usual salads, *tzatziki* and other trimmings. Close to the ferry pier and ideal for a meal or a snack while you await your boat to Kefalloniá or the mainland. Closed Oct–Apr.

KEFALLONIA Patsouras   🍴   €
Ioannou Metaxas 40, Argostóli **Tel** *26710 22779*

Surprisingly cheap taverna in the centre of Kefalloniá's island capital. Patsouras is family-run, and its menu includes all the traditional taverna favourites plus Kefallonian island specialties such as *kreatópita* (meat pie) and *krasáto* (pork in wine sauce). Enjoy lunch or dinner in a pleasant garden setting.

KEFALLONIA Platanos   🍴   €€
Asos **Tel** *26740 51381*

This small taverna is located – as its name suggests – beneath a plane tree in the centre of the pretty, unspoilt fishing village of Asos. Affordable, simple oven-cooked dishes, grilled meats, and salads. If you are staying in Asos, you will probably eat here more than once. Closed Oct–Apr.

KEFALLONIA Tassia   🍴 🍴 🍴   €€€€
Limáni Fiskardou **Tel** *26740 41205*

This internationally-acclaimed restaurant sits right on the harbour and features a good wine list which emphasizes some of Kefalloniá's own vintages, recognised as among the best in Greece. The menu includes fine seafood, simply but well-prepared, as well as pasta dishes, salads, appetizers, and desserts. Closed Oct–Apr.

LEFKADA Sto Molo   🍴 🍴   €€
Golemi 12, Lefkáda town **Tel** *26450 24879*

An attractive restaurant serving an imaginative array of typical small dishes – fish, meat, sausage, cheese, fruit and vegetables – from which you can construct a light snack or settle down for a few hours of sampling, accompanied by ouzo, beer or wine from the cask.

Key to Symbols *see back cover flap*

PAXOS Taka Taka
Gáios **Tel** 26620 32329

A straightforward grill restaurant serving unpretentious meals – mainly fish, but also chicken, lamb, pork chops and beef rissoles, along with salads and a few oven-cooked vegetable dishes. Taka Taka is a long-standing favourite with the locals who come to dine in the delightful, vine-covered garden. Closed lunch; Oct–Feb.

PAXOS Nassos
Longós **Tel** 26620 31604

Located on the main square, this is one of the best places to eat in Longós. The menu usually features octopus (grilled as an appetizer or stewed with wine and onions) along with other seafood dishes, *souvláki* and other grilled meat dishes. Alternative options include casseroles, salads and pasta dishes. Closed Oct–Apr.

ZAKYNTHOS Zakanthi
Kalamáki **Tel** 26950 43586

This lively restaurant and bar is a good place to spend the evening whether you are looking for a light snack with drinks or something more substantial. The menu is eclectic, with a reasonable selection of Greek favourites along with pasta, burgers and other international dishes.

THE ARGO-SARONIC ISLANDS

AIGINA Agora
Fish market, Aígina Town **Tel** 22970 27308

This excellent, old-fashioned fish taverna has been in business for more the 40 years and is undoubtedly the best in town. No fancy dishes, but very fresh seafood – the menu depends on what the boats have brought in that day. Wine is from the barrel. Well worth making an excursion to.

AIGINA Antonis
Pérdikas **Tel** 22970 61443

Popular with Athenian weekend visitors – and justifiably so – Antonis is the best and the most expensive of several fish restaurants on the waterfront of Aígina's most picturesque village. The fish is always fresh and is beautifully prepared and presented. Anyone staying on Aígina should eat here at least once.

KYTHIRA To Korali
Avlémonas **Tel** 27360 34173

This little eating place at Avlémonas, 26 km (15 miles) from Chóra, is everything an island fish taverna should be – tiny, welcoming, with rickety outdoor tables, affordable prices and a choice of fresh grilled seafood and oven-baked taverna dishes, as well as salads and wine from the barrel. An ideal spot for lunch or dinner. Closed Oct–Apr.

KYTHIRA Taverna Magos
Kapsáli **Tel** 27360 33722

This traditional taverna, in the pretty harbour village of Kapsáli, is deservedly popular with people from all over the island, as well as summer visitors. Offerings include freshly-caught seafood and a decent choice of vegetable dishes. Get there early in the evening if you want a table in July or August. Closed Nov–Feb.

POROS O Karavolos
Behind the cinema, Póros Town **Tel** 22980 26158

A traditional, family-run taverna offering typical Greek island cooking and fresh fish at excellent prices. More unusual menu items include snails, for which the restaurant is famous and from which it takes its name. These are much smaller than French escargots, and available only when in season. Closed lunch.

POROS Kathestos
Póros Town **Tel** 22980 24770

On the esplanade, Kathestos has good views from its outside tables and serves typical taverna dishes from the oven and a selection of seafood, some of it fresh off the boat. Good grills include chicken, lamb and pork chops; other offerings include salad and stuffed vegetable dishes. Prices are very affordable by Póros Town standards.

SPETSES Exedra
Palió Limáni, Spétses Town **Tel** 22980 73497

Situated on the waterfront, this traditional Greek taverna offers good local dishes such as shrimps *saganáki* and fish *à la spetsiota* (oven-baked with garlic and tomato sauce). Good value for money although the fish dishes do push the price up. Closed Nov–Feb.

YDRA Gitonikò
Ydra Town **Tel** 22980 53615

Close to the church in Ydra Town, this popular taverna run by genial owners Manòlis and Christina is a bit of an institution. The fare is traditional Greek with oven-cooked dishes such as *pastítsio* and *moussakás*, grilled meat and fresh local fish and seafood. There are also Ydriot specialities. The roof terrace is especially lovely. Closed Dec–Feb.

YDRA Kondylenia

🔲   €€€

Kamíni, just west of Ydra Town **Tel** *22980 53520*

One of this restaurant's biggest selling points is its panoramic sea view across the Gulf to the mountains of the Peloponnese. The menu lives up to the view, with interesting recipes such as spinach, squid and prawn casserole as well as good fresh fish from the grill. A reasonable if not outstanding choice of wines. Closed Oct–Mar.

THE SPORADES AND EVVOIA

ALONISSOS Bambis

📝🔲   €€

Patitíri **Tel** *24240 66184*

This small taverna on the outskirts of Patitíri village has sweeping views of the olive groves and wooded hillsides of Alónissos and across to nearby islands. The menu is traditional Greek taverna cooking at its best, with a leavening of grilled meat and fish dishes as well as a daily choice of hearty oven-cooked dishes. Closed lunch; Oct–May.

EVVOIA Kavo d'Oro

📝🔲   €€

Parodos Sachtouri, Kárystos **Tel** *22240 22326*

Simple and inexpensive old-fashioned taverna which serves good home cooking – most of the meals on offer are very hearty and filling. Rich vegetable and meat stews made with local olive oil, stuffed vine leaves and *papoutsákia* (baked aubergines with cheese, onion and tomato) are all worth sampling. Closed Nov–Feb.

EVVOIA Skýros

📋🔲🎵   €€€

Harbourfront, Kými **Tel** *22220 22624*

A seafront setting and friendly service characterize this excellent taverna. Local grilled shrimps, octopus and whole-fried baby *barboúnia* (red mullet) are on offer as well as oven-baked taverna dishes and salads. Choice of outdoor tables or seating in the air-conditioned interior. Local wines and live entertainment. Closed Nov–May.

EVVOIA Vràhos

📋🔲   €€€

Leofóros Makaríou 4, Chalkida **Tel** *22210 87618*

This traditional taverna enjoys a prime location right on the waterfront. The menu features all the mainstays of Greek and Mediterranean cuisine as well as fresh fish such as sea bream and mackerel which are cooked on the outdoor grill. Salads, vegetables dishes and a good selection of local wines.

SKIATHOS Anemos

🔲   €€€€

Old harbour front, Skiáthos Town **Tel** *24270 21003*

Archetypal waterside eating place with views of the old harbour, the Bourtzi fortress islet, and the bay. Despite its old-fashioned appearance, the menu is modern and influenced by other Mediterranean cuisines, but most visitors still opt for the grilled fish dishes, which are excellent if on the expensive side. Closed lunch; Nov–Apr.

SKIATHOS Karnagio

🔲📋🎵   €€€€

Paraliaki, Skiáthos Town **Tel** *24270 22868*

Karnagio is regularly commended as Skiáthos's most outstanding restaurant, attracting a summer clientele that includes the occasional visiting celebrity. Wide choice of traditional Greek dishes, all very well prepared, and some of the best seafood around. Closed lunch; Oct–May.

SKIATHOS Windmill

🔲   €€€€€

Kotroni Hill, Skiáthos Town **Tel** *24270 24550*

This charming restaurant is housed in an old windmill, up on a hill and with wonderful views over the harbour and Skiáthos Town. Customers dine on a lovely stepped terrace with traditional wooden chairs and tables. The menu here offers mainly Mediterranean cuisine. Closed lunch; Oct–Apr.

SKOPELOS Molos

📝🔲   €€

Paleó Limáni Skópelou **Tel** *24240 22551*

This small taverna, situated beside the pier in Skópelos Town, serves fresh fish and good salads. The more unusual dishes include oven-cooked goat with artichokes. Attractive sea views from the outdoor tables and prices that are pleasantly modest. Closed Nov–Feb.

SKOPELOS Perivóli

🔲🍷   €€€

Skópelos Town **Tel** *24240 23758*

In the centre of town, Perivóli is a lively little restaurant-bar serving drinks, light snacks and full meals. Cooking combines local and international influences, and the wine list is extensive, with some examples of the better new-style Greek wines. No retsina from the barrel here. Closed lunch; Oct–May.

SKYROS O Liakos

📝🔲   €€

Machairas, Skýros Town **Tel** *22220 93509*

This roof garden taverna, with fantastic views over picturesque Skýros Town, serves traditional Greek dishes with imaginative touches. Specialities include *tomatá keftedes* (tomato fritters) and homemade cheese pies. The best time to go is early evening for sunset views and pre-dinner drinks.

Key to Price Guide *see p330* **Key to Symbols** *see back cover flap*

THE NORTHEAST AEGEAN ISLANDS

CHIOS Chotsas
□ □ €€

Georghiou Kondyis 3, Chíos Town **Tel** *22710 42787*

Venerable family-run taverna with a large garden in the centre of Chíos Town. As well as the expected assortment of grilled meats, fish and oven-cooked dishes, the kitchen produces an array of vegetable dishes, dips, fritters and snacks. The proprietor makes his own wine and ouzo, both of which are on tap from the cask. Closed lunch; Sun.

CHIOS Mesaionas
□ □ €€€

Plateía Meston **Tel** *22710 76050*

A highly recommended taverna, attractively situated in the main square of this picturesque village. Serves an interesting array of local island specialties such as aubergine pilaf as well as the usual gamut of grills and casserole dishes. A pleasant place to dine and popular with locals and visitors alike.

CHIOS Pýrgos
□ □ □ €€€€

Kondari, 2 km (1 mile) north of Karfás **Tel** *22710 44740*

The menu here combines Greek and international dishes, with imaginative seafood appetizers. Live traditional music on Wednesday evenings makes Pýrgos a good choice for a special night out. It is also conveniently located for those staying at Karfás, Chíos's most popular beach resort area. Not cheap, nevertheless good value for money.

IKARIA Atsachas
□ □ €€

Mesakti, Armenistís **Tel** *22750 71226*

Simple, family-run taverna perched on a headland between two beaches. The menu is uncomplicated, with a good choice of oven-cooked meat and vegetable dishes and a basic choice of grilled seafood (whitebait, squid, octopus, sea bass and the catch of the day). Local wine from the barrel. Closed Oct–May.

IKARIA Anna
□ □ €€€

Nas **Tel** *22750 71489*

The best of a group of small eating places situated high above the white pebbly cove at Nas (popular with nude sunbathers). The restaurant – which also has rooms to rent – serves fish, grilled meats and oven-cooked dishes, as well as desserts, snacks and local wine. Quiet surroundings and good views. Closed Oct–May.

LESVOS Petra Women's Agrotourism Co-operative
□ €€

Pétra Village **Tel** *22530 41238*

A visit to this sustainable tourism initiative, set up by local women, is an eye-opener for anyone who wants to know more about local food and produce. As well as the more familiar options, the cooks serve up the kind of hearty dishes that tourists rarely encounter. Strong on vegetable dishes, also sticky desserts. Closed Oct–Apr.

LESVOS I Skamnia
□ €€

Harbourfront, Skála Sykaminiás **Tel** *22530 55270*

This restaurant is a favourite with the locals. Sit under the shade of a Mulberry tree in summer and dine on a choice of authentic seafood including sardines and anchovies from the Gulf of Plomári, squid, octopus, stuffed courgette flowers and grilled dishes.

LESVOS Vafeios
□ □ €€€

Dikismos Vafeios, on road to Sykaminiá **Tel** *22530 71752*

An extensive menu of local island dishes, reasonable prices and a lovely terrace setting with sweeping views make this a firm favourite with islanders and with holidaymakers from nearby Mólyvos. The cuisine is rich, with the emphasis on hearty oven-cooked dishes, so come with a big appetite. Closed weekends.

LIMNOS O Glaros
□ €€€

Limáni, Mýrina Town **Tel** *22540 22220*

This classic fish taverna on the harbour front at Mýrina overlooks a flotilla of fishing boats which deliver some of the best and freshest seafood in the Aegean. A wide range of fish from large grouper and *skathari* (bream) which will feed several people to whitebait, prawns, and langouste, and a good assortment of appetizers. Local wines.

SAMOS I Psarades
□ □ €€€

Agios Nikólaos, Kontakaiikia **Tel** *22730 32489*

A wide variety of fish dishes, accompanied by a more limited selection of *mezedés* and salads. Food is served on an attractive terrace with views out to sea. The fish mostly comes from Karlóvasi harbour, 5 km (3 miles) east of the village, and the taverna is at its best in May, when the choice of fish is at its widest. Closed Oct–Apr.

SAMOS Marina
□ €€€

Waterfront, Kokkári **Tel** *22730 92692*

In a village packed with restaurants catering almost entirely to summer tourists, Marina is one of the more authentic options, catering to local diners in low-season. The food is excellent and the menu has a wider choice of local specialities than most of its nearby rivals. Good range of vegetable and oven-cooked meat dishes. Closed Nov–Apr: Mon–Thu.

SAMOTHRAKI 1900

Plateía, Chóra **Tel** *25510 41224*

Of the two restaurants on Chóra's main square, 1900 is the better with a menu that includes imaginative variations on traditional dishes such as stuffed goat along with northern Aegean favourites such *stamnàto* (veal with oven-cooked mushrooms). Views over the sea and the wooded slopes around and above the village. Closed Oct–Mar.

THASOS O Glaros

Alykí Bay **Tel** *25930 31547*

This is the oldest of several tavernas in a pretty hamlet overlooking the sea. Fish, meat and grills, and salads can be enjoyed on a vine-shaded terrace with views across Alyki's twin bays. A pleasant place for lunch within sight of the beach, or for dinner, with friendly service. The owners also have rooms to rent. Closed Oct–May.

THASOS Kraboùsa

Skála Potamiás **Tel** *25930 62312*

Family-run, traditional taverna which now panders mainly to the palates of visiting northern Europeans with fish dishes and simple omelettes and salads. Traditional dishes are also on offer, including Greek mainstays such as *moussakás*, *stifádo* and *pastítsio* (macaroni pie). Set in its own gardens.

THE DODECANESE

ASTYPALAIA Australia

Paralia tou Péra Yaloù, Chóra **Tel** *22430 61275*

A typical island harbour taverna serving a plethora of Cycladic dishes, from courgette fritters to the freshest grilled fish. Meat and vegetables come from their own farm. Justifiably popular, the taverna gets very busy on summer evenings, so book ahead. Affordable dining and a great place to watch fishing boats come and go. Closed Nov–Apr.

CHALKI Althemenis

Nimporió, 85110 **Tel** *22460 31303*

This popular taverna, in Chálki's harbour village, is a reliable option with a menu that includes the usual array of seafood, grilled chicken, pork and lamb, *moussakás*, *pastítsio* (macaroni pie) and other oven-cooked dishes. Can be crowded in the evening in high season. Closed Oct–Apr.

CHALKI Liros

Nimporió, 85110 **Tel** *22460 31264*

This cheap and cheerful family-run restaurant close to the quayside at Nimporió is an adequate choice for lunch or dinner in a location where there is little competition. The menu includes the usual Greek favourites with an attempt at some international dishes too. Pleasant location. Closed Oct–May.

KALYMNOS Drossia

Limanáki Melitsacha, Myrtiès **Tel** *22430 48745*

Beside a tiny, picturesque harbour, this award-winning fish restaurant serves up freshly caught whitebait, squid, prawns, octopus and other fishy appetizers. Main courses include swordfish, sea bass, red mullet and lobster (book couple of days ahead). Fish dishes are grilled, and come with an assortment of dips. Local wine, ouzo and retsina.

KALYMNOS To Limanáki tou Vathý

Rína **Tel** *22430 31333*

This small restaurant, standing above the quay at Rína, is a peaceful spot for lunch or dinner, serving standard taverna samplings including oven-cooked casseroles, grilled meat and grilled and fried fish. A good option, as much for its tranquil location and views as for its food and drink.

KARPATHOS Anoiksis

Diafáni **Tel** *22450 51226*

A classic summer taverna serving old-fashioned island cooking such as oven-cooked meat and vegetable dishes, salads and meat grills – lamb, pork chops and chicken. Occasionally served are fish, stuffed peppers, and kid in red sauce with roast potatoes. Nice location by the harbour at Diafáni.

KASTELLORIZO Lazarakis

Harbour, Kastellórizo **Tel** *22460 49370*

A friendly restaurant with tables either under a vine trellis or on a little jetty in the yacht-filled harbour. Serves perfectly prepared fresh fish, prawns and lobster, good salads and island delicacies such as sea urchin roe. This is one of the classic island tavernas and highly recommended. Closed Nov–Mar.

KOS Limniónas

Limnionas **Tel** *69324 22002*

Traditional fish restaurant beside a small fishing harbour in Limniónas, 48 km (30 miles) from Kos Town. Fresh fish is delivered daily and simply grilled to perfection. Well worth making the journey for, especially for a leisurely lunch beside the sea. Closed Nov–Apr.

Key to Price Guide *see p330* **Key to Symbols** *see back cover flap*

KOS O Makis

Mastichári **Tel** *22420 59061*

One of the best little fish tavernas on the island, O Makis is located next to an attractive little fishing harbour from which comes most of the seafood that it serves. Expect sea-fresh *tsipoura* and *fagri* as well as prawns, squid, octopus and – for special occasions, and worth ordering a couple of days in advance – langouste. Closed lunch.

KOS Plátanos

Plateía Platanou **Tel** *22420 28991*

Located on the square beside the Castle of Knights, this café-restaurant overlooks the ancient agora. More peaceful than Kos Town's parade of harbour-side restaurants, it serves light meals, snacks and drinks, and is reasonably priced compared with many eating places in town. Closed Nov–May.

LEROS Petrino

Lakki **Tel** *22470 24807*

Surprisingly chic taverna with tables inside or outside on the terrace. The menu is traditional, with the emphasis on oven-cooked dishes and grilled meats, but there are imaginative twists to some of the old favourites. Even the chips are more flavoursome than you might expect.

LEROS Da Guisie e Marcello

Alinda (3 km/2 miles from Agia Marina village) **Tel** *22470 24888*

A classic Italian trattoria that takes pride in creating fine classical cooking using only the best locally-sourced organic ingredients. Good choice of vegetarian dishes and authentic Italian pizzas together with a select few Greek wines and an array of organic Italian imported vintages. Good value. Closed Nov–Feb.

LIPSI Kalypso

Waterfront, Lipsí Town **Tel** *22470 41242*

Grilled octopus and roast stuffed kid (in season) are among the better offerings at this lively restaurant with tables shaded by a vine covered trellis on the harbour side. Also on offer are most of the usual Greek favourites and a few more cosmopolitan dishes. Basic wine list.

PATMOS Aspro

Paralía Aspris, Skála **Tel** *22470 32240*

This sophisticated seafood restaurant, on the outskirts of Skála, has an extensive menu of fresh fish dishes (according to season and catch) served from the grill. Also on offer is a good array of accompaniments and starters and an excellent dessert trolley. Closed lunch; Oct–May.

PATMOS Gerovoliés tou Màgou

Skála **Tel** *22470 33226*

Just a stone's throw from Skála port, this traditional Greek taverna does a good turn in the more elaborate grilled meat dishes. The menu features chicken and lamb *exohiko* (stuffed with vegetables and local cheese), *kokoretsi* (a dish of lamb and goat offal traditionally served at Easter) as well juicy steaks. Open until 2am.

RHODES Meltemi

Akti Koundourioti **Tel** *22410 30480*

Newly-renovated, unpretentious family-run restaurant right on the beach. The simple menu has a good choice of hot and cold appetizers, grills and fish dishes – a basket of fresh hot bread is served with every meal. Its terrace with sea views makes it an ideal spot for lunch. Closed Dec–Jan.

RHODES La Casa

28 Mandilara, New Town **Tel** *22410 32926*

La Casa offers a good, basic pizza and pasta menu that also includes a selection of grilled meat dishes and some Greek favourites. Not for those in search of an authentic Greek culinary experience but the portions and prices are reasonable and the service prompt and professional. Closed Nov–Mar.

RHODES Alexis

Sokratous 18, Old Town **Tel** *22410 29347*

This wonderful restaurant in the heart of the picturesque Old Town has been specializing in seafood since it opened in 1957. A host of celebrity guests have come to enjoy the perfectly-grilled fresh fish, good wines and professional but friendly service. Choice of tables on two floors and sunny terrace. Better for dinner than lunch. Closed Nov–Mar.

RHODES Mavrikos

Kentriki Plateía, Lindos **Tel** *22440 31232*

Run by two brothers, this award-winning restaurant attracts a host of celebrity diners. Squid in saffron sauce, skate and pine nuts, diced octopus with nutmeg and bulgur wheat are among the offerings, along with home-made ice cream. Located on the main square of Lindos, with sweeping views from the terrace. Closed Nov–Apr.

SÝMI Mylopetra

Sými **Tel** *22460 72333*

A gourmet restaurant with a passion for Mediterranean fusion cuisine and the best of Greek new-wave wines. The menu and wine list here are superb, the restaunt housed in a sensitively restored Sými mansion with a lovely courtyard. Expensive, but well worth a visit. Closed lunch; Nov–Apr.

TILOS Pavlos
Livádia **Tel** *22460 44011*

Small, friendly, family-run taverna and snack bar overlooking a pebbly beach. On the menu here are traditional Greek dishes as well as some international offerings. All meals are made from fresh, locally-sourced ingredients. This is one of the better places to eat on Tílos. Closed Oct–Apr.

THE CYCLADES

AMORGOS Liotrisi
Chóra **Tel** *22850 71700*

This delightful restaurant in the island capital serves oven-cooked traditional dishes as well as Amorgós specialities such as aubergine (eggplant) stuffed with veal and cheese, rabbit *stifádo* and other typical dishes. Tables inside in the shade or outside under umbrellas .

AMORGOS Mourágio
Katápola **Tel** *22850 71011*

In operation since 1981, Mourágio must be doing something right. Filled with tourists and locals alike, it is a friendly taverna right on the port front. Serving up myriad fresh local fish dishes, the speciality of the house is the delicious lobster spaghetti. Closed Jan–Feb.

ANDROS Nona
Plakoura, old harbour, Chóra **Tel** *22820 23577*

This seasonal fish taverna stands beside the old harbour, below Andros's main village. Served up are all the usual favourites, plus some more off-beat seafood choices such as *fousta* (sea violets) and sea urchin roe. At a stone's throw from the sea you can expect the fish here to be very fresh, and it is. Closed Oct–Apr.

ANDROS Sirocco
Above the harbour, Mpatsí **Tel** *22820 41023*

Housed in a 100-year-old building with a wooden interior, this restaurant offers a surprising array of international dishes as well as the expected Greek favourites. Choices include curry dishes such as chicken biriani, pizzas and pasta dishes. Pleasant seafront location and good service. Closed lunch.

FOLEGANDROS Kritikos
Main square, Chóra **Tel** *22860 41219*

A cheerful, open-air grill house in the centre of the island's picturesque main town. Offerings include grilled chicken, pork chops, lamb and beefburgers, with large salad portions, *tsatsiki* (yoghurt with cucumber) and wine by the bottle or from the barrel. Simple and affordable enough to suit all budgets. Closed Oct–May.

FOLEGANDROS I Melissa
Plateía Kontarini, Chóra **Tel** *22860 41067*

The longest established restaurant on the island, I Melissa has been in business for almost half a century. The food is traditional taverna fare, with oven-cooked dishes such as *moussakás* and *pastítsio*, kid in lemon sauce, salads and fried and grilled meat dishes. Omelettes and rice pudding served in the mornings. Closed Oct–May.

IOS Lordos Byron
Chóra **Tel** *22860 92125*

This is a traditional-style ouzeri and *mezedopoleío* with an interesting choice of snacks and small dishes to sample with your ouzo, raki, beer or retsina. The recorded background music is traditional *rembétika*. Closed lunch; Nov–Mar.

IOS Elpis
Mylopotamós Beach **Tel** *22860 91626*

This long-established cafe-restaurant has a loyal following of guests. Opened in 1962 it has been a fixture on the Ios summer scene ever since, serving traditional meals, snacks and fish dishes. Best during the day, when it makes for the perfect spot to enjoy a long, lazy lunch not far from the beach. Closed Nov–Apr.

KYTHNOS Katerini
Loutrá **Tel** *22810 31418*

The best taverna in Loutrá is found just a little way to the west in the neighbouring bay. It is worth the trip however for the fine sea views and tasty oven-cooked dishes such as kid goat braised in red wine. There are also some local island specialities such as *sfougata* (deep-fried egg and cheese croquettes). Good value for money. Closed Nov–Mar.

KYTHNOS Ostria
Harbour front, Mérichas **Tel** *22810 32263*

This is an excellent harbour-side taverna with good fresh seafood, grills and salads and a menu packed with traditional dishes, well-prepared to meet the tastes of a demanding, mainly Athenian, clientele. Local specialities include *sfougata* (egg and cheese croquettes) and *loukánika* (spicy sausages). Good service. Closed Dec–Feb.

Key to Price Guide *see p330* **Key to Symbols** *see back cover flap*

MILOS Kapetan-Nikola
Apollonía **Tel** 22870 41212

Expensive but atmospheric fish taverna with a good seafood menu complemented by the usual grilled meat dishes. There is also a good selection of imaginatively prepared vegetable dishes. Book ahead, especially on summer weekends when it is very popular with visiting Athenians. Closed Dec.

MILOS Aragosta
Adámantas **Tel** 22870 22292

Considered the best restaurant on Milos, Aragosta serves above-average seafood including lobster and crab and grilled fish dishes. There are spectacular views over the bay of Adámantas from the terrace tables. Attached to the restaurant is an attractive cocktail bar for aperitifs or after-dinner drinks. Closed lunch; Oct–Apr.

MYKONOS Kastro
Kástro area, Little Venice, Mýkonos Town **Tel** 22890 23072

This legendary gay bar, overlooking the Kástro district of Mýkonos Town, is a relaxing place to sip strawberry daiquiris and listen to classical music. Laid-back, popular and a good place in Little Venice from which to watch the spectacular sunsets. Closed lunch; mid-Oct–Apr.

MYKONOS Katrine's
Agios Nikolaos 1, Mýkonos Town **Tel** 22890 22169

Katrine's enjoys a long-standing reputation for fine French and Greek cuisine and has an outstanding list of Greek and imported wines. A very good choice of Greek appetizers and more sophisticated main dishes, attentive service and a dressed-up clientele. Booking is advisable. Closed Nov–Mar.

MYKONOS Sea Satin
Mýloi, Mýkonos Town **Tel** 22890 24676

This lively fish restaurant and traditional grill is located by the sea with open-air tables below the village's famous windmills and with views of Little Venice. The restaurant is a delight for seafood lovers and carnivores alike, this is one of the few places you can get a really good steak. Closed lunch; Nov–Apr.

NAXOS Manolis
Bourgos, Náxos Town **Tel** 22850 25168

This garden taverna, in the heart of the old town, serves traditional Greek cuisine in very pleasant surroundings. Dishes on offer include *melitzánes* (fried aubergines), *skordaliá* (potato and garlic purée) and other well-known favourites at prices to suit even smaller budgets. Closed lunch; Nov–Mar.

NAXOS The Old Inn
Náxos Town **Tel** 22850 26093

For those looking for something different, this restaurant in Náxos's harbour village is an unusual find. A German bar-restaurant serving typically German snacks and dishes such as sausage and black bread, Wiener schnitzel, and more sophisticated international cuisine. Smaller portions available for children. Closed lunch.

NAXOS Il Girasole
Náxos Town **Tel** 22850 23326

"The Sunflower" is a decent Italian restaurant serving a range of around 30 pasta and 40 pizza dishes, plus the usual dessert suspects such as tiramisu and panacotta. Located just 50m from the beach, the seafood spaghetti is unsurprisingly a winner. Good selection of Greek and Italian wines. Closed Nov–Feb.

PAROS Porfyra
Waterfront, Paroikiá **Tel** 22840 23410

This friendly, family-run fish taverna is one of the best bargains in busy Paroikiá and as well as the usual assortment of more familiar fish dishes such as *barboúnia* (red mullet) and sea bream it also offers up some more unusual seafood options such as multi-coloured *fouska* (sea violets) and other shellfish. Closed Jan–Feb.

PAROS Levantis
Paroikiá **Tel** 22840 23613

This smart restaurant serves international dishes and contemporary Greek cuisine using the best local produce. The menu changes seasonally to ensure freshness and the emphasis is on light, clean-tasting food. Exotic offerings from around the Mediterranean include North African dishes such as couscous. Closed Oct–Apr.

SANTORINI Nikolas
Erithrou Stavrou, Firá **Tel** 22860 24550

A long-established traditional taverna in the heart of Firá with an authentic Greek menu that makes few concessions to 21st-century influences. Fish, a good choice of local cheeses and salads, and island wines from the barrel. Good views and air-conditioning compensate for the lack of outdoor seating. Closed lunch Sun.

SANTORINI Iliovasilema
Ammoudi, Oia **Tel** 22860 71614

A delightfully simple quayside restaurant where the fish comes straight off the boat and onto the grill. The menu here mainly depends on the catch of the day which can be enjoyed at rickety wooden tables beside a tiny pebbly beach. Extremely good value. Closed Nov–Apr.

SANTORINI 1800

Main street, Oía **Tel** *22860 71485*

Book ahead for an evening at Oía's swankiest restaurant where the sophisticated menu blends Greek and European influences. There is an extensive wine list with a good sampling of Santorini's more interesting labels and vintages. Not too formal, but not ideal for children. Closed lunch; Nov–Apr.

SERIFOS Takis

Waterfront, Livádi **Tel** *22810 51159*

Takis is a sophisticated eating place with a large menu featuring plenty of excellent seafood dishes, salads and organic vegetarian dishes. There is also an extensive wine list, with some of the better wines imported from the islands and the Greek mainland, as well as good local wine from the barrel. Closed Dec–Feb.

SIFNOS Artemónas

Agiou Kostantínou 3, Artemónas **Tel** *22840 31303*

Located in the Artemónas hotel, this restaurant offers good home-made dishes using organic vegetables. Dine in the charming traditional interior with wooden furniture and blue and white tablecloths or sit in the hotel's peaceful garden.

SIFNOS Odos Oneiron

Apollonía **Tel** *22840 33389*

This is Sífnos's classiest and most expensive restaurant, offering a wide choice of Mediterranean fusion-influenced cooking alongside an array of Greek favourites including grilled seafood. The restaurant is housed in a renovated 19th century island mansion and is a delightful place for a romantic evening. Closed lunch; Oct–Apr.

SYROS Archontariki

Emmanouil Roidi 8, Ermoúpoli **Tel** *22810 86771*

Down a side street off the main square in Ermoúpoli, this rustic taverna is a deservedly popular spot for locals and tourists alike. There's live music, a pretty shaded roof terrace and a wide range of traditional dishes and local specialities such as louza sausages. Good local wines also served.

SYROS Iliovasilema

Galissás **Tel** *22810 43325*

This aptly named fish taverna at Galissás, 9km (6 miles) from Ermoúpoli, has an extensive menu of mainly fish dishes On offer are some interesting seafood pasta recipes, and dishes such as stuffed squid, but the menu also features a reasonable choice of grills and traditional oven-cooked dishes. Closed Oct–Apr.

TINOS Paleá Palláda

Kontogiorgi 1, Tínos Town **Tel** *22830 23516*

Large, old-fashioned taverna with indoor and outdoor tables in the main market square of Tínos Town. Serves traditional taverna cuisine with plenty of oven-cooked dishes, grills, salads and *loukánika* (sausages). The local wine from the barrel is better than average, and a strong local clientele lends authenticity. Closed Dec–end Feb.

TINOS Metaxi mas

Tínos Town **Tel** *22830 25945*

A recent addition to Tínos's portfolio of eating places, this classy *mezedopoleío* offers a good choice of snacks and larger meals, an array of classic ouzos to choose from and wine from the barrel or by the glass. Good assortment of dishes for vegetarians as well as meat-eaters, and a very lively atmosphere, especially in the evening. Closed Nov–Dec.

CRETE

AGIA ROUMELI To Farangi

Main square **Tel** *28250 91225*

This family-run taverna is the best place to satisfy your appetite after a hike in the Samariá gorge. Specialities include *sfakaniés pítes* (pastries topped with honey, a favourite of the Sfakiá region). The menu also includes stuffed vegetables, vine leaves, grilled meat dishes and fresh fish from the nearby Libyan Sea. Closed Oct–May.

AGIOS NIKOLAOS To Koutouki

Lassithíou 4A **Tel** *28410 26877*

Affordable *mezedopoleío-ouzerí* offering a wide range of *mezédes* and snacks that change seasonally. Staples include octopus, cheeses, sausage, shrimps and vegetable dips. Traditional drinks include ouzo and raki, and wine and retsina from the barrel or bottle. Good for a light lunch on a hot afternoon.

AGIOS NIKOLAOS Itanos

Plateía Venizélou **Tel** *28410 25340*

Popular taverna in a lane just off Agios Nikólaos's main square. The menu includes spit-roasted chicken, lamb and goat, oven-baked dishes, and an assortment of stuffed and stewed vegetables. Good value and a decent choice of local wines. Seats inside or out.

Key to Price Guide *see p330* **Key to Symbols** *see back cover flap*

CHANIA Tholos €€

Agion Déka 36, 73100 **Tel** *28210 46725*

This reasonably-priced restaurant enjoys a good reputation and uses ingredients sourced from the best local produce. Housed in a 14th-century building with tables set in a courtyard surrounded by old stone walls. Extensive wine list and very busy during the summer months. Closed Nov–Mar.

CHANIA To Pigadi tou Tourkou (The Well of the Turk) €€

1–3 Kallinikou Sarpaki, Splantzia, 73100 **Tel** *28210 54547*

This cosy restaurant is housed in the arched stone cellar of an ancient Chaniá mansion, with candle-lit tables and Cretan and Levantine background music. The menu is more imaginative than most – try the aubergine meatballs, or the spicy stuffed squid – and uses local produce. Closed lunch; Wed; mid-Nov–Feb.

CHANIA Dínos €€€

8 Akti Enoseos, 73100 **Tel** *28210 57448*

One of Chaniá's best seafood restaurants, Dínos is on the quayside of Chaniá's delightful inner harbour. Sit at traditional blue wooden tables and chairs with white tablecloths and order from an extensive choice of seafood, from octopus and whitebait to langouste and sea bass. Closed Nov–Mar.

CHANIA To Karnagio €€€

8 Katehaki, 73100 **Tel** *28210 53366*

This old-style taverna on the harbour continues to serve good fish and solid traditional favourites including *moussakás, pastítsio*, grilled lamb, pork, chicken and large salads with feta cheese. Wine is served from the barrel. Good for a hearty meal on a limited budget. Closed Nov–Apr.

CHANIA Ela €€€€

42 Kondylaki, 73100 **Tel** *28210 74128*

Ela's is located within the stone-paved courtyard of an attractively restored Venetian mansion in the centre of Chaniá's old quarter. A wide menu which includes the run-of-the-mill Greek offerings and a sprinkling of authentic Cretan dishes along with a range of options that owe much to its multi-national clientele.

CHANIA Tamam €€€€€

49 Zambeliou, 73100 **Tel** *28210 96080*

Tamam serves a wide range of mainstream Greek, Cretan and eastern Mediterranean-influenced dishes. Lots of dips – hummus, *fáva, tzatzíki* – and cooked vegetable dishes (courgette fritters, fried chickpea balls) make this one of the few decent options for vegetarians in Crete.

ELOUNTA Argo €€€€€

Eloúnta Gulf Villas, Eloúnta **Tel** *28410 90300*

This is the smart restaurant of the upmarket Eloúnta Gulf Villas resort, serving some outstanding and imaginative dishes with real Mediterranean flair, such as lobster and crayfish risotto. More traditional dishes, such as meat in filo pastry with yoghurt, are also on offer. Service is professional and the wine list good. Closed Nov–Mar.

ELOUNTA The Old Mill €€€€€

Eloúnta Mare Hotel **Tel** *28410 41102*

Located in one of the island's grandest hotels, The Old Mill is the most luxurious restaurant in eastern Crete. The menu combines the best of Greek flavours and island produce with international cuisine, and the extensive wine list allows sampling of some of the finest vintages from Crete and elsewhere in Greece. Booking essential. Closed Nov–Mar.

IERAPETRA Odeion €

Lasthenous 18 **Tel** *28420 27429*

This stylish *mezedopoleío* and café-bar is housed in an elegant Neo-Classical mansion. Younger locals and summer visitors come for the range of cocktails, imported beers and local wines, as well as for the good selection of *mezédes* and light meals. Tables inside and out. Closed lunch; Sun in winter.

IRAKLEIO O Kyriákos €

Leofóros Dimokratías 53 **Tel** *28102 22464*

You will be beckoned into the kitchen of this old-fashioned restaurant to choose your meal from bubbling pots or glass cases filled with fish, vegetables and cuts of meat. O Kyriakós is an Irákleio institution, unpretentious and catering to a loyal local following. Wine from the barrel.

IRAKLEIO Loukoulos €€€€

Korai 5 **Tel** *28102 24435*

A mainly Italian menu in up-market surroundings – with white linen tablecloths, candles in the evening, and a dining room decorated with antique prints and paintings. With its excellent food, Loukoulos has been regarded for years as one of the best restaurants in Irákleio.

MAKRYGIALOS Spilia tou Drakoú €€

Paralía Kaló Neró **Tel** *28430 51494*

Above the beach at Kaló Neró, this is one of the better places to eat in Makrýgialos. Sofía and Vasíli Drakoú serve hearty, traditional grills and oven-cooked casserole dishes, as well as a choice of fish dishes and lighter snacks. Friendly atmosphere, and nice views of the sea and the surrounding hills.

MARATHI Pandelis
Marathi Beach **Tel** *22470 32609*

A simple beach taverna by a small sandy cove, and run by a Greek-Australian couple. There are tables among lots of shady greenery, and an imaginative menu – try the octopus with caper leaves, spit-roasted goat or the vegetable fritters. Also has several simple rooms to rent and serves breakfast, lunch and dinner daily. Closed Nov–Apr.

PALAICHORA Oriental Bay
Anatoliko Paralia **Tel** *28230 41322*

A little family restaurant beneath the palm trees of Palaiochora's long, sandy eastern beach. On the menu is a reasonable choice of Cretan dishes, Greek standards and international favourites, all at affordable prices. This is a good spot for lunch beside the sea. Choice of Cretan and Greek wines. Closed Oct–Apr.

RETHYMNO Kyria Maria
20 Moschovitou, 74100 **Tel** *28310 29078*

This colourful little taverna on a narrow lane in Rethymno's old quarter has a small menu of plain, village-style cretan dishes. Its busy during the evenings (get there early to find a table) but quieter at lunchtime. Friendly service, good food and very good value. Closed Nov–Mar.

RETHYMNO Avlí
22 Xanthoudidou, 74100 **Tel** *28310 26213*

This charming restaurant offers dining in vaulted rooms or in the courtyard of an old mansion. The menu consists mainly of grilled and roast meat and there is an extensive wine list featuring some of the best Greek wines. Attentive service and spotless interior.

RETHYMNO Veneto
Epimenidou 4 **Tel** *28310 56634*

The fine dining restaurant of one of Rethymno's smartest boutique hotels is the place to sample some of Greece's best wines. The elegantly prepared meals include lamb in vine leaves, cuttlefish-ink pasta, prawns in saffron sauce, and Byzantine spicy meat balls. Tables indoors, in a cool, stone-vaulted cellar, or in a courtyard. Closed Nov–Apr.

SITEIA Zorbas
Harbour front, Siteia **Tel** *28430 22689*

This traditional-style taverna on the waterfront is favoured as much for its attractive location as for its menu. On offer is the usual array of grilled meat and fish dishes, oven-baked *moussakás*, beans and salads. There are also more expensive dishes such as grilled lobster, octopus and *barboúnia* (red mullet). Adequate choice of local wines.

VAMOS Sterna tou Bloumofisi
Vamos **Tel** *28250 22932*

Perhaps the best place in Crete to sample the real Cretan diet, with dishes such as snails, shredded broad beans, wild artichokes, cuttlefish in fennel, rooster in wine, and oven-cooked rabbit, goat and chicken. Set in an authentic taverna dating from the 19th century. Fine selection of local wines. Closed Dec–Feb.

ATHENS

EKALI Wine Bar 103
Thisseos 103, 14565 **Tel** *21062 60400*

Modern, French-influenced cooking is the keynote of Avenue 103, the trendy restaurant of Athens's newest "style" hotel in the up-and-coming suburb of Ekali. The menu here is influenced by the Greek seasons and there is also an excellent choice of Greek and imported wines. Closed Sun & Mon evenings.

GAZI Brothel
Orpheos 33 & Dekeleon, 11853 **Tel** *21034 70505*

This happening bar-restaurant in the lively former industrial quarter of Gazi (gasworks), attracts a lively crowd of young Athenians. The décor, with the emphasis on red and black, is somewhat Gothic, and this is no place for a quiet evening. Imaginative menu. Closed Jun–Aug.

KIFISSIA Vardis
Pendelikón Hotel, Diligiannis 66, Kifissiá, 14562 **Tel** *21062 30650*

You will probably want to dress up if you plan to eat at this grand hotel restaurant in the upmarket suburb of Kifissiá. Vardis is one of the elite few Michelin-starred eating places in Greece. Its menu focuses on contemporary French cooking with a Greek twist, using the best regionally sourced ingredients. Closed lunch; Sun.

KOLONAKI Le Grand Balcon
St George Lycabettus Hotel, 10675 **Tel** *21072 90711*

It is worth visiting Le Grand Balcon, the St George Lycabettus Hotel's rooftop restaurant-bar, if only for the magnificent view of Athens at night. The food is good too and the wine list features some of the better wines that have begun emerging from Greek vineyards in recent years. Good cocktails.

Key to Price Guide *see p330* **Key to Symbols** *see back cover flap*

YKAVITTOS Orizontes Lykavittou   €€€€€

Lofos Lykavittou, 10673 **Tel** *21072 27065*

A mainly Athenian clientele come to savour the modern cuisine and good wine list of this excellent restaurant. The service is attentive (and multi-lingual) and the food is modern Mediterranean. Wonderful views over the city from its location on Lykavittós Hill.

MAKRIGIANNI Strofi   €€€€

25 Rovertou Galli, 11742 **Tel** *21092 14130*

The view here is outstanding, especially at night, when your table has a prospect of the floodlit columns of the Acropolis. Good, solid, traditional Greek dishes and rapid service. Reservations recommended, especially during summer evenings. Closed lunch; Sun.

MONASTIRAKI Ydria   €€€

Andrianoú 68, corner of Aiólou 10555 **Tel** *21032 51619*

Just around the corner from the Tower of the Winds, this sprawling café is the best choice in a square packed with other cafés. The outside seating is especially lovely and shaded by Mulberry trees. The menu offers an extended grill menu but it's the fine coffees, cakes and cocktails that make it a favourite hangout with young Athenians.

MONASTIRAKI To Koutí   €€€€

Adrianou 23, 10555 **Tel** *21032 13229*

A few steps from the flea-market area, To Koutí is a great place for a meal, a snack, or just coffee, with tables outside and inside an attractive old building with mellow painted walls and marble tables. This restaurant is good for salads and grilled meat dishes, with a laid-back atmosphere during the day.

OMONOIA Ideal   €€€

Panepistemiou 46, 10678 **Tel** *21033 03000*

A grand, Art-Deco interior and a menu that includes international standard dishes as well as Greek classics such as milk-fed veal with aubergine, stuffed courgettes and artichokes in lemon juice. Ideal has been an Omonoia favourite since it opened in 1922. Closed Sun.

PANGRATI Spondi   €€€€€

Pyrronos 5, 11636 **Tel** *21075 20658*

This Michelin-star restaurant combines Greek and French produce and influences, with a fusion à la carte menu that changes at least twice a year. It also boasts a remarkable wine list. Tables outdoors during summer in a paved courtyard, or indoors in a cool, vaulted cellar. Very stylish. Reservations recommended. Open for dinner only.

PLAKA Scholarchio   €

Tripodon 14, 10558 **Tel** *21032 47605*

A cheap and cheerful, split-level *ouzeri-mezedopoleio* located on a pedestrian street below the Acropolis. The food is uncomplicated and served in traditional fashion – you pick each dish from the waiter's tray. Eat inside, or on a shaded terrace with fine views.

PLAKA Klimataria   €€

Klepsydras 5, 10558 **Tel** *21032 11215*

Just a stone's throw from the Acropolis, Klimataria is a friendly, rustic Greek taverna with authentic 1960s aesthetics. It offers typical dishes such as *moussaka*, *tzatziki* and slow-cooked meat dishes and breads made in its own bread oven. At weekends there is live guitar and accordion music. Don't be surprised if your neighbour starts to sing along!

PLAKA O Damigos   €€

Kydathinaion 41, 10558 **Tel** *21032 25084*

A Plaka institution, this venerable basement restaurant (hidden away beneath a bar) is famous for cod and chips, Athenian-style, served with *skordaliá* (mashed garlic) and washed down with pungent retsina from the barrel. Inside, ancient stone pillars support its ceiling. Closed lunch.

PLAKA Daphne's   €€€€€

Lysikratous 4, 10557 **Tel** *21032 27971*

A grand, ornate restaurant with an impressive list of famous visitors, including several heads of state. Walls are graced by colourful frescoes, and there are tables in a choice of inner rooms or an outdoor courtyard. Stick to the simpler dishes such as grilled fish or meat. Extensive list of Greek wines. Closed lunch.

THISEIO Abibagio   €€

Iraklidon 3, 11851 **Tel** *21034 26794*

One of the capital's best *mezedopoleio* restaurants where you can sample countless small dishes ranging from half a dozen kinds of cheese, salted and pickled anchovies and herrings, vegetable dishes, and dips such as *taramasaláta*, hummus, *fáva* and *tsatsíki*. Ouzo, beer and retsina are the favoured drinks.

THISEIO Pil Poul   €€€€

Corner of Apostolou Pavlou & Poulopoulou, 11851 **Tel** *210 342 3665*

Housed in a 1920s Neo-Classical mansion, Pil Poul serves fashionable Mediterranean cooking with a strong French influence. The magnificent views of the Acropolis from its rooftop terrace draw the crowds to this busy and expensive restaurant. Smart dress and reservations essential. Closed lunch; Sun; May.

SHOPPING IN GREECE

Shopping in the Greek islands can be an entertaining pastime, especially when you buy directly from the producer. This is often the case in the smaller villages, where crafts are a major source of income. Embroiderers and lace makers can

Honey from Evvoia

often be seen sitting outside their houses, and potters can be found in their workshops. Apart from these industries, and the food and drink produced locally, most other goods are imported to the islands and therefore carry a heavy mark-up.

Olive-wood bowls and other souvenirs from Corfu Old Town

VAT AND TAX FREE SHOPPING

Usually included in the price, FPA (*Fóros Prostitheménis Axías*) – the equivalent of VAT or sales tax – is about 18 per cent in Greece.

Visitors from outside the EU staying less than three months may claim this money back on purchases over 117 euros. A "Tax-Free Cheque" form must be completed in the store, a copy of which is then given to the customs authorities on departure. You may be asked to show your receipt or goods as proof of purchase.

OPENING HOURS

Allowing for plenty of exceptions, shops and boutiques are generally open on Monday, Wednesday and Saturday from 9am to 2:30pm, and on Tuesday, Thursday and Friday from 9am to 2pm and 5pm to 8pm. Supermarkets, found in all but the smallest communities, are often family-run and tend to stay open longer hours, typically Monday to Saturday from 8 or 9am to 8 or 9pm. Sunday shopping is possible in most tourist resorts. The corner *períptero* (street kiosk), found in nearly every town, is open

from around 7am to 11pm or midnight, selling everything from aspirins to ice cream.

MARKETS

Most towns in the Greek islands have their weekly street market (*laïki agorá*), a colourful jumble of the freshest and best-value fruit and

Basket of herbs and spices from a market stall in Irákleio, Crete

vegetables, herbs, fish, meat and poultry – often juxtaposed with a miscellany of shoes and underwear, fabrics, household items and sundry electronic equipment.

In larger towns, the street markets are in a different neighbourhood each day, usually opening early and packing up by about 1:30pm, in time for the afternoon siesta. Prices are generally cheaper than in the supermarkets, and a certain amount of bargaining is also acceptable, at least for non-perishable items.

FOOD AND DRINK

Culinary delights to look out for in the shops and markets of the Greek islands include honey, pistachios, olives, herbs and spices. Good cheeses include the salty feta, and the sweet *anthótyro* from Crete; for something sugary, try the numerous pastries and biscuits (cookies) of the *zacharoplasteío*.

Greece is also well known for several of its wines and spirits. These include brandy, ouzo (an aniseed-flavoured spirit), retsina (a resinated wine) and, from Crete, the firewater known as raki.

SIZE CHART

Women's dresses, coats and skirts

Greek		44	46	48	50	52	54	(size)
GB/Australian		10	12	14	16	18	20	(size)
US		8	10	12	14	16	18	(size)

Men's suits, shirts and jumpers

Greek	44	46	48	50	52	54	56	(size)
GB/US	34	36	38	40	42	44	46	(inches)
Australian	87	92	97	102	107	112	117	(cm)

Women's shoes

Greek	36	37	38	39	40	41	(size)
GB	3	4	5	6	7	8	(size)
US/Australian	5	6	7	8	9	10	(size)

Men's shoes

Greek	40	41	42	43	44	45	(size)
GB/Australian	7	7½	8	9	10	11	(size)
US	7½	8	8½	9½	10½	11½	(size)

What to Buy in Greece

Traditional handicrafts, though not particularly cheap, do offer the most genuinely Greek souvenirs. These cover a range of items from finely wrought gold reproductions of ancient Minoan pendants to rustic pots, wooden spoons and handmade sandals. Leatherwork is particularly noted on the island of Crete, where the town of Chaniá *(see p252)* hosts a huge leather market. Among the islands renowned for their ceramics are Crete, Lésvos and Sífnos. Many villages throughout the Greek islands produce

Rug from Anógeia, Crete

brightly coloured embroidery *(kéntima)* and wallhangings, which are often hung out for sale. You may also see thick *flokáti* rugs. They are handwoven from sheep or goat's wool, but are more often produced in the mountainous regions of mainland Greece than on the islands themselves. In the smaller island communities, crafts are often cottage industries, which earn the entire family a large chunk of its annual income during the summer. There is usually room for some bartering when buying from the villagers.

Gold jewellery *is sold mainly in larger towns. Modern designs are found in jewellers such as Lalaounis, and reproductions of ancient designs in museum gift shops.*

Icons *are generally sold in shops and monasteries. They range from very small portraits to substantial pictures. Some of the most beautiful, and expensive, use only age-old traditional techniques and materials.*

Ornate utensils, *such as these wooden spoons, are found in traditional craft shops. As here, they are often hand-carved into the shapes of figures and produced from the rich-textured wood of the native olive tree.*

Kombolói, *or worry beads, are a traditional sight in Greece; the beads are counted as a way to relax. They are sold in souvenir shops and jewellers.*

Kitchenware *is found in most markets and in specialist shops. This copper coffee pot (mpríki) is used for making Greek coffee.*

Leather goods *are sold throughout Greece. The bags, backpacks and sandals make useful and good-value souvenirs.*

Ornamental ceramics *come in many shapes and finishes. Traditional earthenware, often simple, functional and unglazed, is frequently for sale on the outskirts of Athens and the larger towns of the islands.*

SPECIALIST HOLIDAYS AND OUTDOOR ACTIVITIES

If you feel you want more of a focus to your holiday in the Greek islands, there are many organized tours and courses available that cater to special interests. You can visit ancient archaeological sites with a learned academic as your guide, you can improve your writing skills, paint the Greek landscape or learn the Greek language, learn to cook Greek food and appreciate Greek wines, or develop your spirituality. All kinds of

Moped in Rhodes

walking tours, as well as botanical and bird-watching expeditions, are available in the islands. So too are golf, tennis, cycling and horse-riding holidays. If you prefer to be pampered or rejuvenated, Greek spas now rival the best in Europe, and there is even a naturist hotel on Crete for help with the all-over tan. Information on sailing and watersports, and advice on choosing the perfect beach are covered on pages 348–9.

Visitors at the ancient theatre at Delos (see pp218–19)

ARCHAEOLOGICAL TOURS

For those interested in Greece's glorious ancient past, a tour to some of the famous archaeological sites, accompanied by qualified archaeologists, can make for a fascinating and memorable holiday. In addition to visiting ruins, many tours take in Venetian fortresses, Byzantine churches, caves, archaeological museums and monasteries along the way.

Filoxenia organizes tours with Minoan, Roman and

Tourists visiting caves near Psychró, in Crete

medieval interests. The island of Crete is one of their main destinations, taking in sites at Réthymno (see pp258–9), as well as the Minoan palace at Knosós (see pp272–3). History and archaeology specialists **Andante Travels** also operate tours of Minoan Crete, with two specialist guest lecturers accompanying the group.

CREATIVE HOLIDAYS

With their vivid landscapes and renowned quality of light, the Greek islands are an inspirational destination for artistic endeavour. Courses in creative writing, and drawing and painting, are available at all levels.

The Skyros Centre (see p116), on the island of the same name, offers two locations – one at the main town and another at the

remote village of Atsítsa – for self-development and therapeutic holidays, including themes directed towards writing and painting as well as yoga. **Filoxenia** also offers various creative holidays on several of the islands. **AegeanScapes** runs painting and Raku pottery holidays on Pátmos and Páros. **Simpson's Greece** organises drawing and painting holidays focusing on the landscape of Crete. Their courses cater for beginners through to advanced level. For the flexible, yoga holidays on Paxós and Corfu are available through **Travel à la Carte**.

GREEK LANGUAGE COURSES

Immersing yourself in a language is the best and most enjoyable way to learn. Greek language courses at all levels are available in Límni on the island of Evvoia and on Sýros. The courses can be booked through the Greek company of **Omilo** in Athens, who offer a variety of courses from two weeks up to eight weeks.

NATURE HOLIDAYS

The Greek islands are rich in natural beauty, and you need not be a fanatical botanist or ornithologist to enjoy the stunning wild flowers and variety of birdlife. Spring is the best time to explore the countryside, when the colourful flowers are in bloom, especially on lush islands such as Corfu and on mountainous Crete. It is also a good time to see the influx of migrating birds, which rest and feed in Greece on their journeys between Africa and Europe.

The **Hellenic Ornithological Society** details further information on wild birds and their habitats as well as related activities and events. **Limosa Holidays** is a specialist tour operator offering trips centred around bird-watching and botany. They have established tours to several islands, as has **The Travelling Naturalist. Island Holidays** offers similarly themed tours on Crete. **Simpson's Greece** also explore the wildlife of Crete on their specialist walking tours. The tours are adapted to the needs and abilities of

A chameleon, found mainly on Crete

the group but generally operate at a relaxed pace.

More information on the wildlife of Crete and other specialist tour operators is given on pages 246–7. Note that these types of holidays also incorporate into the tours visits to nearby historical and archaeological sites.

WALKING AND TREKKING

The hills of the Greek islands are a walker's paradise, particularly between March and June, when the countryside is verdant, the sun is not too hot and wild flowers abound. Many of the islands provide fine locations and scenery in which to walk, and the lack of too many organised trails gives a greater sense of freedom and discovery.

Trekking Hellas arranges walking holidays in the White Mountains of Crete, and on Andros and Tínos in the Cyclades. **Sherpa Expeditions** leads tours through the mountainous interior of western Crete, including the Samariá Gorge *(see pp254–5)*, and **Ramblers Holidays** offers

Walkers climbing Mount Idi in central Crete

walking throughout the Greek islands, including some of the lesser-visited islands such as Nísyros and Ikaría.

Simply Crete arranges walking and trekking tours on Crete, with a professional tour leader, while **Inntravel** features walking tours of Crete, Lésvos and Sámos. Walking tours of Crete to see the spring flowers are available through **Freelance Holidays** and from **Simpson's Greece. Explore** organises walks along the Corfu Trail, a walking holiday in Crete, and other trips including visiting several of the Aegean and Cyclades islands, while **Travelsphere** has walking in Crete and **Waymark Holidays** has walking in both Crete and Lésvos.

For the independent trekker, guides such as *The Mountains of Greece: A Walker's Guide* (Cicerone Press), and the various Sunflower Guides dealing with the Greek islands are invaluable sources of infor-

mation. If you are not one for the hardy mountain hike, there are plenty of less strenuous options too.

Trails in Greece are not generally marked as well as in many other countries in Europe, with exceptions such as the excellent Corfu Trail. The Greek way is much simpler than signposts: they put a blob of red paint on rocks and walls, to indicate the path. Needless to say, these do not always work as well as they should.

On the positive side, many of the islands have locally published booklets or leaflets containing walks, which can be bought in shops, though some of them are available for free, supported by local walking groups and organisations. In the Ionian Islands, for example, there is a wide range of excellent walking leaflets published by **Friends of the Ionian**. These leaflets are widely available.

CRUISES AND BOAT TRIPS

Greece's unique combination of natural beauty and fascinating history makes a cruising holiday both relaxing and stimulating. Greek cruises run between April and October, and there are a variety of options available, ranging from a full luxury cruise to short boat trips.

Odyssey Sailing Greece provide information on a wide range of available options,

Daytrip boats in Mandráki Harbour, Rhodes

from economy cabin cruises to fully crewed VIP motor yachts. Operators such as **Swan Hellenic Cruises, Travelsphere** and **Voyages of Discovery** in the UK, **Metro Tours** and **Hellenic Holidays** in the US, offer all-inclusive holidays onboard large luxury liners, with guest speakers versed on a range of subjects from archaeology to marine biology. Such cruises tend to incorporate the Greek islands into extensive routes from Italy to the Middle East, or to the Black Sea.

Explore runs week- and fornight-long cruises aboard a traditional Greek caïque. A more informal option is to take a trip on one of the graceful tall ships operated by **Star Clippers**, who have various routes linking Athens with Venice or Istanbul, or through the Cycladic islands.

There are also less extensive boat trips to nearby islands and places of interest. Organized locally, these trips are best booked on the spot.

CYCLING AND MOPEDS

Freewheeling cyclists can hire bikes at most holiday resorts, including the latest mountain bikes, but more organised options are available on Crete from **Simpson's Greece** and on Kefalloniá with **Explore**. Even the smallest resorts will also have moped, scooter and perhaps motor-bike rental agencies.

Mopeds are a cheap and easy way of getting about, but holidaymakers are advised to use them with caution, especially if you do not normally drive one when at home. In fact some tour operators discourage their clients from renting them. Island roads can have many rough patches, with sudden potholes or patches of loose gravel, causing mopeds and scooters to skid and frequently come off the road. Greek car drivers also drive aggressively, some with little regard for vul-nerable moped users. Accidents are so commonplace that anyone rents a scooter or moped at their own risk.

HORSE RIDING

Those who prefer horse riding are also well catered for, with **Unicorn Trails** organising trips to Kefalloniá, including the chance to swim with your horse in the sea, and to Crete, with its mountainous terrain. Riding in Corfu features in the programme of **Equitour**, based at the Vassilika Stables in the Rópa Plain to the south *(see p82)*.

GOLF AND TENNIS

Tucked away on the Rópa Plain on Corfu is one of the best courses in Europe, where you can play as a guest if you happen to be there on holiday, or you can organise a special tour out there with golfing specialists such as **Bill Goff Golf Tours** and **3D Golf**. Golf courses on the islands are not widespread – apart from mini-golf and crazy-golf! But **Golf Afandou** at Afántou on Rhodes has an 18-hole course, and there are two courses on Crete: the **Crete Golf Club** in Chersónisos and the Porto Elounda Golf Course located within the **Porto Elounda Resort**. For information on these and other golf courses on the mainland, contact the **Hellenic Golf Federation**.

Tennis players would be advised to book a holiday at one of the bigger hotels, many of which have their own tennis courts. Municipal clubs and private clubs do exist but tend not to be as good. The Portomyrina Hotel on Límnos has three courts and two tennis coaches available, with special tennis holidays bookable through **Neilson Active Holidays**.

NATURISM

Nude sunbathing is only allowed in Greece on desig-nated nudist beaches, but in practice people strip off on quiet beaches all over the islands. As long as the beaches are reasonably private and you do not offend local people, there is seldom a problem.

There is one licensed naturist hotel in the Greek islands: the **Vritomartis Hotel**

near Sfakiá on the south coast of Crete. It is a delightful hotel, and also welcomes non-naturist guests as naturism is only practiced around the swimming pool and at the beach, and not in any indoor areas.

SPAS

Greece is well endowed with natural hot springs – a result of volcanic activity – and several islands have developed these as spas, offering such treatments as hydrotherapy, physiotherapy and hydromassage.

The main centres are listed on the EOTs (Greek Tourist Offices) information sheet *Spas in Greece*, and include Kos and Nísyros in the Dodecanese, Ikaría, Lésvos and Límnos in the Northeast Aegean group, Zákynthos in the Ionians and Kýthnos in the Cyclades.

Some of the large resort hotels also have excellent spa facilities, most notably around Eloúnta *(see p278)* on Crete and on upmarket islands such as Mýkonos and Santoríni.

FOOD AND WINE

In medieval times Greece produced the best wine in Europe and after a long lull, when a lot of Greek wine was barely drinkable, today's wine makers are redis-covering their skills. A cruise which visits several of the country's leading vineyards can be booked through UK wine tour specialists, **Arblaster and Clarke**.

There is a greater interest in Greek cuisine recently too, and cookery holidays on the island of Sými are available with **Simply Crete**.

Also offering holidays that combine the culture, food and wine of Crete are **Simpson's Greece**. Their specialised tours take you into the homes of ordinary Cretans, where you can savour traditional home-cooked food and local wines. The tours are orga-nised thematically and look at local activities, such as organic olive farming and wine making.

DIRECTORY

ARCHAEOLOGICAL TOURS

Andante Travels
The Old Barn, Old Road, Alderbury, Salisbury, SP5 3AR, UK.
Tel 01722 713800.
www.andantetravels.co.uk

Filoxenia
Inntravel, Castle Howard, York, YO60 7JU, UK.
Tel 01653 617755.
www.filoxenia.co.uk

CREATIVE HOLIDAYS

Aegean Scapes
Karali 79, Papagou 15669 Athens, Greece.
Tel 210 641 0972.
www.aegeanscapes.com

Filoxenia
(See Archaeological Tours.)

Simpson's Greece
44 Sheen Lane, East Sheen, London, SW14 8LP, UK.
Tel 0845 811 6502.
www.simpsontravel.com

Travel à la Carte
The White House, Drove Lane, Thatcham, RG18 9NL, UK.
Tel 01635 201250.

GREEK LANGUAGE COURSES

Omilo
PO Box 61070, 15101 Maroussi, Athens.
Tel 210 612 2896.
www.omilo.com

NATURE HOLIDAYS

Hellenic Ornithological Society
Vasileos Irakleiou 24, 10682 Athens.
Tel 210 822 7937.
www.ornithologiki.gr/en/enmain.htm

Island Holidays
PO Box 26317, Comrie, Crieff PH6 2YL, UK.
Tel 01764 670107.
www.islandholidays.co.uk

Limosa Holidays
Suffield House, Northrepps, Norfolk, NR27 0LZ, UK.
Tel 01263 578143.
www.limosaholidays.co.uk

Simpson's Greece
(See Creative Holidays.)

The Travelling Naturalist
PO Box 3141, Dorchester, Dorset, DT1 2XD, UK.
Tel 01305 267994.
www.naturalist.co.uk

WALKING AND TREKKING

Explore
55 Victoria Road, Farnborough, Hants, GU14 7PA, UK.
Tel 0870 333 4001.
www.explore.co.uk

Freelance Holidays
www.freelance-holidays.co.uk

Friends of the Ionian
www.foi.org.uk

Inntravel
Castle Howard, York, YO60 7JU, UK.
Tel 01653 617906.
www.inntravel.co.uk

Ramblers Holidays
Lemsford Mill, Lemsford Village, Welwyn Garden City, Herts, AL8 7TR, UK.
Tel 01707 331133.
www.ramblersholidays.co.uk

Sherpa Expeditions
131a Heston Road, Hounslow, Middlesex, TW5 0RF, UK.
Tel 0208 577 2717.
www.sherpaexpeditions.com

Simply Crete
www.simplytravel.co.uk

Travelsphere
Compass House, Rockingham Road, Market Harborough, Leicestershire, LE16 7QD, UK.
Tel 0800 191418.
www.travelsphere.co.uk

Trekking Hellas
Filellinou 7, 10557 Athens.
Tel 210 331 0323.
www.trekking.gr

Waymark Holidays
44 Windsor Road, Slough, Berkshire, SL1 2EJ.
Tel 01753 516477.
www.waymarkholidays.com

CRUISES AND BOAT TRIPS

Explore
(See Walking & Trekking.)

Hellenic Holidays
1501 Broadway, Suite 1512, New York, NY 10036, USA.
Tel 212 944 8288.
www.hellenicholidays.com

Metro Tours
484 Lowell St, Peabody, MA 01960, USA.
Tel 800 221 2810.
www.metrotours.com

Odyssey Sailing Greece
Antonopoulo 158D, 38221 Volos.
Tel 24210 36676/58116.
www.odysseysailing.gr

Star Clippers
Crown House, Crown Street, Ipswich IP1 3HS, UK.
Tel 01473 292029
www.starclippers.co.uk

Swan Hellenic Cruises
Richmond House, Perminus Terrace, Southampton SO14 3PN, UK.
Tel 0845 355 5111.
www.swanhellenic.com

Travelsphere
(See Walking & Trekking.)

Voyages of Discovery
1 Victoria Way, Burgess Hill, W. Sussex, RH15 9NF, UK.
Tel 0845 612 9149.
www.voyagesofdiscovery.com

CYCLING AND MOPEDS

Explore
(See Walking & Trekking.)

Simpson's Greece
(See Creative Holidays.)

HORSE RIDING

Equitour
Petershold Farm, Minstead, Lyndhurst, Hants, SO43 7GJ, UK.
Tel 0800 043 7942.
www.equitour.co.uk

Unicorn Trails
2 Acorn Centre, Chestnut Avenue, Biggleswade, Beds, SG18 0RA, UK.
Tel 01767 600606.
www.unicorntrails.com

GOLF AND TENNIS

Bill Goff Golf Tours
Unit 9, Bankside, The Waterfront, Gateshead NE11 9SY, UK.
Tel 0870 4012020.
www.billgoff.com

Crete Golf Club
PO Box 106, 70014 Hersonissos, Crete.
Tel 28970 26000.
www.crete-golf.com

Golf Afandou
Afántou Bay, Rhodes.
Tel 22410 51451.
www.afandougolfcourse.com

Hellenic Golf Federation
www.hgf.gr

Neilson Active Holidays
Locksview, Brighton Marina, Brighton BN2 5HA. *Tel 0870 333 3356.*
www.neilson.co.uk

Porto Elounda Resort
Elounda, Crete 72053.
Tel 28410 6800.
www.portoelounda.com

3D Golf
61 New Road, Ayr, Scotland KA8 8DA, UK.
Tel 01292 293001.
www.3dgolf.com

NATURISM

Vitromatris Hotel
Chora Sfakion, Crete.
Tel 282 509 1112.
www.vritomartis.gr

FOOD AND WINE

Arblaster and Clarke
Farnham Road, West Liss, Hants, GU33 6JQ, UK.
Tel 01730 8933440.
www.winetours.co.uk

Simpson's Greece
(See Creative Holidays.)

BEACHES AND WATERSPORTS

With hundreds of islands, crystal-clear seas and beaches of every kind, it is not surprising that so many water-lovers are attracted to Greece. Although people swim most of the year round, the main season for watersports is from late May to early November. All kinds of watersports can be

Beach parasol

enjoyed, especially in the larger and more developed resorts, and rental fees are still quite reasonable compared with other Mediterranean destinations. But if you prefer a more leisurely vacation, you can always choose from the many beautiful and tranquil beaches to be found on the islands.

Holiday company flags flying on "Golden Beach", Páros

BEACHES

Beaches vary greatly in the Greek islands, offering everything from shingle and volcanic rock to gravel and fine sand. The Cyclades and Ionian Islands are where the sandy beaches tend to be, and of these the best are usually on the south of the islands. Crete's beaches are also mostly sandy, but not exclusively. The Northeast Aegean, Dodecanese and Sporades are a mixture of sandy and pebbly beaches. Some islets, such as Chálki and Kastellórizo, have few or no beaches at all. But, in compensation, they often have very clear seas, which can be good for snorkelling.

Swimmers diving off the boards at a pool by the beach on Rhodes

Any beach with a Blue Flag (awarded annually by the Hellenic Society for the Protection of Nature, in conjunction with the European Union) is guaranteed to have its water tested every 15 days for cleanliness and purity, as well as meeting over a dozen other environmental criteria. These beaches tend to be among the best, and safest for children, though they can be very crowded.

Also worth trying out are beaches recommended in the headings for each entry in this guide. Occasionally the main beach near the port of an island is run by the EOT (Greek Tourist Office). There will be a charge for its use, but it will be kept clean and often have the added benefit of showers. Topless bathing is widespread, though nude bathing is still officially forbidden, except on a few designated beaches; it is never allowed within sight of a church.

The Greek seas are generally safe and delightful to swim in, though lifeguards are almost non-existent in Greece. Every year there are at least a few casualties, especially on

windy days when the sea is rough and there are underwater currents. Sharks and stingrays are rare around beaches, but more common are sea urchins and jellyfish. Both can be painful, but are not particularly dangerous.

WATERSPORTS

With so much coastline, facilities catering for watersports are numerous. Windsurfing has become very popular, and waters recommended for this include those around Corfu, Lefkáda and Zákynthos in the Ionian islands, Lésvos and Sámos in the Northeast Aegean, Kos in the Dodecanese, Náxos in the Cyclades and the coast around Crete. The **Hellenic Water-ski Federation** can offer the best advice. For a little more money you could take up water-skiing or jet-skiing; and at the larger resorts parasailing is also available. If you need instruction, you will find that many of the places that rent equipment also provide tuition.

Holiday-makers learning the skills of windsurfing in coastal waters

Hire centre for watersports equipment, Rhodes

SCUBA AND SNORKELLING

The amazingly clear waters of the Mediterranean and Aegean reveal a world of submarine life and archaeological remains. Snorkelling (see pp24–5) can be enjoyed almost anywhere along the coasts, though scuba diving is severely restricted. Designated areas for diving are around Crete, Rhodes, Kálymnos and Mýkonos, and also around most of the Ionian Islands. A complete list of places where it is permissible to dive with oxygen equipment can be obtained from the EOT, or by mail from the **Department of Underwater Archaeology** in Athens. Wherever you go snorkelling or diving, it is strictly forbidden to remove any antiquities you see, or even to photograph them.

SAILING HOLIDAYS

Sailing vacations can be booked through yacht charter companies in Greece or abroad. The season runs from April to the end of October or early November, and itineraries are flexible. Charters fall into four main categories. Bareboat charter is without a skipper or crew and is available to those with previous sailing experience (contact the **Hellenic Professional and Bareboat Yacht Owners' Association**). Crewed charters range from the modest services of a skipper, assistant or cook to a yacht with a full crew. Sailing within a flotilla, typically in a group of around 6 to 12 yachts, provides the opportunity of independent sailing with the support of a lead boat, contactable by radio. **Thomas Cook** and **Sunsail**

Learning the techniques of sailing

both offer sailing holidays in a flotilla. They also offer the popular "combined vacation". This type of vacation mixes cruiser sailing with the added interest of coastal pursuits, such as shore-based dinghy sailing and windsurfing.

DIRECTORY

USEFUL ORGANIZATIONS

Department of Underwater Archaeology
Kallispéri 30, 11742 Athens.
Tel 210 924 7249.

Greek Yacht Brokers' and Consultants' Association
Marina Zeas, 185 36 Piraeus.
Tel 210 453 3134.

Hellenic Professional and Bareboat Yacht Owners' Association
Marina Zeas, 185 36 Piraeus.
Tel 210 452 6335
or 210 428 0465.
Hellenic Water-ski Federation
Leofóros Possidónos,
16777 Athens.
Tel 210 994 4334.

Hellenic Yachting Federation
Possidónos 51, Piraeus.
Tel 210 940 4828.

YACHT CHARTER COMPANIES

Sunsail
Port House, Port Solent,
Portsmouth PO6 4TH,
England.
Tel 02392 222222.
@ sales@sunsail.com

Tenrag Yacht Charters
Tenrag House, Freepost CU986,
Preston, Canterbury, Kent CT3
1EB, England.
Tel 01227 721874.
@ info@tenrag.com

Variety Cruises
Papada 2, 11525 Athens.
Tel 210 691 9191.

Thomas Cook Holidays
Tel 08705 666222 (direct booking).

Ulysses Tours Inc.
645 Fifth Ave, New York,
NY 10022.
Tel 1-800-431-1424.
www.ulyssestours.com

Sailing aboard a yacht in the Greek seas

SURVIVAL
GUIDE

PRACTICAL INFORMATION

Greece's appeal is both cultural and hedonistic. Its physical beauty, hot climate and warm seas, together with the easy-going outlook of its people, are all conducive to a relaxed holiday. It does pay, however, to know something about the nuts and bolts of Greek life to avoid unnecessary frustrations – when to visit, what to bring, how to get around and what to do if things go wrong. Greece is no longer the cheap holiday destination it once was, though public transport, vehicle hire, eating out and hotel accommodation are still relatively inexpensive compared with most other European countries. Tourist information is available through the many EOT offices *(see p354)*, which offer plenty of advice on the practical aspects of your stay.

Soldier in ceremonial dress

WHEN TO VISIT

High season in the Greek islands – from late June to early September – is the hottest *(see p51)* and most expensive time to visit, as well as being very crowded. December to March are the coldest and wettest months everywhere, with reduced public transport, and many hotels and restaurants closed throughout the winter.

Spring (from late April to May) is one of the loveliest times to visit the islands – the weather is sunny but not yet debilitatingly hot, there are relatively few tourists about, and the countryside is ablaze with brightly coloured wild flowers, against a backdrop of fresh, verdant vegetation.

WHAT TO BRING

Most of life's comforts are available in Greece, but a few items that are advisable to take include: a good map of the area in which you intend to stay *(see p370)*; an AC adaptor for your electrical gadgetry *(see p355)*; sunglasses and a sun hat, mosquito repellent, any medical supplies you might need and a high-factor suntan lotion.

Apart from swimwear, light clothing is all you need for most of the year, although a sweater or light jacket for the evening is also recommended, and is essential either side of high season, in May and October. During winter and spring, rainwear should be taken, as well as warm clothes.

Many religious buildings have dress codes (usually signposted) that should be adhered to *(see p355)*.

ΕΛΕΓΧΟΣ ΔΙΑΒΑΤΗΡΙΩΝ
PASSPORT CONTROL

Passport control sign at a Greek airport

VISA REQUIREMENTS

Visitors from EU countries, the US, Canada, Australia and New Zealand need only a valid passport for entry to Greece (no visa is required), and can stay for a period of up to 90 days. For longer stays a resident's permit must be obtained from the **Aliens' Bureau** in Athens, or the local police in remoter areas.

Non-EU citizens planning to work or study in Greece should contact their local Greek consulate a few months in advance about visa requirements and work permits.

CUSTOMS

Visitors entering Greece from within the EU are no longer subject to any customs controls or other formalities. Limits for duty-paid goods have been similarly relaxed in recent years, though anything valuable should be recorded in your passport upon entry if it is to be re-exported. Visitors coming from non-EU countries may be subject to the occasional spot check on arrival in Greece.

The unauthorized export of antiquities and archaeological artifacts from Greece is treated as a serious offence, and penalties range from hefty fines to prison sentences.

Any prescription drugs that are brought into the country should be accompanied by a copy of the prescription for the purposes of the customs authorities *(see p357)*.

Restrictions on the import and export of money are covered on page 359.

On 30 June 1999, the intra-EU Duty and Tax Free Allowances, better known as Duty Free and mainly affecting such luxury items as alcohol, perfumes and tobacco, were abolished. EU residents can now import greater amounts of these goods, as long as they are for personal use.

Visitors on the beach in high summer

◁ **Fisherman at Skála Sykaminiás on Lésvos**

A family arriving at a Greek airport

TRAVELLING WITH CHILDREN

Children are much loved by the Greeks and welcomed just about everywhere. Baby-sitting facilities are provided by most hotels on request, though check before you book in *(see p299)*.

Concessions of up to 50 per cent are offered on most forms of public transport for children aged 10 and under, but in some cases it is 8 and under.

Swimming in the sea is generally safe for kids, but keep a close eye on them as lifeguards are rare in Greece. Also be aware of the hazards of overexposure to the sun.

WOMEN TRAVELLERS

Greece is a very safe country and foreign women travelling alone are usually treated with respect, especially if dressed modestly *(see p355)*. However, in tourist areas lone women may draw unwanted attention from young Greek men. Hitchhiking alone in Greece is not advisable.

STUDENT AND YOUTH TRAVELLERS

Within Greece itself, no concessions are offered on ferry, bus or train travel, except to students actually studying in Greece. However, there are plenty of deals to be had getting to Greece, especially during low season. There are scores of agencies for student and youth travel, including **STA Travel**, which has 120 offices worldwide. IYHF (International Youth Hostel Federation) membership cards are rarely asked for in Greek hostels, but to be on the safe side it is worth joining before setting off.

Most state-run museums and archaeological sites are free to EU students holding a valid International Student Identity Card (ISIC); non-EU students with an ISIC card are usually entitled to a 50 per cent reduction. There are no youth concessions available for these entrance fees, but occasional discounts are possible with a "Go 25" card, which can be obtained from any STA office by travellers who are under 26.

International student identity card

FACILITIES FOR THE DISABLED

There are few facilities in Greece for assisting the disabled – sights with wheelchair access are indicated for entries in this guide. In the UK, organizations such as **Tourism For All** and **Door to Door** *(see p301)* give advice. In the US, **SATH** also has useful information. Agencies such as **The Assisted Travel Service** organize holidays for the disabled. Sailing holidays are offered by **Odyssey Sailing** *(see p345)*.

A sign directing access for wheelchairs at a Greek airport

DIRECTORY

GREEK TOURIST OFFICES (EOT)

Greek National Tourist Board Internet Site
www.gnto.gr

Athens
Tsocha 24, Ambelokipi, 11521 Athens. *Tel 210 870 7000.*
Amalias 26. *Tel 210 331 0716.*

Australia
37-49 Pitt St, Sydney, NSW 2000.
Tel (612) 9241 1663.

Canada
91 Scollard St, 2nd Floor, Toronto, Ontario M5R 1G4.
Tel (416) 968-2220.

United Kingdom and Republic of Ireland
4 Conduit St, London W1S 2DJ.
Tel 020-7495 9300.

USA
Olympic Tower, 645 Fifth Ave, New York, NY 10022.
Tel (212) 421-5777.

USEFUL ADDRESSES

Aliens' Bureau
Antigónis 99, Kolonos, Athens.
Tel 210 510 2813.

Hosteling International USA
8401 Colesville Road, Suite 600, Silver Spring, MD 20910.
Tel (301) 495-1240.

Pacific Travel
Nikis 26, 10557 Athens.
Tel 210 324 1007
or *210 322 3213.*

STA Travel
11 Goodge St, London W1T 2PF.
Tel 08701 600599.
www.statravel.co.uk

10 Downing St, New York, NY 10014.
Tel (212) 627-3111.

SATH (Society for the Advancement of Travel for the Handicapped)
347 Fifth Ave, Suite 610, New York, NY 10016.
Tel (212) 447-7284.
www.sath.org

The Assisted Travel Service
1 Tank Lane, Purfleet, Essex RM19 1TA.
Tel 01708 863198.
www.assistedholidays.com

Holiday Essentials

The EOT's Greek tourism emblem

For a carefree holiday in Greece, it is best to adopt the philosophy *sigá, sigá* (slowly, slowly). Within this principle is the ritual of the afternoon siesta, a practice that should be taken seriously, particularly during the hottest months when it is almost a physiological necessity. Almost everything closes for a few hours after lunch, reopening later in the day when the air cools and Greece comes to life again. The shops reopen their doors, the restaurants start filling up and, at seafront locales, practically everyone partakes in the *vólta*, or evening stroll – a delightful Greek institution.

TOURIST INFORMATION

Tourist information is available in many towns and villages in Greece, either in the form of government-run EOT offices (Ellinikós Organismós Tourismoú, or the Hellenic Tourism Organization), municipally run tourist offices, the local tourist police *(see p356)*, or privately owned travel agencies. The EOT publishes an array of tourist literature, including maps, brochures and leaflets on transport and accommodation – be aware though that not all of their information is up to date and reliable. The addresses and phone numbers of the EOT and municipal tourist offices, as well as the tourist police, are listed throughout this guide.

GREEK TIME

Greece is always 2 hours ahead of Britain (GMT), 1 hour ahead of European countries on Central European Time (such as France), 7 hours ahead of New York, 10 hours ahead of Los Angeles and 8 hours behind Sydney.

As Greece is now part of the EU, it follows the rule that all EU countries must put their clocks forward to summertime, and back again to wintertime on the same days, in order to avoid any confusion when travelling between countries. This should lessen the chance of missing a ferry or flight due to confusion over the time!

Entry ticket to an archaeological site

OPENING HOURS

Opening hours tend to be vague in Greece, varying from day to day, season to season and place to place. It is therefore advisable to use the times given in this book as rough guidelines only and to check with local information centres for accurate times.

State-run museums and archaeological sites generally open from around 8:30am to 2:45pm (the major ones stay open as late as 8 or 9pm in the summer months).

Mondays and main public holidays *(see p50)* are the usual closing days for most tourist attractions. Locally run and private museums may be closed on additional public holidays and also on local festival days.

A *periptero*, or kiosk, with a wide array of papers and periodicals

Monasteries and convents are open during daylight hours, but will close for a few hours in the afternoon.

Opening times for shops are covered on page 342, pharmacies on page 357, banks on page 358, post offices on page 361 and OTE (telephone) offices on page 360.

Most shops and offices are closed on public holidays and local festival days, with the exception of some shops within tourist resorts.

The dates of major local festivals are included in the Visitors' Checklists in each main town entry in this guide.

ADMISSION CHARGES

Most state-run museums and archaeological sites charge an entrance fee of between 1.5 and 6 euros. Reductions are available, however, ranging from around 25 per cent for EU citizens aged 60 years and over (use your passport as proof of age) to 50 per cent for non-EU students armed with an international student identity card (ISIC) *(see p353)*.

Though most museums and sites are closed on public holidays, the ones that do remain open are free of charge.

EVENTS

The english-language paper *Athens News* has a What's On column, gazetting events all over the city and also those of special interest to children. The tourist office in Amerikis Street has a free monthly English-language magazine, *Now in Athens*, which has details of cultural events and entertainment in Athens, as does the weekly Time Out and Greek-language *Athinorama*.

A list of Greek festivals and cultural events is given on pages 46–50, but it is worth asking your nearest tourist office about what's happening locally. Other forms of entertainment include the outdoor cinema in summer, which is very popular with the Greeks; most films are in English with Greek subtitles. There are also

A typical sign about dress codes at a monastery

bars, discos and nightclubs in the resorts, as well as tavernas and *kafeneía* (coffee shops), found in every village and often the centre of social life.

RELIGION

Greece is almost entirely-Greek Orthodox. The symbols and rituals of the religion are deeply rooted in Greek culture and are visible everywhere. Saints' days are celebrated throughout Greece *(see p50)*, both locally and nationally.

The largest religious minorities are the Muslims of Rhodes and Kos, and the Catholics of Sýros and Tínos, though they constitute less than 1 per cent of the country's total population. Most other non-Orthodox places of worship are situated in Athens.

ETIQUETTE

Like anywhere else, common courtesy and respect is appreciated in Greece, so try speaking a few words of the language, even if your vocabulary only extends as far as the basics *(see pp404–8)*.

Though formal attire is rarely needed, modest clothing (trousers for men and skirts for women) is *de rigueur* for visits to churches and monasteries.

Topless sunbathing is generally tolerated, but nude bathing is officially restricted to a few designated beaches.

In restaurants, the service charge is always included in the bill, but tips are still appreciated – the custom is to leave between 10 and 15 per cent. Public toilet attendants should also be tipped. Taxi drivers do not expect a tip, but they are not averse to them either; likewise hotel porters and chambermaids.

PHOTOGRAPHY

Photographic film is readily available in Greece, though it is often quite expensive in tourist areas and close to the major sights.

Taking photographs inside churches and monasteries is officially forbidden; within museums photography is usually permitted, but flashes and tripods are often not. In most cases where a stills camera is allowed, a video camera will also be fine, but you may have to pay an extra fee.

At sites, museums or religious buildings it is best to gain permission before using a camera, as rules do vary.

A Greek priest

ELECTRICAL APPLIANCES

Two-pin adaptor, for use with all British appliances when in Greece

Greece, like other European countries, runs on 220 volts/50 Hz AC. Plugs have two round pins, or three round pins for appliances that need to be earthed. The adaptors required for British electrical appliances are difficult to find in Greece so bring one with you. Similarly, transformers are needed for North American equipment.

CONVERSION CHART

Greece uses the metric system, with two small exceptions: sea distances are expressed in nautical miles and land is measured in *strémmata*, the equivalent of about 0.1 ha (0.25 acre).

Imperial to Metric
1 inch = 2.54 centimetres
1 foot = 30 centimetres
1 mile = 1.6 kilometres
1 ounce = 28 grams
1 pound = 454 grams
1 pint = 0.6 litres
1 gallon = 4.6 litres

Metric to Imperial
1 millimetre = 0.04 inches
1 centimetre = 0.4 inches
1 metre = 3 feet 3 inches
1 kilometre = 0.64 miles
1 gram = 0.04 ounces
1 kilogram = 2.2 pounds
1 litre = 1.8 pints

DIRECTORY

EMBASSIES IN GREECE

Australia
Kifissias and Alexandras,
11523 Athens.
Tel 210 870 4000.

Canada
Gennadíou 4, 11521 Athens.
Tel 210 727 3400.

Republic of Ireland
Vassiléos Konstantínou 7,
10674 Athens.
Tel 210 723 2771/2.

New Zealand
Kifissias 76, 11232 Athens.
Tel 210 692 4136.

United Kingdom
Ploutárchou 1,
10675 Athens.
Tel 210 727 2600.

USA
Vasilíssis Sofías 91,
10160 Athens.
Tel 210 721 2951.

Personal Health and Security

Fire service emblem

Greece is one of the safest European countries to visit, with a time-honoured tradition of honesty that still survives despite the onslaught of mass tourism. But, like travelling anywhere else, it is still advisable to take out a comprehensive travel insurance policy. One place where danger is ever present, however, is on the road. Driving is a volatile matter in Greece, which now has the highest accident rate in Europe. Considerable caution is recommended, for drivers and pedestrians.

PERSONAL SECURITY

The crime rate in Greece is very low compared with other European countries. Nevertheless, a few precautions are worth taking, like keeping cars and hotel rooms locked, watching your handbag in public, and not keeping all your documents together in one place. If you do have anything stolen, contact the police or tourist police.

POLICE

Greece's police are split into three forces: the regular police, the port police and the tourist police. The tourist police are the most useful for vacationers, combining normal police duties with tourist advice. Should you suffer a theft, lose your passport or have cause to complain about shops, restaurants, tour guides or taxi drivers, your case should first be made to them. As every tourist police office claims to have at least one English speaker, they can act as interpreters if the case needs to involve the local police. Their offices also offer maps, brochures, and advice on finding accommodation.

LEGAL ASSISTANCE FOR TOURISTS

European consumers' associations together with the European Commission have created a programme, known as **EKPIZO**, to inform tourists of their rights. Its aim is specifically to help vacationers who experience problems with hotels, campsites, travel

A policeman giving directions to vacationers

agencies and so forth. They will furnish tourists with the relevant information and, if necessary, arrange legal advice from lawyers in English, French or German. Contact the Crete office for their telephone numbers on the other islands.

MEDICAL TREATMENT AND INSURANCE

British and EU citizens are entitled to free emergency medical care in Greece on presentation of a European Health Insurance Card (available from the UK Department of Health and most UK post offices). Be aware, however, that public health facilities are limited on the islands and private clinics are expensive. Visitors are strongly advised to take out comprehensive travel insurance (available from travel agents, banks and insurance brokers) covering both private medical treatment and loss or theft of personal possessions. Be sure, too, to read the small print: not all policies, for instance,

will cover you for activities of a "dangerous" nature, such as motorcycling and trekking; not all policies will pay for doctors' or hospital fees direct, and only some will cover you for ambulances and emergency flights home. Paying for your flight with a credit card such as Visa or American Express will also provide limited travel insurance, including reimbursement of your air fare if the agent happens to go bankrupt.

HEALTH PRECAUTIONS

It costs little or nothing to take a few sensible precautions when travelling abroad, and certain measures are essential if vacationing in the extreme heat of high summer. The most obvious thing to avoid is overexposure to the sun, particularly for the fair-skinned: wear a hat and good-quality sunglasses, as well as a high-factor suntan lotion. If you do burn, calamine lotion or aloe gel are soothing. Heat stroke is a real hazard for which medical attention should be sought immediately; heat exhaustion and dehydration (made worse by alcohol consumption) are also serious.

Be sure to drink plenty of water, even if you don't feel thirsty, and if in any doubt invest in a packet of electrolyte tablets (a mixture of potassium salts and glucose) available at any Greek pharmacy, to avoid dehydration and replace lost minerals.

Port policeman's uniform

City policeman's uniform

An ambulance with the emergency number emblazoned on its side

Fire engine

Police car

Always go prepared with an adequate supply of any medication you may need while away, as well as a copy of the prescription with the generic name of the drug – this is useful not only in case you run out, but also for the purposes of customs when you enter the country. Also be aware that codeine, a painkiller commonly found in headache tablets, is illegal in Greece.

Tap water in Greece is generally safe to drink, but in remote communities it is a good precaution to check with the locals. Bottled spring water is for sale throughout the islands, and often has the advantage of being chilled.

However tempting the sea may look, swimming after a meal is not recommended for at least two hours, since stomach cramps out at sea can be fatal. Underwater hazards to be aware of are weaver fish, jellyfish and sea urchins. The latter are not uncommon and are extremely unpleasant if trodden on. If you do tread on one, the spine will need to be extracted using olive oil and a sterilized needle. Jellyfish stings can be relieved by vinegar, bicarbonate of soda, or by various remedies sold at Greek pharmacies. Though a rare occurrence, the sand-dwelling weaver fish has a powerful sting, its poison causing extreme pain. The immediate treatment is to immerse the affected area in very hot water to dilute the venom's strength.

No inoculations are required for visitors to Greece, though tetanus and typhoid boosters may be recommended.

PHARMACIES

Greek pharmacists are highly qualified and can not only advise on minor ailments, but also dispense medication not usually available over the counter back home. Their premises, *farmakeía*, are identified by a red or green cross on a white background. Pharmacies are open from 8:30am to 2pm, but are usually closed in the afternoon and on Saturday mornings.

Pharmacy sign

However, in larger towns there is often a rota system to maintain a service throughout the day and also late at night. Details are posted in pharmacy windows, in both Greek and English.

EMERGENCY SERVICES

In case of emergencies the appropriate services to call are listed in the directory below. For accidents or other medical emergencies, a 24-hour ambulance service only operates within Athens. Outside Athens, in rural towns and on the islands, ambulances are rarely on 24-hour call. But, if necessary, patients can be transferred from island ESY (Greek National Health Service) hospitals or surgeries to a main ESY hospital in Athens by ambulance and ferry, or helicopter.

A complete list of ESY hospitals, private hospitals and clinics is available from the tourist police.

DIRECTORY

NATIONWIDE EMERGENCY NUMBERS

Police
Tel 100.

Ambulance
Tel 166.

Fire
Tel 199.

Road assistance
Tel 10400.

Coastguard patrol
Tel 108.

ATHENS EMERGENCY NUMBERS

Tourist police
Tel 171.

Doctors
Tel 1016.

Pharmacies
For information on 24-hour pharmacies:
📱 14944.

Poison treatment centre
Tel 01 779 3777.

EKPIZO BUREAU

Athens branch
Valtetsíou 43–45, 10681 Athens.
Tel 210 330 4444.
Fax 210 330 0591.
www.ekpizo.org.gr

Banking and Local Currency

Greece has now converted to the common European currency, the euro, which replaces the former drachma. Changing money from other currencies into euros is straightforward and can be done at banks or post offices. Even in small towns and resorts you can expect to find a car hire firm or travel agency that will change cash and travellers' cheques – albeit with a sizeable commission. Larger towns and tourist centres all have the usual banking facilities, including a growing number of cash machines (ATMs) that make it easy to access funds abroad.

Visitors changing money at a foreign exchange bureau

BANKING HOURS

All banks are open from 8am to 2pm Monday to Thursday, and from 8am to 1:30pm on Friday. In the larger cities and tourist resorts there is usually at least one bank that reopens its exchange desk for a few hours in the evening and on Saturday mornings during the summer season.

Cash machines, though seldom found outside the major towns and resorts, are in operation 24 hours a day. All banks are closed on public holidays *(see p50)* and may also be closed on any local festival days.

BANKS AND EXCHANGE FACILITIES

There are banks in all major towns and resorts, as well as exchange facilities at post offices (which tend to charge lower commissions and are found in the more remote areas of Greece), travel agents, hotels, tourist offices and car hire agencies. Always take your passport with you when cashing travellers' cheques, and check exchange rates and commission charges beforehand, as they vary greatly. In major towns and tourist areas you may find a foreign exchange machine

Foreign exchange machine

for changing money at any time of day or night. These operate in several languages, as do the ATMs.

CARDS, CHEQUES AND EUROCHEQUES

Visa, mastercard (Access), American Express and Diners Club are the most widely accepted credit cards in Greece. They are the most convenient way to pay for air tickets, international ferry journeys, car hire, some hotels and larger purchases. Cheaper tavernas, shops and hotels as a rule do not accept credit cards.

You can get a cash advance on a foreign credit card at some banks, though the minimum amount is 44 euros, and you will need to take your passport with you as proof of identity. A credit card can be used for drawing local currency at cash machines. At a bank or ATM, a 1.5 per cent processing charge is usually levied for Visa, but none for other cards.

Cirrus and Plus debit card systems operate in Greece. Cash can be obtained using the Cirrus system at National Bank of Greece ATMs and the Plus system at Commercial Bank ATMs.

Travellers' cheques are the safest way to carry large sums of money. They are refundable if lost or stolen, though the process can be time-consuming. American Express and Travelex are the best-known brands of travellers' cheques in Greece. They usually incur two sets of commissions: one when you buy them (1–1.5 per cent) and another when you cash them. Rates for the latter vary considerably, so shop around

before changing your money. Travellers' cheques can be cashed at large post offices *(see p361)* – an important consideration if you are travelling to a rural area or remote island.

Eurocheques, available only to holders of a European bank account in the form of a chequebook, are honoured at banks and post offices throughout Greece, as well as many hotels, shops and travel agencies. There is no commission charged when cashing Eurocheques, though there is an annual fee of about £8 for holding a European account and a fee of about 2 per cent for each cheque used. All fees are debited directly from the account.

(see p361)

DIRECTORY

To report a lost or stolen credit card call the following numbers collect from Greece:

American Express
Tel 00 44 1273 696933.

Diners Club
Tel 00 44 1252 513500.

MasterCard
Tel 00 800 11887 0303.

Visa
Tel 00 800 11638 0304.

To report lost or stolen travellers' cheques call the following free-phone numbers from Greece:

American Express
Tel 00 800 44 127569.

Travelex
Tel 00 800 44 131409 (toll free).

Visa
Tel 00 800 44 128366

THE EURO

Twelve member states of the EU have replaced their traditional currencies with a single European currency, the Euro. These are Austria, Belgium, Finland, France, Germany, Greece, Ireland, Italy, Luxembourg, the Netherlands, Portugal and Spain; the UK, Denmark and Sweden stayed out, with an option to review their decision. The euro was introduced in most countries, but only for banking purposes, on 1 January, 1999. Greece adopted it on 1 January 2001. In all countries, a transition period saw euros and local currency used simultaneously.

In Greece, euro notes and coins came into circulation on 1 January 2002 and became the sole legal tender at the end of March 2002.

Bank Notes

Euro banknotes have seven denominations. The 5-euro note (grey in colour) is the smallest, followed by the 10-euro note (pink), 20-euro note (blue), 50-euro note (orange), 100-euro note (green), 200-euro note (yellow) and 500-euro note (purple). All notes show the stars of the European Union.

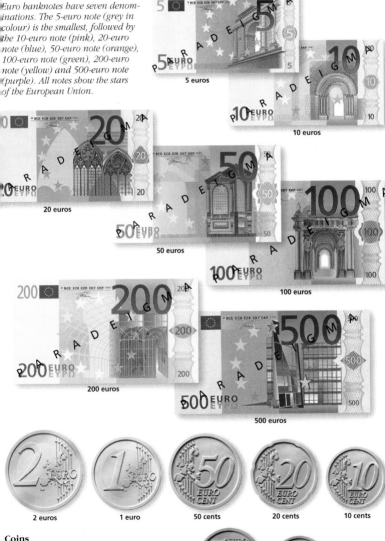

5 euros

10 euros

20 euros

50 euros

100 euros

200 euros

500 euros

2 euros

1 euro

50 cents

20 cents

10 cents

Coins

The euro has eight coin denominations: 2 euros and 1 euro (both silver-and-gold); 50 cents, 20 cents, 10 cents (all gold); 5 cents, 2 cents and 1 cent (all bronze). The reverse of each coin is the same in all Eurozone countries; the obverse is different in each country.

5 cents

2 cents

1 cent

Communications

Post office logo

The Greek national telephone company is the OTE (Organismós Tilepikoinonión Elládos). Telecommunications have improved dramatically in recent years, and now there are direct lines to all major countries. These are often better than local lines, but the rates are among the highest in Europe. Greek post is reasonably reliable and efficient, especially from the larger towns and resorts; faxes are also easy to send and receive. The Greeks are avid newspaper readers, and in addition to a vast array of Greek publications, there are also a few good English-language papers and magazines.

TELEPHONES AND FAXES

Public telephones can be found in many locales – hotel foyers, telephone booths, street kiosks, or the local OTE office. Long-distance calls are best made in a telephone booth using a phonecard – available at any kiosk. Alternatively, calls can be made at a metered phone in an OTE office, where you can also make reverse-charge calls. OTE offices are open daily from 7am to 10pm or midnight in the larger towns; or until around 3pm in smaller communities. Call charges are variable, but in general local calls are inexpensive, out-of-town

domestic calls are surprisingly expensive, and long-distance calls are extortionate. You can ring the operator first for specific rates, as well as information about peak and cheap times, which vary depending on the country you are phoning.

Ship-to-shore and shore-to-ship calls can be made through INMARSAT; for information on this service call the marine operator from Greece on 158.

Faxes can be sent from OTE offices, hotels, a few city post offices, and some car hire and travel agencies, though expect to pay a heavy surcharge wherever you go.

The easiest way to receive a fax is to become friendly with your nearest car hire or travel agency – both will usually keep faxes aside for you.

INTERNET ACCESS

Most large hotels offer Internet access to guests. You will find at least one Internet café in the major towns on almost all islands, where it costs around 3–6 euros per hour. Some travel agencies also offer Internet access for a small charge.

RADIO AND TV

With three state-owned radio channels and a plethora of local stations, the airwaves are positively jammed in Greece, and reception is not always dependable. There are many Greek music stations to listen to, as well as classical music stations such as ER-3, one of the three state-run channels, which can be heard on 95.6 FM. Daily news summaries are broadcast in English, French and German, and with a shortwave radio you will be able to pick up the BBC World Service in most parts of Greece. Its frequency varies, but in the Greater Athens area it can be heard on 107.1 FM. Galaxy on 92.1 FM has CNN news summaries in English every two hours.

Greek TV is broadcast by two state-run, and several privately run, channels, plus a host of cable and satellite stations from across Europe. Most Greek stations cater to popular taste, with a mix of dubbed foreign soap operas, game shows, sport and films, Fortunately for visitors, foreign language films tend to be subtitled rather than dubbed.

USING A PHONECARD TELEPHONE IN GREECE

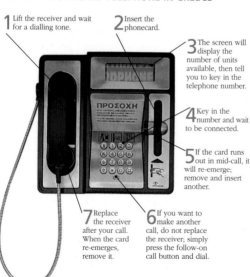

1 Lift the receiver and wait for a dialling tone.

2 Insert the phonecard.

3 The screen will display the number of units available, then tell you to key in the telephone number.

4 Key in the number and wait to be connected.

5 If the card runs out in mid-call, it will re-emerge; remove and insert another.

6 If you want to make another call, do not replace the receiver; simply press the follow-on call button and dial.

7 Replace the receiver after your call. When the card re-emerges, remove it.

A public phone

A Greek telephone card

The standard bright yellow post box

Satellite stations CNN and Euronews televise international news in English round the clock. Guides that give details of the coming week's television programmes are published in all the English-language papers.

NEWSPAPERS AND MAGAZINES

The trusty corner *períptera* (kiosks), bookshops in larger towns and tourist shops in the resorts often sell day-old foreign newspapers and magazines, though the mark-up is substantial. Much cheaper, and also widely available, is the English-language paper published in Athens, *Athens News*, which is published every day except Monday. The *Odyssey,* a bi-monthly, glossy magazine, is available in most of the resorts as well as the capital. These two publications are excellent sources of information on local entertainment, festivals and cultural goings-

on, while also providing coverage of domestic and international news. The most popular Greek language newspapers are *Eleftherotypía, Eléftheros Týpos, Kathemeriní* and *Ta Nea.*

POST

Greek post offices *(tachy-dromeía)* are generally open from 7:30am to 2pm Monday to Friday. Some main branches in the larger towns stay open as late as 8pm (main branches occasionally open for a few hours at the weekend as well).

All post offices are closed on public holidays *(see p50).* Those with an "Exchange" sign will change money in addition to the usual services.

Post boxes are usually bright yellow; those with two slots are marked *esoterikó,* meaning domestic, and *exoterikó,* meaning overseas. Bright red post boxes are reserved for express mail, for both domestic and overseas destinations. Express is a little more expensive, but cuts delivery time by a few days.

Stamps *(grammatósima)* can be bought over the counter at post offices and also at *períptera;* the latter usually charge a 10 per cent commission.

Airmail letters from the Greek islands to most European countries take between three and six days, and anywhere from five days to a week or more to North America, Australia and New Zealand. Postcards always take a little longer,

so if you are sending them, allow an additional couple of days to reach any destination.

The poste restante system – whereby mail can be sent to, and picked up from a post office – is widely used in Greece. Mail should be clearly marked "Poste Restante", with the recipient's surname underlined so that it gets filed in the right place. A passport, or some other proof of identity, is needed when collecting the post, which is kept for a maximum of 30 days before being returned to the sender.

If you are sending a parcel from Greece to a non-EU country, do not seal it before going to the post office. The contents will need to be inspected by security before it is sent, and if the package is sealed they will not hesitate to unwrap it.

DIRECTORY

Domestic Calls

Tel 129 (domestic operator).

Tel 11888 (for all calls within Greece).

International Calls

Tel 139 (international operator and directory assistance).

International Calls from Greece

Dial 00, the country code (a list is given below), the local area code (minus the initial 0) and then the number itself.

Australia 61.

Ireland 353.

New Zealand 64.

UK 44.

USA & Canada 1.

International Calls to Greece from Abroad

Dial the international access code (a list is given below), 30 (country code), the area code and then the number itself.

Australia 0011.

Ireland, UK & New Zealand 00.

USA & Canada 011

Athenians reading newspapers on a clothes line at a street kiosk

TRAVEL INFORMATION

Reliably hot, sunny weather makes Greece an extremely popular destination for vacationers. During the warmer months (May to October), countless charter flights bring millions to the Greek islands, although it is also possible to reach Greece by car, rail or coach, and continue to the islands by ferry. While many of the larger islands are accessible by plane, the ferry network reaches even the remotest islands. This is matched by the bus service, which has frequent

Olympic Airways passenger aeroplane

services on all major routes and local buses to the tiniest communities. Travelling around by car or motorcycle offers the most flexibility on larger islands, allowing the traveller to reach places that are inaccessible by public transport. But the roads in remoter parts can be rough and potentially dangerous (see p370). If, however, you do not wish to rent a car, taxis provide another inexpensive option, and on many islands, taxi boats sail around coasts, offering pick-up and drop-off points along the way.

GETTING TO GREECE BY AIR

The main airlines operating direct scheduled flights from London to Athens and Thessaloníki are **Olympic Airways** (the Greek national airline) and **British Airways**. Athens now has a new airport, Elefthérios Venizélos, which handles all international and domestic flights. The old airport (Hellinikon) is no longer used.

From Europe, there are around 20 international airports in Greece that can be reached directly. Only Crete, Rhodes and Corfu among the islands, and Athens and Thessaloníki on the mainland, handle both charter and scheduled flights. The other airports can only be reached directly by charter flights.

Travellers with airport shopping

From outside Europe, all scheduled flights to Greece arrive in Athens, although only a few airlines offer direct flights – most will require changing planes, and often airlines, at a connecting European city. There are direct flights daily from New York operated by Olympic and **Delta**. From Australia, Olympic Airways operates flights out of Sydney, Brisbane and Melbourne. These necessitate a stop-off in Southeast Asia or Europe. Flights from New Zealand are also via Melbourne or Sydney. Other carriers with services from Australasian cities to Athens include Qantas, **Singapore Airlines** and **KLM**. At present, none of these airlines operate direct flights between Australia and Greece.

Check-in desks at Athens' new Elefthérios Venizélos Airport

CHARTERS AND PACKAGE DEALS

Charter flights to Greece are nearly all from within Europe, and mostly operate between May and October. Tickets are sold by travel agencies either as part of an all-inclusive package holiday or as a flight-only deal.

Although they tend to be the cheapest flights available, charters do carry certain restrictions: departure dates cannot be changed once booked and there are usually minimum and maximum limits to one's stay (typically between three days and a month). Another consideration if you plan to visit Turkey from Greece is that charter passengers can only go for a day trip; if you stay any longer you will forfeit the return portion of your air ticket.

Athens' new international airport

Booking agency in Athens

FLIGHT TIMES

Flying to Athens from London or Amsterdam takes about 3.5 hours; the journey time from Paris and Berlin is around 3 hours – the trip from Berlin being a little quicker. From Madrid it takes just over 4 hours and from Rome a little under 2 hours. There are direct flights to Athens from New York, which take 10 hours, although a non-direct flight can take more than 12. From Los Angeles the flight's duration is from 17 to 19 hours, depending on the European connection. From Sydney, via Bangkok, the flight takes around 19 hours.

AIR FARES

Fares to Greece are generally at their highest from June to September, but how much you pay will depend more on the type of ticket you decide to purchase. Charters are usually the cheapest option during peak season, though discounted scheduled flights are also common and worth considering for longer visits or during the low season, when there are few charters available. Reasonable savings can also be made by booking an APEX (Advance Purchase Excursion) ticket well in advance but, like charters, these are subject to minimum and maximum limits to one's stay and other

restrictions. Budget travellers can often pick up bargains through agents advertising in the national press, and cheap last-minute deals are also advertised on Teletext and Ceefax in the UK. Whoever you book through, be sure that the company is a fully bonded and licensed member of ABTA (the Association of British Travel Agents) or ASTA (the American Society of Travel Agents); this will ensure that you get home should the company go bankrupt during your stay, and should guarantee compensation. Note that domestic flights in Greece are subject to an airport tax *(see p364)*.

ATHENS' AIRPORT

Greece's largest and most prestigious infrastructure development project for the new millennium opened to air traffic in 2001. Located at Spáta, 27 km (17 miles) northeast of the city centre, Athens' airport now handles all the city's passenger and cargo flights. It has two runways, designed for simultaneous, round-the-clock operation, and a Main Terminal Building for all arrivals and departures. Arrivals are located on the ground floor (level 1) and departures on the first floor (level 2). The smaller Satellite Building is accessed along an underground corridor with moving walkways. The airport has been designed to allow for a 45-minute connection time between two scheduled flights.

Departure gate symbol

The airport's modern business and service facilities include a shopping mall, restaurants and cafés in the Main Terminal Building and a four-star hotel within the airport complex. Car rental firms, banks, bureaux de change and travel agencies are located in the arrivals area.

TRANSPORT FROM ATHENS AIRPORT

A new six-lane highway links the airport to the Athens City Ring Road. From the airport, the E95 bus runs to and

Ultramodern interior of Athens' new airport

from Plateía Syntágmatos in the city centre every 15 minutes with a journey time of about one hour. Bus E96 runs to and from Piraeus every 20 minutes, taking about 100 minutes. Tickets for both journeys cost around 3 euros. These tickets are in effect one-day travel cards and can also be used to travel around the city *(see p295)*. A taxi-ride into town costs about 35 euros.

Athens' metro and rail networks were extended to the airport in time for the 2004 Olympic Games.

One of the smaller planes in Olympic's fleet, for short-haul flights

Athens' new airport, designed in the blue and white national colours

FLIGHT CONNECTIONS IN GREECE

As well as having the largest number of international flights in Greece, Athens also has the most connecting air services to the islands. Both international and domestic flights now arrive at and depart from the main terminal at the city's Elefthérios Venizélos airport. Thessaloníki also handles scheduled flights, but only from within Europe. Greece's other international airports are served by charters only, again mostly from within Europe.

DOMESTIC FLIGHTS

Greece's domestic airline network is extensive. **Olympic Airways** and its affiliate, **Olympic Aviation**, operate most internal flights, though there are also a Number of private companies, such as **Aegean Airways** and **Interjet**, providing services between Athens and some of the major island destinations. Fares for domestic flights are at least double the equivalent bus journey or deck-class ferry trip. Tickets and timetables for Olympic flights are available from any Olympic Airways office in Greece or abroad, as well as from most major travel agencies. Reservations are essential in peak season.

Olympic Airways operates direct flights from Athens to over two dozen islands, and from Thessaloníki, there are direct Olympic flights to nine of the islands. A number of inter-island services operate during the summer, and about a dozen of these fly throughout the year.

A small airport departure tax is charged on domestic flights of between 62 and 460 air miles. For "international" flights (that is, those over 460 air miles) the tax is doubled.

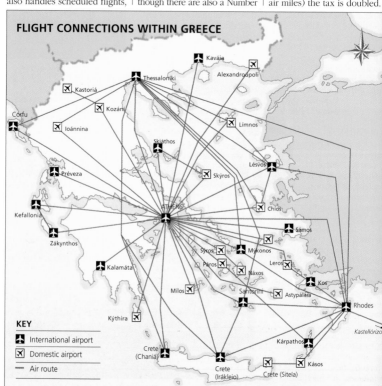

FLIGHT CONNECTIONS WITHIN GREECE

Kavála
Alexandroúpoli
Thessaloníki
Kastoriá
Kozáni
Corfu
Ioánnina
Límnos
Skiáthos
Lésvos
Skýros
Préveza
ATHENS
Chíos
Kefallonía
Sámos
Zákynthos
Sýros
Mýkonos
Leros
Kalamáta
Páros
Náxos
Milos
Santoríni
Astypálaia
Kýthira
Kos
Rhodes
Kastellórizo
Crete (Chaniá)
Kárpathos
Crete (Irákleio)
Crete (Siteía)
Kásos

KEY

✈ International airport

☒ Domestic airport

— Air route

ISLAND	DISTANCE	FLYING TIME	ISLAND	DISTANCE	FLYING TIME
Corfu	381 km (237 miles)	40 minutes	Crete (Chaniá)	318 km (198 miles)	45 minutes
Rhodes	426 km (265 miles)	45 minutes	Santoríni	228 km (142 miles)	40 minutes
Skýros	128 km (80 miles)	40 minutes	Kos	324 km (201 miles)	45 minutes
Skiáthos	135 km (84 miles)	30 minutes	Mýkonos	153 km (95 miles)	30 minutes
Límnos	252 km (157 miles)	45 minutes	Páros	157 km (98 miles)	35 minutes

DIRECTORY

ATHENS AIRPORT

Elefthérios Venizélos – Athens International Airport
5th km Spáta-Loutsa Ave., 10904 Spáta.
Tel 210 353 0000.
Fax 210 369 8883.
www.aia.gr

OLYMPIC AIRWAYS

Arrivals and Departures
Tel 210 353 0000.
🔲 1440.
Tel 801 11 44444
(reservations).

Athens Office
Syngroú 96, 11741 Athens.
Tel 210 926 9111.

Crete Office
Plateía Eleftherías, Irákleio.
Tel 2810 288073.

Rhodes Office
Tel 22410 24571-5.

Thessaloníki Office
Kountouriótou 3, Thessaloníki.
Tel 2310 368 311.
www.olympic airlines.com

OTHER AIRLINES

Aegean Airlines
Leof. Vouliagménis 572, 16451 Athens.
Tel 801 11 20000.
www.aegeanair.com

Air Canada
Syngroú 5, 11741 Athens.
Tel 210 900 6081.
www.aircanada.ca

Air France
Leof. Vouliagménis 18, 16674 Athens.
Tel 210 960 1100.
www.airfrance.com

British Airways
Vouliagménis & Themistokleous 1, 16674 Athens.
Tel 801 11 56000.
www.ba.com

Delta Airlines
Elefthérios Venizélos Airport.
Tel 210 331 1678.
www.delta.com

EasyJet
Tel 210 353 0300.
www.easyjet.com

KLM
Tel 210 911 0000.
www.klm.com

Singapore Airlines
Xenofóntos 9, 10557 Athens.
Tel 210 356 8120.
www.singapore air.com

PRIVATE AIRLINES (FOR DOMESTIC TRAVEL)

Aviator
Syngroú 179, 17564 Athens.
Tel 210 353 3713.
www.aviator.gr

Interjet
Eleftheríou Venizélou 32, 16675 Athens.
Tel 210 961 2050.
www.interjet.gr

Olympic Aviation
Syngroú 96, 11741 Athens.
Tel 210 356 9111.
(reservations),
(or via Olympic Airways).

TRAVEL AGENCIES IN ATHENS

American Express Travel Services
Mesoghion 318, 15341 Athens.
Tel 210 324 4975.

Blue Star Ferries
Attica Premium, Attica Posidonos, Piraeus.
Tel 210 891 9550.

Dolphin Hellas Travel
Syngroú 16, 11742 Athens.
Tel 210 922 7772/3.
www.dolphin-hellas.gr

Ginis Vacances
3rd floor, Ermoú 23–25, 10563 Athens.
Tel 210 325 0401.

Oxygen Travel
Eslin 4, Athens.
Tel 210 641 0881.
info@oxygentravel.gr

OLYMPIC AIRWAYS OFFICES ABROAD

Australia
37–49 Pitt Street, Suite 303, Level 3, Underwood House, Royal Exchange, Sydney, NSW 2001.
Tel (02) 9251 1047.

Canada
80 Bloor Street, Suite 502, Toronto, Ontario M5S F2V1.
Tel (416) 964 2720.

UK
11 Conduit Street, London W15 2LP.
Tel 0870 606 0460
(reservations).

USA
Satellite Airlines Terminal, 125 Park Avenue, New York, NY 10017.
Tel (212) 867 0970.
Tel 800 223 1226
(reservations).

Travelling by Sea

Greece has always been a nation of seafarers and,
with its hundreds of islands and thousands of miles of
coastline, the sea has played an important part in the
history of the country and continues to do so today. It is
now a major source of revenue for Greece, with millions
of vacationers choosing the idyllic Greek islands for
their break. The network of ferries is a lifeline for the
islanders, and for the tourist an enjoyable and relaxing
way of island-hopping or reaching a single destination.

TRAVELLING TO GREECE BY SEA

There are regular year-round
ferry crossings from the
Italian ports of Ancona, Bari
and Brindisi to the Greek
ports of Igoumenítsa in Epirus
and Pátra in the Peloponnese.
During the summer, there are
additional sailings from Venice
and Trieste. Journey times
and fares vary considerably,
depending on the time of
year, point of embarkation,
ferry company and type of
ticket. There are also
reductions possible for
students, travellers under 26
and railcard holders.

Other year-round ferry services include the route from
Haifa in Israel, via Cyprus, to
Rhodes and Piraeus (with a
stop-off at Crete in summer).
From Turkey's Aegean coast
ferries operate year round
between Kusadasi and Sámos,
and Çeşme and Chíos, with
additional summer sailings
between Bodrum and Kos,
Marmaris and Rhodes, and
Ayvalık and Lésvos.

If you are transporting your
car into Greece by ferry, you
will require a vehicle registra-
tion document and, in summer,

Car ferry leaving from Mandráki harbour, on Nísyros

will need to reserve ahead.
Addresses and numbers of
agents for advance bookings
are given on page 369.

GREEK FERRY SERVICE

The reputation of the Greek
ferry service suffered as a
result of the disaster in September 2000 when more than
80 people were drowned off
the coast of Páros after their
ferry sank. The government
has now implemented stricter

checks on all ferries, which
should result in higher safety
standards in future.

The smaller ports have
limited services, so check the
timetable on arrival to see if
you can get a ferry on the day
and for the destination you
want. The larger ports have
many more services. Piraeus,
the port of Athens, is Greece's
busiest port and has many
routes emanating from its
harbour. The hub of activity is
at Plateía Karaïskáki, where
the majority of ticket agents
reside, as well as the port
police. A number of competing
companies run the ferry
services, each with its own
agents handling bookings and
enquiries. This makes the task
of finding out when ferries
sail, and from which dock, a
more challenging one. The
ferries are approximately
grouped by destination, but
when the port is busy ferries
dock wherever space permits.
So, finding your ferry usually
involves studying the agency's
information board or asking
the port police *(limenarcheío)*.

In this guide, we show the
direct ferry routes in high
season on the individual
island maps, pictorial maps
for each island group, and
the back endpaper for the
country-wide network; high
season is from June to
August. In low season, expect
all services to be significantly
reduced and some routes to
be suspended altogether.
The routes on these maps
should be taken as guidelines
only – check local sources
for the latest information.

The Greek Tourist Office's
weekly schedules can serve
as a useful guideline to departure times. Visit www.gtp.
gr for information.
Alternatively, ask at a local
travel agency. Some of the
English-language papers also
print summer ferry schedules.
Hydrofoils, catamarans, caïques
and taxi boats supplement the
ferry services *(see pp368–9)*.

FERRY TICKETS

Tickets for all ferry journeys
can be purchased from the
shipping line office, any
authorized travel agency,

Motorcyclists waiting for a ferry at the port of Piraeus

uise ship sailing towards the harbour at Zákynthos

On major routes, ferries have essentially three classes, ranging from deck class to de luxe – the latter costing almost as much as flying.

First class usually entitles you to a two-bunk exterior cabin with bathroom facilities. A second-class ticket costs around 25 per cent less and gives you a three- or four-bunk cabin with washing facilities, such as a basin. Second-class cabins are usually within the interior of the vessel. A deck-class ticket gives you access to most of the boat, including a lounge with reclining seats. But during the summer, on a warm, starry night, the deck is often the best place to be.

n the quayside, or on the rry itself. All fares except first ass are set by the Ministry of ransport, so a journey should ost the same amount egardless of which shipping ne you choose. Advance ooking is essential for a car in igh season. For motorbikes and cars a supplement is also payable. Cars can cost as much as three or four times the passenger fare.

Children under two travel free, those aged from two to nine pay half fare, and once over the age of ten, children must pay the full adult fare.

ERRY COMPANY FUNNELS

he funnels of each company's fleet are bold and brightly oloured, and serve as beacons for travellers searching the arbour for their ferry. In the busiest port, Piraeus, ferries ften dock wherever there is space and even in high eason each company is unlikely to have more than two or three boats in dock at a time. Targeting the funnel, therefore, is often the easiest way to find your ferry.

Ventouris Sea Lines

ANEK Lines **GA Ferries** **Minoan Lines**

NEL Lines

PIRAEUS PORT MAP

This map shows the layout of the main harbour, with a guide to the quay you are likely to need for various destinations.

Piraeus Port Authority
Tel 1441.
Coastal Service Timetables
Tel 1441.

KEY TO DEPARTURE POINTS

- Argo-Saronic Islands
- Northeast Aegean Islands
- Dodecanese
- Cyclades
- Crete
- International ferries
- Hydrofoils and catamarans

For key to symbols see back flap

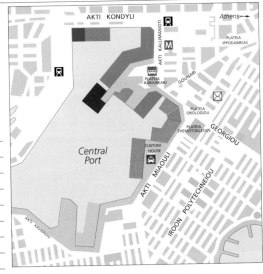

Hydrofoil, known as a "Flying Dolphin"

HYDROFOILS AND CATAMARANS

Some of the islands can be reached by Greece's 60 or so hydrofoils. The main operators are **Hellenic Seaways** and **Dodecanese Hydrofoils**, though there are many smaller companies running "Flying Dolphins", as they are known locally. They are twice as fast as a ferry but, as a consequence, are double the price.

The major drawback of hydrofoils is that most vessels only function in the summer months and are often cancelled if weather conditions are poor. In fact, on seas that are anything other than calm, hydrofoils are quite slow, and can prove a bad idea for those prone to seasickness.

Hydrofoils can accommodate around 140 passengers, but have no room for cars or motorcycles. Advance booking is often essential, and it is as well to book as early as possible during high season. Tickets are bought from an agent or on the quayside, but rarely on board the vessel itself. Routes are around the mainland and Peloponnese coasts, and to island groups close to the mainland – the Argo-Saronic group, Evvoia and the Sporades, and to several islands within the Cyclades. There are also routes between Rhodes, in the Dodecanese, and Sámos, at the southern end of the Northeast Aegean.

Catamaran [departu...]

Catamarans are a more recent innovation in Greece, offering an airline-type service in terms of seating, bar facilities and on-board television. They are also better designed for handicapped passengers. There are services around the Ionian Islands, and about half a dozen catamarans operating in the Aegean, mostly between the mainland port of Rafína and the islands of Andros, Tínos and Mýkonos. Costs are on a par with hydrofoils, and tickets should be bought from a travel agency a few days prior to sailing. But, if seats are available, they can be purchased on board.

TOURIST EXCURSIONS

Many holiday resorts put on small excursion boats that take groups of tourists to out-of-the-way beaches and caves, or on day cruises and picnics. Routes and times are dictated by local conditions.

Passengers on a departing ferry

Up-to-date information and booking arrangements are available on arrival in the islands at any local travel agency or information centre.

An excursion caïque on the Dodecanese island of Sými

LOCAL INTER-ISLAND FERRIES

In addition to the large ferries that cover the main routes, there are smaller ferries making inter-island crossings in the summer. Local ferries, regardless of size, are subject to government price controls, but boats chartered by tourist agencies can charge what they like, and often prove expensive. These boats do, however, provide direct connections, which shortcut circuitous routes via mainland ports.

Taxi boat travelling around the coast of Spétses

TAXI BOATS

Taxi boats (or caïques) are even more ad hoc, sailing along coastlines and making short trips between adjacent islands. They are usually only available during high season and, as the smallest vessels, are most prone to cancellation in adverse sea conditions. They tend to be more expensive than ferries, given the short distances involved, but often provide a route where few or no others are available.

Routes and itineraries are at the discretion of the boat owners, and the only place to determine if one is going your way is at the quayside.

DIRECTORY

UK & US AGENTS

Hellenic Holidays
1501 Broadway, Suite 1512, New York, NY 10036.
Tel (212) 944-8288.

Viamare Travel Ltd
Graphic House, 2 Sumatra Rd, London NW6 1PU.
Tel 020-7431 4560.
www.viamare.com

HYDROFOIL AND CATAMARAN SERVICES

Blue Star Ferries
Attica Premium, Attica Posidonos, Piraeus.
Tel 210 891 9550.
www.bluestarferries.com

Dodecanese Hydrofoils
Australias 3, 85100 Rhodes.
Tel 22410 78052.

Hellenic Seaways
6 Astiggos, Karaiskaki Square, Piraeus.
Tel 210 419 9000 (booking line).
www.hellenicseaways.gr

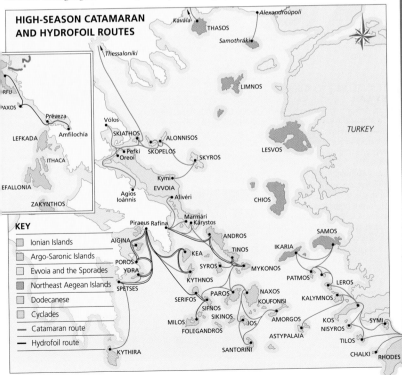

HIGH-SEASON CATAMARAN AND HYDROFOIL ROUTES

KEY
- Ionian Islands
- Argo-Saronic Islands
- Evvoia and the Sporades
- Northeast Aegean Islands
- Dodecanese
- Cyclades
- — Catamaran route
- — Hydrofoil route

Travelling by Road and Rail

There has been much upgrading of the roads on the islands but, particularly in remote areas, they can still be rough, and in some cases suitable only for four-wheel drive vehicles. Cars and motorcycles are easily rented though, and the extensive bus network is complemented by many taxis. Maps from local travel agents are less than reliable, however, and visitors are advised to bring their own: GeoCenter and Freytag & Berndt are both good.

You have priority

You have right of way

Do not use car horn

Wild animals crossing

Hairpin bend ahead

Roundabout (traffic circle)

TRAVELLING TO GREECE BY CAR

The most direct overland route to Greece from the UK, via the former Yugoslavia, is currently not recommended to motorists. The alternative route is through France, Switzerland and Italy, and from there to Greece by ferry. Motoring organizations, such as the **AA**, **RAC** or **ELPA** (the Automobile and Touring Club of Greece), offer advice on routes and regulations. You need a full, valid national driving licence, and insurance cover (at least third party insurance is compulsory).

RULES OF THE ROAD

Driving is on the right in Greece and, with the exception of some rural back-roads, road signs conform to European norms. The speed limit on national highways is 120 km/h (75 mph) for cars; on country roads it is 90 km/h (55 mph) and in towns 50 km/h (30 mph). Seat belts are required by law and children under ten are not allowed in the front seat.

CAR HIRE

Scores of car rental agencies in all main resorts offer a range of vehicles from small cars to minibuses. International companies such as **Avis** and **Budget** tend to be considerably more expensive than their local counterparts, though the latter are generally as reliable. Third party is the minimum insurance required by law, but personal accident insurance is strongly recommended. A full licence that has been held for at least one year is needed, and the minimum age requirement ranges from 21 to 25 years.

MOTORCYCLE, MOPED AND BICYCLE RENTAL

Motorcycles and mopeds are readily available to rent on the islands. The latter are ideal for short distances

Dual-language road sign

on flattish terrain, but for anything mountainous a motorcycle is a must. Make sure that the vehicle is in good condition before you set out and that the price includes adequate insurance cover; also check that your own travel insurance covers motorcycle accidents, as many do not.

The speed limit on national highways is 70 km/h (45 mph) for bikes up to 100 cc, and 90 km/h (55 mph) for larger bikes; helmets are compulsory.

Bicycles can also be rented in some resorts, though the steep mountainous terrain and hot sun can be deterrents to even the toughest enthusiast. Bicycles can, however, be transported free on most Greek ferries and buses.

PETROL (GAS) STATIONS

Petrol stations are plentiful in towns, though less so in rural areas – always set out with a full tank to be on the safe side. Fuel is sold by the litre, and there are usually three or four grades available: super (95 octane), unleaded, super unleaded and diesel, which is confusingly called *petrélaio*. Filling stations set their own working hours, but generally they are open seven days a week from 7 or 8am to between 7 and 9pm.

Sign for petrol stat

Rack of bikes for hire, at the beach in Kos town

Passengers aboard a taxi truck on the island of Lipsí

TAXIS

Taxis provide a very reasonably priced way of getting around on the islands. All taxis are metered, but for longer journeys a price can usually be negotiated per diem, or per trip. Drivers are generally amenable to dropping you off and picking you up a few hours later. Most rural villages have at least one taxi, and the best place to arrange for one is at the local *kafeneío* (café). Taxi trucks often take several passengers, each paying for their part of the journey.

TRAVELLING BY BUS AND COACH

International buses connect Greece with the rest of Europe, though fares are not as cheap as charter flights during the vacation season.

Greece's domestic bus system is operated by **KTEL** (Koinó Tameío Eispráxeon Leoforeíon), a syndicate of privately run companies that provides almost every community with services of some sort. In remote rural villages the bus might call once or twice a day, usually at the local taverna or *kafeneío*, while services between the larger centres are frequent and efficient. You can also usually rely on there being a bus service between the port and main town of any island, if the latter is situated inland.

On many of the larger islands travel agents offer a wide range of excursions on air-conditioned coaches, accompanied by qualified guides. These include trips to major archaeological and historical sites, other towns and resorts, popular beaches, areas for established walks, such as the Samariá Gorge in Crete, and organized events, such as an evening out in a "typical Greek taverna".

Front view of a local bus on the island of Nísyros

TRAVELLING BY TRAIN

Travelling to greece by train from London takes over three days. The journey is via France, Switzerland and Italy, crossing by ferry to Corfu and the mainland port of Pátra.

Within Greece, the network is run by the **OSE** (Organismós Sidirodrómon Elládos). The system is restricted to the mainland, but there are useful routes out of Athens to Pátra, Vólos (for ferries to Skiáthos and Skópelos) and up the Attic coast to Evvoia.

The distinctive front end of an express train

General Index

Acknowledgments

Dorling Kindersley would like to thank the following people whose contributions and assistance have made the preparation of this book possible.

Main Contributor
Marc Dubin is an American expatriate who divides his time between London and Sámos. Since 1978 he has travelled in every province of Greece. He has written or contributed to numerous guides to Greece, covering such diverse topics as trekking and contemporary Greek music.

Stephanie Ferguson, a freelance journalist and travel writer, has hopped around almost 50 Greek islands. She became bewitched by Greece after a holiday 20 years ago and since then has contributed to eight guide books and written travel features on Greece for several national publications.

Mike Gerrard is a travel writer and broadcaster who has written several guides to various parts of Greece, which he has been visiting annually since 1964.

Andy Harris is a travel and food journalist based in Athens. He is the author of *A Taste of the Aegean*.

Tanya Tsikas is a Canadian writer and travel guide editor. Married to a Greek, she has spent time in Crete and currently lives in Oxford.

Deputy Editorial Director
Douglas Amrine

Deputy Art Director Gillian Allan

Managing Editor Georgina Matthews

Managing Art Editor Annette Jacobs

Additional Illustrations
Richard Bonson, Louise Boulton, Gary Cross, Kevin Goold, Roger Hutchins, Claire Littlejohn.

Design and Editorial Assistance
Emma Anacootee, Claire Baranowski, Sonal Bhatt, Hilary Bird, Tony Clark, Elspeth Collier, Michelle Crane, Michele Crawford, Catherine Day, Jim Evoy, Robin Gauldie, Emily Green, Emily Hatchwell, Leanne Hogbin, Kim Inglis, Maria Kelesidi, Lorien Kite, Esther Labi, Felicity Laughton, Andreas Michael, Ella Milroy, Lisa Minsky, Robert Mitchell, Adam Moore, Jennifer Mussett, Tamsin Pender, Eva Petrou, Marianne Petrou, Pollyanna Poulter, Jake Reimann, Ellen Root, Simon Ryder, Collette Sadler, Rita Selvaggio, Liz Sharp, Ellie Smith, Claire Stewart, Claire Tennant-Scull, Amanda Tomeh, Dora Whitaker, Andy Wilkinson.

Dorling Kindersley would also like to thank the following for their assistance: The Greek Wine Bureau, Odysea.

Additional Research
Anna Antoniou, Garifalia Boussiopoulou, Anastasia Caramanis, Michele Crawford, Magda Dimouti, Shirley Durant, Panos Gotsi, Zoi Groummouti, Peter Millett, Tasos Schizas, Garifalia Tsiola, Veronica Wood.

Artwork Reference
Ideal Photo S.A., The Image Bank, Melissa Publishing House, Tony Stone Worldwide.

Additional Photography
Jane Burton, Frank Greenaway, Derek Hall, Nigel Hicks, Dave King, Neil Lucas, National History Museum, Ian O'Leary, Stephen Oliver, Roger Philips, Kim Sayer, Clive Steeter, Harry Taylor, Kim Taylor, Mathew Ward, Jerry Young.

Photography Permissions
Dorling Kindersley would like to thank the following for their assistance and kind permission to photograph at their establishments:

Nelly Dimoglou Folk Dance Theatre, Rhodes; Museum of Greek Folk Art, Athens; Karpathos Museum; Markos Vamvakaris Museum, Syros; Kymi Folk Museum, Evvoia; Stavros Kois's House, Syros. Also all other cathedrals, churches, museums, hotels, restaurants, shops, galleries, and sights too numerous to thank individually.

Picture Credits
t = top; tl = top left; tlc = top left centre; tc = top centre; trc = top right centre; tr = top right; cla = centre left above; ca = centre above; cra = centre right above; cl = centre left; c = centre; cr = centre right; clb = centre right below; cb = centre below; crb = centre right below; bl = bottom left; b = bottom; bc = bottom centre; bcl = bottom centre left; br = bottom right; d = detail.

Works of art have been reproduced with the permission of the following copyright holders: © ADAGP, Paris and DACS, London 1997 *The Kiss* Constantin Brancusi 211br. The work of art *Three Standing Figures,* Henry Moore (1947) 211bl is reproduced by permission of the Henry Moore Foundation.

The publisher would like to thank the following individuals, companies and picture libraries for permission to reproduce their photographs:

AISA ARCHIVO ICONGRAFICO, Barcelona: Museo Archeologique, Bari 59tr; Museo Archeologique, Florence 56tl; AKG, London: 190b, 324t; Antiquario Palatino 55bl; British Museum 289b; Erich Lessing Akademie der Bildenden Künste, Vienna 56c; Musée du Louvre 55tl; Naples Archaeological Museum 145b; National Archeological Museum, Athens 28–9(d), 29t; Staatliche Kunstsammlungen, Albertinum, Dresden 33crb, Liebighaus, Frankfurt/Main 35c; Staatliche Antikensammlungen und Glyptotek, München 54b; Mykonos Museum 57t; alamy images: Jan Baks 10tc; Peter Horree 327tl; Imagebroker / Egmont Strigl 10br; IML Image Group Ltd 10cl, 11br; naglestock.com 11cr; Robert Harding World Imagery 327c; Terry Harris just greece photo library 12bl; ANCIENT ART AND ARCHITECTURE: 31ca, 32t, 36t, 36ca, 37cl, 39t, 58cb, 58b(d), 99b; ANTIKENMUSEUM BASEL UND SAMMLUNG LUDWIG: 62–3; APERION: John Hios 47c; ARGYROPOULOS PHOTO PRESS: 47t, 49cr, 50t, 50cb; Athens International Airport: 362cr/b, 363cr, 364; ATHENS URBAN TRANSPORT ORGANIZATION (OASA): 292cla, 293c.

BENAKI MUSEUM: 27b, 38ca, 41t, 41c, 43ca, 291b; PAUL BERNARD: 35t; BIBLIOTHEQUE NATIONAL, Paris: Caoursin folio 175 4c(d), 40–41(d), Caoursin folio 33 189bl, Caoursin folio 79 189br; BODLEIAN LIBRARY, Oxford: MS Canon Misc 378 170v 36cb; BRIDGEMAN ART LIBRARY, London: Birmingham City Museums and Art Galleries *Pheidias Completing the Parthenon Frieze,* Sir Lawrence Alma-Tadema 60t; Bibliothèque Nationale, Paris *The Author Guillaume Caoursin, Vice Chancellor of the Order of St John of Jerusalem Dedicating his Book to Pierre d'Aubusson, Grand Master of the Order of St John of Jerusalem who is Seated Surrounded by High Dignitaries of the Order* (1483), illustrated by the Master of Cardinal of Bourbon Lat 6067 f 3v 26(d); British Museum, London *Cup, Tondo,*

with Scene of Huntsmen Returning Home 31cb, *Greek Vase Showing Diver About to Enter the Sea in Search of Sponges* (c.500 BC) 169br; Fitzwilliam Museum, University of Cambridge *Figurine of Demosthenes* Enoch Wood of Burslem (c.1790) (lead glazed earthenware) 59tl, *Attic Red-figured Pelike: Pigs, Swineherd and Odysseus,* Pig Painter (470–60BC) 87bl; Freud Museum, London *Figure of Artemis from Myrina,* Greek, Hellenistic Period (2nd century BC) 218tl; Giraudon/Louvre *Alexander the Great, Portrait Head* (3rd century BC) Greek (marble) 34tr; House of Masks, Delos *Mosaic of Dionysus riding a Leopard* (c.AD 180) 37t; Kunsthistorisches Museum, Vienna *Elizabeth of Bavaria, Wife of Emperor Franz Joseph I of Austria,* Franz Xavier Winterhalter 83c(d); Lauros-Giraudon/Louvre, Paris *Rhodes Winged Victory of Samothrace* (early 2nd century BC) 132c; Louvre, Paris *Double Bust of Aristophanes and Sophocles* (15th century) 58t; National Archaeological Museum, Athens *Bronze Satue of Poseidon* (c.460–450 BC) photo Bernard Cox 54c; Private Collection *Two-tiered Icon of the Virgin and Child and Two Saints,* Cretan School (15th century) 40cl; Victoria and Albert Museum, London *Corfu,* Edward Lear 81b; © THE BRITISH MUSEUM: 28clb, 29cb, 30tl, 32cb, 33clb, 55tr(d), 55br, 57c(d), 62tl, 63tl, 63tr, 139cra.

CAMERA PRESS, London: ANAG 45tl, 45bl; Christopher Simon Sykes 77t; Wim Swaan 219t; BRUCE COLEMAN LTD.: Philip van de Berg 247cl; Luiz Claudio Marigo 91t; Natalio Feneck 115ca; Gordon Langsbury 247bl; Andrew J Purcell 24ca; Kim Taylor 115bl; World Wildlife Fund for Nature 115t; Konrad Wothe 247br; CORBIS: John Heseltine 326cl; Gail Mooney 11tl.

C M DIXON PHOTO RESOURCES: 29ca; Glyptotek, Munich 34tr; MARC DUBIN: 22t, 23bla, 23bra, 48tl, 174t, 199t, 199b, 223b, 233c, 235b, 252b.

ECOLE FRANÇAISE D'ATHÈNES: 218tr; ECOLE NATIONALE SUPERIEURE DES BEAUX ARTS, Paris: *Delphes Restauration du Sanctuaire Envoi,* Tournaire (1894) 32–3; EKDOTIKI ATHINON: 3, 28crb, 164b, 187b(d), 188bl(d); ELIA: 109b; JANICE ENGLISH: 192t, 193t, 193b, 194c, 194b; ET ARCHIVE: National Archaeology Museum, Naples 34ca; EUROPEAN COMMISSION: 359;

Mary Evans Picture Library: 87cr, 87bc, 87br, 133b, 156b.

Ferens Art Gallery: Hull City Museums and Art Galleries and Archives *Electra at the Tomb of Agamennon* (1869), Lord Frederick Leighton 57b.

Giraudon, Paris: Chateau Ecouen *Retour d'Ulysse* Ecole Siennoise 87cl; Louvre Paris 62c, *Scène de Massacres de Scio,* Eugene Delacroix 42ca(d), 147b(d); Musée Nationale Gustave Moreau *Hesiode et Les Muses,* Gustave Moreau 58ca; Musée d'Art Catalan, Barcelona 289cb; Nicholas P Goulandris Foundation Museum of Cycladic and Ancient Greek Art: 211tl, 211tc, 211tr, 211cl, 211cr, 284b, 291t; Ronald Grant Archive: *Zorba the Greek,* 20th Century Fox 276b.

Greek National Tourism Organization: 354tl.

Robert Harding Picture Library: David Beatty 46br; Tony Gervis 46ca, 46bl; Adam Woolfitt 48c; Helio Photo: 98ca; Hellenic Post Service: 45cla, 360tl, 361tl; Hellenic War Museum, Athens: 251t; Historical Museum of Crete, Irákleio: *Landscape of the Gods-Trodden Mount Sinai,* El Greco 269tr; michael holford: British Museum 34cb, 54t; Hulton Getty Collection: 43cb(d); Central Press Photo 44clb(d).

Ideal Photo SA: T Dassios 301c; A Pappas 235t; C Vergas 47clb, 47br, 48tr, 83b, 117b, 161t; Images Colour Library: 94t, 270bl; Impact Photos: Jeremy Nicholl 344b; Caroline Penn 47bl.

Carol Kane: 198br; Gulia Klimi: 49t; Kostos Kontos: 28cr, 44cra, 44cla, 47crb, 50ca, 163b, 164c, 188br, 293b, 357t, 349c.

Frank Lane Pictures: Eric and David Hoskings 246bl; Ilias Lalaounis: 343cla.

Magnum Photos Ltd.: Constantine Manos 46t; Mansell Collection: 54–5.

National Gallery of Victoria, Melbourne: *Greek by the Inscriptions Painter Challidian* Felton Bequest (1956) 56b; National Historical Museum: 40t, 42t, 42–3(d), 43t, 44b; Natural Image: Bob Gibbons 247tl; Peter Wilson 254tl; Nature Photographers: Brinsley Burbridge 247tr; Robin Bush 247cr; Michael J Hammett 24tr; Paul Sterry 115br, 247tc; Antonis

Nicolopoulos: 348c, 349b.

Olympic Airways: 362t; Oronoz Archivo Fotografico: Biblioteca National Madrid *Invasions Bulgares Historia Matriksiscronica FIIIV* 38cb(d); Charlottenberg, Berlin 56tr; El Escorial, Madrid *Battle of Lepanto,* Cambiaso Luca 40cr(d); Museo Julia 55c(d); Musée du Louvre 62b; Museo Vaticano 59b, 63bl; Oxford Scientific Films: Paul Kay 25tr.

Romylos Parisis: City of Athens Museum 42cb; Pictor International: 48b; Pictures: 46crb, 348br; Planet Earth Pictures: Wendy Dennis 246c; Jim Greenfield 25cb; Ken Lucas 115cb; Marty Snyderman 25tl; Private Collection: 281c; Popperfoto: 44crb, 45cra, 45br.

Rex Features: Sipa Press/C Brown 45tr.

Scala, Florence: Gallerie degli Uffizi 30cb; Museo Archeologico, Firenze 31t; Museo Mandralisca Cefalu 32ca; Museo Nationale Tarquinia 63br; Museo de Villa Giulia 30–31, 62tr; Spectrum Colour Library: 254tl; Maria Stefossi 20bl.

Tap (Service Archaeological Receipts Fund) Hellenic Republic Ministry of Culture: 1st Epharat of Antiquities 44cb, 60br, 288cb, 288b, 289t, 289ca; Acropolis Museum 290tl, 290tr; Andros Archaeological Museum 208cl; Agios Nikolaos Archaeological Museum 278c; 2nd Epharat of Antiquities 66bl, 98t, 98cb, 98b, 99t, 99c; Chalkida Archaeological Museum 120c; Chania Archaeological Museum 253c; Corfu Archaeological Museum 69t, 79c; Eretreia Archaeological Museum 5t, 119b, 121c; 5th Epharat of Byzantine Antiquities 39c; 14th Epharat of Byzantine Antiquities 137t, 140t, 144b; 4th Epharat of Byzantine Antiquities 18c, 165tr, 165c, 165b, 199c; 18th Epharat of Antiquities 128t, 129t, 129b; 19th Epharat of Antiquities 132b, 133ca, 133cb; Ikia Varelizidenas 125c; Irakleio Archaeological Museum 266bl, 270t, 270ca, 270cb, 271t, 271c, 272br; 20th Epharat of Antiquities 135b; 21st Epharat of Antiquities 67t, 156t, 156c, 218b, 219ca, 219bl, 219br, 227c, 230cl, 240cb, 241c, 344ca; 22nd Epharat of Antiquities 171b, 172t, 172c, 180t, 186tl, 186tr, 186c, 186b, 187t, 187c, 196t, 196cr; 23rd Epharat of Antiquities 249t, 263c, 264c, 264b, 265 all, 266t, 266c, 266br, 267 all, 272t, 272c,

272bl, 273tl, 273tr, 273b, 274t, 274c, 274b, 275b, 221t; 24th Epharat of Antiquities 277c; Kos Archaeological Museum 172b; Milos Archaeological Museum 236t; Mykonos Archaeological Museum 214tr; National Archeological Museum, Athens 28t, 30tr, 241b, 286c; Naxos Archaeological Museum 230t; Nea Moni Archaeological Museum 67ca; Numismatic Museum of Athens 275c; Rhodes Archaeological Museum 184cb; Vathy Archaeological Museum, Samos 154cr; 2nd Epharat of Byzantine Antiquities 213c, 221t, 224t, 227b, 251c; 7th Epharat of Byzantine Antiquities 109t; 6th Epharat of Byzantine Antiquities 89c; Thessaloniki Archaeological Museum 33t; Thira Archaeological Museum 238tl; 3rd Epharat of Byzantine Antiquities 144b, 152tl, 153c, 155b; 13th Epharat of Byzantine Antiquities 150–51 all, 261cra, 277b, 279t; Tinos Archaeological Museum 212t; 3rd Epharat of Antiquities 287b; TERIADE MUSEUM:

Dafnis and Chloe, Marc Chagall ©ADAGP, Paris and DACS, London 1997 138br; travel ink: Nigel Bowen-Morris 242–3; travel library: Faltaits Museum 116tr; yannis tsarouchis foundation: Private Collection *Barber Shop in Marousi,* Yannis Tsarouchis (1947) 44tr. WERNER FORMAN ARCHIVE: Thessaloniki Archaeological Musem 38t; LORRAINE WILSON: 66br; PETER WILSON: 20c, 61bl, 67b, 288ca; BRIAN WOODYATT: 18b.

JACKET
Front – ALAMY IMAGES: Jon Arnold Images / Gavin Hellier main image; DK IMAGES: clb; Matthew Ward clb. Back – ALAMY IMAGES: Andre Jenny bl; DK IMAGES: Joe Cornish tl; Rupert Horrox cla, Rob Reichenfeld clb. Spine – ALAMY IMAGES: Jon Arnold Images / Gavin Hellier t; DK IMAGES: Linda Whitwam b.

All other images © Dorling Kindersley. For further information see: www.dkimages.com

---

SPECIAL EDITIONS OF DK TRAVEL GUIDES

DK Travel Guides can be purchased in bulk quantities at discounted prices for use in promotions or as premiums. We are also able to offer special editions and personalized jackets, corporate imprints, and excerpts from all of our books, tailored specifically to meet your own needs.

To find out more, please contact:
(in the United States) **SpecialSales@dk.com**
(in the UK) **travelspecialsales@uk.dk.com**
(in Canada) DK Special Sales at **general@tourmaline.ca**
(in Australia) **business.development@pearson.com.au**

Phrase Book

There is no universally accepted system for representing the modern Greek language in the Roman alphabet. The system of transliteration adopted in this guide is the one used by the Greek Government. Though not yet fully applied throughout Greece, most of the street and place names have been transliterated according to this system. For Classical names this guide uses the k, os, on and f spelling, in keeping with the modern system of transliteration. In a few cases, such as Socrates, the more familiar Latin form has been used. Classical names do not have accents. Where a well-known English form of a name exists, such as Athens or Corfu, this has been used. Variations in transliteration are given in the index.

GUIDELINES FOR PRONUNCIATION

The accent over Greek and transliterated words indicates the stressed syllable. In this guide the accent is not written over capital letters nor over monosyllables, except for question words and the conjunction ή (meaning "or"). In the right-hand "Pronunciation" column below, the syllable to stress is given in bold type.

On the following pages, the English is given in the left-hand column with the Greek and its transliteration in the middle column. The right-hand column provides a literal system of pronunciation and indicates the stressed syllable in bold.

THE GREEK ALPHABET

Α α	A a	**arm**
Β β	V v	**v**ote
Γ γ	G g	**y**ear (when followed by e and i sounds) **no** (when followed by ξ or γ)
Δ δ	D d	**th**at
Ε ε	E e	**egg**
Ζ ζ	Z z	**zoo**
Η η	I i	bel**ie**ve
Θ θ	Th th	**think**
Ι ι	I i	bel**ie**ve
Κ κ	K k	**kid**
Λ λ	L l	**land**
Μ μ	M m	**man**
Ν ν	N n	**no**
Ξ ξ	X x	ta**xi**
Ο ο	O o	**fox**
Π π	P p	**port**
Ρ ρ	R r	**room**
Σ σ	S s	**sorry** (zero when followed by μ)
ς	s	(used at end of word)
Τ τ	T t	**tea**
Υ υ	Y y	bel**ie**ve
Φ φ	F f	**fish**
Χ χ	Ch ch	lo**ch** in most cases, but **h**e when followed by a, e or i sounds
Ψ ψ	Ps ps	ma**ps**
Ω ω	O o	**fox**

COMBINATIONS OF LETTERS

In Greek there are two-letter vowels that are pronounced as one sound:

Αι αι	Ai ai	**egg**
Ει ει	Ei ei	bel**ie**ve
Οι οι	Oi oi	bel**ie**ve
Ου ου	Ou ou	l**u**te

There are also some two-letter consonants that are pronounced as one sound:

Μπ μπ	Mp mp	**b**ut, sometimes nu**mb**er in the middle of a word
Ντ ντ	Nt nt	**d**esk, sometimes u**nd**er in the middle of a word
Γκ γκ	Gk gk	**g**o, sometimes bi**ng**o in the middle of a word
Γξ γξ	nx	a**nx**iety
Τζ τζ	Tz tz	han**ds**
Τσ τσ	Ts ts	i**t's**
Γγ γγ	Gg gg	bi**ng**o

In an Emergency

Help!	Βοήθεια! / Voítheia	vo-**ee**-theea
Stop!	Σταματήστε! / Stamatíste	sta-ma-**tee**-steh
Call a doctor!	Φωνάξτε ένα γιατρό / Fonáxte éna giatró	fo-**nak**-steh **e**-na ya-**tro**
Call an ambulance/ the police/the fire brigade!	Καλέστε το ασθενοφόρο/την αστυνομία/την πυροσβεστική / Kaléste to asthenofóro/tin astynomía/tin pyrosvestikí	ka-**le**-steh to as-the-no-**fo**-ro/teen a-sti-no-**mia**/teen pee-ro-zve-stee-**kee**
Where is the nearest telephone/hospital/ pharmacy?	Πού είναι το πλησιέστερο τηλέφωνο/νοσοκο-μείο/φαρμακείο; / Poú eínai to plisiéstero tiléfono/ nosoko-meío/farmakeío?	poo **ee**-ne to plee-see-**e**-ste-ro tee-le-pho-no/no-so-ko-**mee**-o/far-ma-**kee**-o?

Communication Essentials

Yes	Ναι / Nai	neh
No	Όχι / Ochi	**o**-chee
Please	Παρακαλώ / Parakaló	pa-ra-ka-**lo**
Thank you	Ευχαριστώ / Efcharistó	ef-cha-ree-**sto**
You are welcome	Παρακαλώ / Parakaló	pa-ra-ka-**lo**
OK/alright	Εντάξει / Entáxei	en-**dak**-zee
Excuse me	Με συγχωρείτε / Me synchoreíte	me seen-cho-**ree**-teh
Hello	Γειά σας / Geiá sas	**yeea** sas
Goodbye	Αντίο / Antío	an-**dee**-o
Good morning	Καλημέρα / Kaliméra	ka-lee-**me**-ra
Good night	Καληνύχτα / Kalinýchta	ka-lee-**neech**-ta
Morning	Πρωί / Proí	pro-**ee**
Afternoon	Απόγευμα / Apógevma	a-**po**-yev-ma
Evening	Βράδυ / Vrádi	**vrath**-i
This morning	Σήμερα το πρωί / Símera to proí	**see**-me-ra to pro-**ee**
Yesterday	Χθές / Chthés	**chthes**
Today	Σήμερα / Símera	**see**-me-ra
Tomorrow	Αύριο / Avrio	**av**-ree-o
Here	Εδώ / Edó	ed-**o**
There	Εκεί / Ekeí	e-**kee**
What?	Τι; / Τi?	**tee**?
Why?	Γιατί; / Giatí?	ya-**tee**?
Where?	Πού; / Poú?	**poo**?
How?	Πώς; / Pós?	**pos**?
Wait!	Περίμενε! / Perímene!	pe-**ree**-me-neh

Useful Phrases

How are you?	Τί κάνεις;	tee ka-nees
	Tí kááneis?	
Very well, thank you	Πολύ καλά, ευχαριστώ	po-lee ka-la, ef-cha-ree-sto
	Polý kalá, efcharistó	
How do you do?	Πώς είστε;	pos ees-te?
	Pós eíste?	
Pleased to meet you	Χαίρω πολύ	che-ro po-lee
	Chaíro polý	
What is your name?	Πώς λέγεστε;	pos le-ye-ste?
	Pós légeste?	
Where is/are...?	Πού είναι;	poo ee-ne?
	Poú eínaï?	
How far is it to...?	Πόσο απέχει...;	po-so a-pe-chee?
	Póso apéchei...?	
How do I get to?	Πώς μπορώ να πάω....;	pos bo-ro-na pa-o?
	Pós mporó na páo...?	
Do you speak English?	Μιλάτε Αγγλικά;	mee-la-te an-glee-ka?
	Miláte Angliká?	
I understand	Καταλαβαίνω	ka-ta-la-ve-no
	Katalavaíno	
I don't understand	Δεν καταλαβαίνω	then ka-ta-la-ve-no
	Den katalavaíno	
Could you speak slowly?	Μιλάτε λίγο πιο αργά παρακαλώ;	mee-la-te lee-go pyo ar-ga pa-ra-ka-lo?
	Miláte lígo pio argá parakaló?	
I'm sorry	Με συγχωρείτε	me seen-cho-ree teh
	Me synchoreíte	
Does anyone have a key?	Έχει κανένας κλειδί;	e-chee ka-ne-nas klee-dee?
	Echei kanénas kleidí?	

Useful Words

big	Μεγάλο	me-ga-lo
	Megálo	
small	Μικρό	mi-kro
	Mikró	
hot	Ζεστό	zes-to
	Zestó	
cold	Κρύο	kree-o
	Krýo	
good	Καλό	ka-lo
	Kaló	
bad	Κακό	ka-ko
	Kakó	
enough	Αρκετά	ar-ke-ta
	Arketá	
well	Καλά	ka-la
	Kalá	
open	Ανοιχτά	a-neech-ta
	Anoichtá	
closed	Κλειστά	klee-sta
	Kleistá	
left	Αριστερά	a-ree-ste-ra
	Aristerá	
right	Δεξιά	dek-see-a
	Dexiá	
straight on	Ευθεία	ef-thee-a
	Eftheía	
between	Ανάμεσα / Μεταξύ	a-na-me-sa/me-tak-see
	Anámesa / Metaxý	
on the corner of....	Στη γωνία του...	stee go-nee-a too
	Sti gonía tou...	
near	Κοντά	kon-da
	Kontá	
far	Μακριά	ma-kree-a
	Makriá	
up	Επάνω	e-pa-no
	Epáno	
down	Κάτω	ka-to
	Káto	
early	Νωρίς	no-rees
	Norís	
late	Αργά	ar-ga
	Argá	
entrance	Η είσοδος	ee ee-so-thos
	I eísodos	
exit	Η έξοδος	eee-kso-dos
	I éxodos	
toilet	Οι τουαλέτες /WC	ee-too-a-le-tes
	Oi toualétes / WC	
occupied/engaged	Κατειλημμένη	ka-tee-lee-me-nee
	Kateiliméni	

unoccupied/vacant	Ελεύθερη	e-lef-the-ree
	Eléftheri	
free/no charge	Δωρεάν	tho-re-an
	Doreán	
in/out	Μέσα/Έξω	me-sa/ek-so
	Mésa/ Exo	

Making a Telephone Call

Where is the nearest public telephone ?	Πού βρίσκεται ο πλησιέστερος τηλεφωνικός θάλαμος;	poo vrees-ke-teh o plee-see-e-ste-ros tee-le-fo-ni-kos tha-la-mos?
	Poú vrísketai o plisiésteros tilefonikós thálamos?	
I would like to place a long-distance call	Θα ήθελα να κάνω ένα υπεραστικό τηλεφώνημα	tha ee-the-la na ka-no e-na ee-pe-ra-sti-ko tee-le-fo-nee-ma
	Tha íthela na káno éna yperastikó tilefónima	
I would like to reverse the charges	Αα ήθελα να χρεώσω το τηλεφώνημα στον παραλήπτη	tha ee-the-la na chre-o-so to tee-le-fo-nee-ma ston pa-ra-lep-tee
	Tha íthela na chreóso to tilefónima ston paralípti	
I will try again later	Θα ξαναπλε φωνήσω αργότερα	tha ksa-na-tee-le-fo-ni-so ar-go-te-ra
	Tha xanatilefoníso argótera	
Can I leave a message?	Μπορείτε να του αφήσετε ένα μήνυμα;	bo-ree-te na too a-fee-se-teh e-na mee-nee-ma?
	Mporeíte na tou afísete éna mínyma?	
Could you speak up a little please?	Μιλάτε δυνατότερα, παρακαλώ;	mee-la-teh dee-na-to-te-ra, pa-ra-ka-lo
	Miláte dynatótera, parakaló	
Local call	Τοπικό τηλεφώνημα	to-pi-ko tee-le-fo-nee-ma
	Topikó tilefónima	
Hold on	Περιμένετε	pe-ri-me-ne-teh
	Periménete	
OTE telephone office	Ο ΟΤΕ / Το τηλεφωνείο	o O-TE / To tee-le-fo-nee-o
	Ο ΟΤΕ / Το tilefoneío	
Phone box/kiosk	Ο τηλεφωνικός θάλαμος	o tee-le-fo-ni-kos tha-la-mos
	Ο tilefonikós thálamos	
Phone card	Η τηλεκάρτα	ee tee-le-kar-ta
	I tilekárta	

Shopping

How much does this cost?	Πόσο κάνει;	po-so ka-nee?
	Póso kánei?	
I would like....	Αα ήθελα...	tha ee-the-la...
	Tha íthela...	
Do you have....?	Έχετε...;	e-che-teh
	Echete...?	
I am just looking	Απλώς κοιτάω	a-plos kee-ta-o
	Aplós koitáo	
Do you take credit cards/travellers' cheques?	Δέχεστε πιστωτικές κάρτες/ travellers' cheques;	the-ches-teh pee-sto-tee-kes kar-tes/ travellers' cheques
	Déch: στε pistotikés kártes/travellers' cheques?	
What time do you open/close?	Ποτέ ανοίγετε/ κλείνετε;	po-teh a-nee-ye-teh/ klee-ne-teh?
	Póte anoígete/ kleínete?	
Can you ship this overseas?	Μπορείτε να το στείλετε στο εξωτερικό;	bo-ree-teh na to stee-le-teh sto e-xo-te-ree ko?
	Mporeíte na to steílete sto exoterikó?	
This one	Αυτό εδώ	af-to e-do
	Aftó edó	
That one	Εκείνο	e-kee-no
	Ekeíno	

expensive	Ακριβό Akrivó	a-kree-vo
cheap	Φθηνό Fthinó	fthee-no
size	Το μέγεθος To mégethos	to me-ge-thos
white	Λευκό Lefkó	lef-ko
black	Μαύρο Mávro	mav-ro
red	Κόκκινο Kókkino	ko-kee-no
yellow	Κίτρινο Kítrino	kee-tree-no
green	Πράσινο Prásino	pra-see-no
blue	Μπλε Mple	bleh

Types of Shop

antique shop	Μαγαζί με αντίκες Magazí me antíkes	ma-ga-zee me an-dee-kes
bakery	Ο φούρνος O foúrnos	o foor-nos
bank	Η τράπεζα I trápeza	ee tra-pe-za
bazaar	Το παζάρι To pazári	to pa-za-ree
bookshop	Το βιβλιοπωλείο To vivliopoleío	to vee-vlee-o-po-lee-o
butcher	Το κρεοπωλείο To kreopoleío	to kre-o-po-lee-o
cake shop	Το ζαχαροπλαστείο To zacharoplasteío	to za-cha-ro-pla-stee-o
cheese shop	Μαγαζί με αλλαντικά Magazí me allantiká	ma-ga-zee me a-lan-dee-ka
department store	Πολυκάταστημα Polykatástima	Po-lee-ka-ta-stee-ma
fishmarket	Το ιχθυοπωλείο/ ψαράδικο To ichthyopoleío/ psarádiko	to eech-thee-o-po-lee-o /psa-rá-dee-ko
greengrocer	Το μανάβικο To manáviko	to ma-na-vee-ko
hairdresser	Το κομμωτήριο To kommotírio	to ko-mo-tee-ree-o
kiosk	Το περίπτερο To períptero	to pe-reep-te-ro
leather shop	Μαγαζί με δερμάτινα είδη Magazí me dermátina eídi	ma-ga-zee me ther-ma-tee-na ee-ethee
street market	Η λαϊκή αγορά I laïkí agorá	ee la-ee-kee a-go-ra
newsagent	Ο εφημεριδοπώλης O efimeridopólis	O e-fee-me-ree-tho-po-lees
pharmacy	Το φαρμακείο To farmakeío	to far-ma-kee-o
post office	Το ταχυδρομείο To tachydromeío	to ta-chee-thro-mee-o
shoe shop	Κατάστημα υποδημάτων Katástima ypodimáton	ka-ta-stee-ma ee-po-dee-ma-ton
souvenir shop	Μαγαζί με "souvenir" Magazí me "souvenir"	ma-ga-zee meh "souvenir"
supermarket	Σουπερμάρκετ/ Υπεραγορά "Supermarket"/ Yperagorá	"Supermarket" / ee-per-a-go-ra
tobacconist	Είδη καπνιστού Eídi kapnistoú	Ee-thee kap-nees-too
travel agent	Το ταξειδιωτικό γραφείο To taxeidiotikó grafeío	to tak-see-thy-o-tee-ko gra-fee-o

Sightseeing

tourist information	Ο ΕΟΤ O EOT	o E-OT
tourist police	Η τουριστική αστυνομία I touristikí astynomía	ee too-rees-tee-kee a-stee-no-mee-a
archaeological	αρχαιολογικός archaiologikós	ar-che-o-lo-yee-kos

art gallery	Η γκαλερί I gkalerí	ee ga-le-ree
beach	Η παραλία I paralía	ee pa-ra-lee-a
Byzantine	βυζαντινός vyzantinós	vee-zan-dee-nos
castle	Το κάστρο To kástro	to ka-stro
cathedral	Η μητρόπολη I mitrópoli	ee mee-tro-po-lee
cave	Το σπήλαιο To spílaio	to spee-le-o
church	Η εκκλησία I ekklisía	ee e-klee-see-a
folk art	λαϊκή τέχνη laïkí téchni	la-ee-kee tech-nee
fountain	Το συντριβάνι To syntriváni	to seen-dree-va-nee
hill	Ο λόφος O lófos	o lo-fos
historical	ιστορικός istorikós	ee-sto-ree-kos
island	Το νησί To nisí	to nee-see
lake	Η λίμνη I límni	ee leem-nee
library	Η βιβλιοθήκη I vivliothíki	ee veev-lee-o-thee-kee
mansion	Η έπαυλις I épavlis	eee-pav-lees
monastery	Μονή moní	mo-ni
mountain	Το βουνό To vounó	to voo-no
municipal	δημοτικός dimotikós	thee-mo-tee-kos
museum	Το μουσείο To mouseío	to moo-see-o
national	εθνικός ethnikós	eth-nee-kos
park	Το πάρκο To párko	to par-ko
garden	Ο κήπος O kípos	o kee-pos
gorge	Το φαράγγι To farángi	to fa-ran-gee
grave of....	Ο τάφος του... O táfos tou...	o ta-fos too
river	Το ποτάμι To potámi	to po-ta-mee
road	Ο δρόμος O drómos	o thro-mos
saint	άγιος/άγιο/αγία /αγίες ágios/ágioi/agía/agíes	a-yee-os/a-yee-ee/a-/yee-a/a-yee-es
spring	Η πηγή I pigí	ee pee-yee
square	Η πλατεία I plateía	ee pla-tee-a
stadium	Το στάδιο To stádio	to sta-thee-o
statue	Το άγαλμα To ágalma	toa-gal-ma
theatre	Το θέατρο To théatro	to the-a-tro
town hall	Το δημαρχείο To dimarcheío	To thee-mar-chee-o
closed on public holidays	κλειστό τις αργίες kleistó tis argíes	klee-sto tees aryee-es

Transport

When does the .... leave?	Πότε φεύγει το ....; Póte févgei to...?	po-teh fev-yee to...?
Where is the bus stop?	Πού είναι η στάση του λεωφορείο υ; Poú eínai i stási tou leoforeíou?	poo ee-neh ee sta-see too le-o-fo-ree-oo?
Is there a bus to...?	Υπάρχει λεωφορείο για....; Ypárchei leoforeío gia...?	ee-par-chee le-o-fo-ree-o yia...?
ticket office	Εκδοτήρια εισιτηρίων Ekdotíria eisitiríon	Ek-tho-tee-reea ee-see-tee-ree-on
return ticket	Εισιτήριο με επιστροφή Eisitírio me epistrofí	ee-see-tee-ree-o meh e-pee-stro-fee
single journey	Απλό εισιτήριο Apló eisitírio	a-plo ee-see-tee-reeo

bus station	Ο σταθμός λεωφορείων Ο stathmós leoforeíon	o stath-mos leo-fo-ree-on
bus ticket	Εισιτήριο λεωφορείου Eisitírio leoforeíou	ee-see-tee-ree-o leo-fo-ree-oo
trolley bus	Το τρόλλεϋ Το trólley	to tro-le-ee
port	Το λιμάνι Το limáni	to lee-ma-nee
train/metro	Το τρένο Το tréno	to tre-no
railway station	σιδηροδρομικός σταθμός sidirodromikós stathmós	see-thee-ro-thro-mee-kos stath-mos
moped	Το μοτοποδήλατο / το μηχανάκι Το motopodílato / Το michanáki	o mo-to-po-thee-la-to/to mee-cha-na-kee
bicycle	Το ποδήλατο Το podílato	to po-thee-la-to
taxi	Το ταξί Το taxí	to tak-see
airport	Το αεροδρόμιο Το aerodrómio	to a-e-ro-thro-mee-o
ferry	Το φερυμπότ Το "ferry-boat"	to fe-ree-bot
hydrofoil	Το δελφίνι / Το υδροπτέρυγο Το delfíni / Το ydroptérygo	to del-fee-nee / To ee-throp-te-ree-go
catamaran	Το καταμαράν Το katamarán	to catamaran
for hire	Ενοικιάζονται Enoikiázontai	e-nee-kya-zon-deh

Staying in a Hotel

Do you have a vacant room?	Έχετε δωμάτια; Echete domátia?	e-che-teh tho-ma-tee-a?
double room with double bed	Δίκλινο με διπλό κρεβάτι Díklino me dipló kreváti	thee-klee-no meh thee-plo kre-va-tee
twin room	Δίκλινο με μονά κρεβάτια Díklino me moná krevátia	thee-klee-no meh mo-na kre-vat-ya
single room	Μονόκλινο Monóklino	mo-no-klee-no
room with a bath	Δωμάτιο με μπάνιο Domátio me mpánio	tho-ma-tee-o meh ban-yo
shower	Το ντους Το douz	To dooz
porter	Ο πορτιέρης Ο portiéris	o por-tye-rees
key	Το κλειδί Το kleidí	to klee-dee
I have a reservation	Έχω κάνει κράτηση Echo kánei krátisi	e-cho ka-nee kra-tee-see
room with a sea view/balcony	Δωμάτιο με θέα στη θάλασσα/μπαλκόνι Domátio me théa stí thálassa/mpalkóni	tho-ma-tee-o meh the-a stee tha-la-sa/bal- ko-nee
Does the price include breakfast?	Το πρωινό συμπεριλαμβάνεται στην τιμή; Το proïnó symperi- lamvánetai stin timí?	to pro-ee-no seem-be-ree-lam-va-ne-teh steen tee-mee?

Eating Out

Have you got a table?	Έχετε τραπέζι; Echete trapézi?	e-che-te tra-pe-zee?
I want to reserve a table	Θέλω να κρατήσω ένα τραπέζι Thélo na kratíso éna trapézi	the-lo na kra-tee-so é-na tra-pe-zee
The bill, please	Τον λογαριασμό, παρακαλώ Ton logariazmó parakaló	ton lo-gar-yas-mo pa-ra-ka-lo
I am a vegetarian	Είμαι χορτοφάγος Eímai chortofágos	ee-meh chor-to-fa-gos
What is fresh today?	Τί φρέσκο έχετε σήμερα; Tí frésko échete símera?	tee fres-ko e-che-teh see-me-ra?

waiter/waitress	Κύριε / Γκαρσόν / Κυρία (female) Kýrie/Garsón"/Kyría	Kee-ree-eh/Gar-son/Kee-ree-a
menu	Ο κατάλογος Ο katálogos	o ka-ta-lo-gos
cover charge	Το κουβέρ Το "couvert"	to koo-ver
wine list	Ο κατάλογος με τα οινοπνευματώδη Ο katálogos me ta oinopnevmatódi	o ka-ta-lo-gos meh ta ee-no-pnev-ma-to-thee
glass	Το ποτήρι Το potíri	to po-tee-ree
bottle	Το μπουκάλι Το mpoukáli	to bou-ka-lee
knife	Το μαχαίρι Το machaíri	to ma-che-ree
fork	Το πηρούνι Το piroúni	to pee-roo-nee
spoon	Το κουτάλι Το koutáli	to koo-ta-lee
breakfast	Το πρωινό Το proïnó	to pro-ee-no
lunch	Το μεσημεριανό Το mesimerianó	to me-see-mer-ya-no
dinner	Το δείπνο Το deípno	to theep-no
main course	Το κυρίως γεύμα Το kyríos gévma	to kee-ree-os yev-ma
starter/first course	Τα ορεκτικά Τα orektiká	ta o-rek-tee-ka
dessert	Το γλυκό Το glykó	to ylee-ko
dish of the day	Το πιάτο της ημέρας Το piáto tis iméras	to pya-to tees ee-me-ras
bar	Το μπαρ Το "bar"	To bar
taverna	Η ταβέρνα I tavérna	ee ta-ver-na
café	Το καφενείο Το kafeneío	to ka-fe-nee-o
fish taverna	Η ψαροταβέρνα I psarotavérna	ee psa-ro-ta-ver-na
grill house	Η ψησταριά I psistariá	ee psee-sta-rya
wine shop	Το οινοπωλείο Το oinopoleío	to ee-no-po-lee-o
dairy shop	Το γαλακτοπωλείο Το galaktopoleío	to ga-lak-to-po-lee-o
restaurant	Το εστιατόριο Το estiatório	to e-stee-a-to-ree-o
ouzeri	Το ουζερί Το ouzerí	to oo-ze-ree
meze shop	Το μεζεδοπωλείο Το mezedopoleío	To me-ze-do-po-lee-o
take away kebabs	Το σουβλατζίδικο Το souvlatzídiko	To soo-vlat-zee-dee-ko
rare	Ελάχιστα ψημένο Eláchista psiméno	e-lach-ees-ta psee-me-no
medium	Μέτρια ψημένο Métria psiméno	met-ree-a psee-me-no
well done	Καλοψημένο Kalopsiméno	ka-lo-psee-me-no

Basic Food and Drink

coffee	Ο καφές Ο Kafés	o ka-fes
with milk	με γάλα me gála	me ga-la
black coffee	σκέτος skétos	ske-tos
without sugar	χωρίς ζάχαρη chorís záchari	cho-rees za-cha-ree
medium sweet	μέτριος métrios	me-tree-os
very sweet	γλυκύς glykýs	glee-kees
tea	τσάι tsái	tsa-ee
hot chocolate	ζεστή σοκολάτα zestí sokoláta	ze-stee so-ko-la-ta
wine	κρασί krasí	kra-see
red	κόκκινο kókkino	ko-kee-no
white	λευκό lefkó	lef-ko
rosé	ροζέ rozé	ro-ze

raki	Το ρακί Το rakí	to ra-ke**e**
ouzo	Το ούζο Το oúzo	to **oo**-zo
retsina	Η ρετσίνα I retsína	ee ret-see-na
water	Το νερό Το neró	to ne-**ro**
octopus	Το χταπόδι Το chtapódi	to chta-**po**-dee
fish	Το ψάρι Το psári	to psa-ree
cheese	Το τυρί Το tyrí	to tee-**ree**
halloumi	Το χαλούμι Το chaloúmi	to cha-**loo**-mee
feta	Η φέτα I féta	ee **fe**-ta
bread	Το ψωμί Το psomí	to pso-me**e**
bean soup	Η φασολάδα I fasoláda	ee fa-so-**la**-da
houmous	Το χούμους Το houmous	to choo-moos
halva	Ο χαλβάς Ο chalvás	o chal-**vas**
meat kebabs	Ο γύρος Ο gýros	o yee-ros
Turkish delight	Το λουκούμι Το loukoúmi	to loo-koo-mee
baklava	Ο μπακλαβάς Ο mpaklavás	o bak-la-**vas**
klephtiko	Το κλέφτικο Το kléftiko	to klef-tee-ko

Numbers

1	ένα éna	**e**-na
2	δύο dýo	thee-o
3	τρία tría	tree-a
4	τέσσερα téssera	**te**-se-ra
5	πέντε pénte	**pen**-deh
6	έξι éxi	**ek**-si
7	επτά eptá	ep-**ta**
8	οχτώ ochtó	och-**to**
9	εννέα ennéa	e-**ne**-a
10	δέκα déka	**the**-ka
11	έντεκα énteka	**en**-de-ka
12	δώδεκα dódeka	th**o**-the-ka
13	δεκατρία dekatría	de-ka-tree-a
14	δεκατέσσερα dekatéssera	the-ka-**te**s-se-ra
15	δεκαπέντε dekapénte	the-ka-**pen**-de
16	δεκαέξι dekaéxi	the-ka-**ek**-si
17	δεκαεπτά dekaeptá	the-ka-ep-**ta**
18	δεκαοχτώ dekaochtó	the-ka-och-**to**
19	δεκαεννέα dekaennéa	the-ka-e-**ne**-a
20	είκοσι eíkosi	ee-ko-see
21	εικοσιένα eikosiéna	ee-ko-see-**e**-na
30	τριάντα triánta	tree-**an**-da
40	σαράντα saránta	sa-**ran**-da
50	πενήντα penínta	pe-**neen**-da
60	εξήντα exínta	ek-**seen**-da
70	εβδομήντα evdomínta	ev-tho-**meen**-da

80	ογδόντα ogdónta	og-th**on**-da
90	ενενήντα enenínta	e-ne-**neen**-da
100	εκατό ekató	e-ka-**to**
200	διακόσια diakósia	thya-**kos**-ya
1,000	χίλια chília	cheel-ya
2,000	δύο χιλιάδες dýo chiliádes	thee-o cheel-**ya**-thes
1,000,000	ένα εκατομμύριο 	**e**-na e-ka-to-mee-ree-o

Time, Days and Dates

one minute	ένα λεπτό éna leptó	**e**-na lep-**to**
one hour	μία ώρα mía óra	mee-a **o**-ra
half an hour	μισή ώρα misí óra	mee-see **o**-ra
quarter of an hour	ένα τέταρτο éna tétarto	**e**-na te-tar-to
half past one	μία και μισή mía kai misí	mee-a keh mee-see
quarter past one	μία και τέταρτο mía kai tétarto	mee-a keh te-tar-to
ten past one	μία και δέκα mía kai déka	mee-a keh the-ka
quarter to two	δύο παρά τέταρτο dýo pará tétarto	thee-o pa-**ra** te-tar-to
ten to two	δύο παρά δέκα dýo pará déka	thee-o pa-**ra** the-ka
a day	μία μέρα mía méra	mee-a me-ra
a week	μία εβδομάδα mía evdomáda	mee-a ev-tho-**ma**-tha
a month	ένας μήναςénas mínas	**e**-nas mee-nas
a year	ένας χρόνος énas chrónos	**e**-nas chro-nos
Monday	Δευτέρα Deftéra	thef-te-ra
Tuesday	Τρίτη Tríti	tree-tee
Wednesday	Τετάρτη Tetárti	te-**tar**-tee
Thursday	Πέμπτη Pémpti	pemp-tee
Friday	Παρασκευή Paraskeví	pa-ras-ke-ve**e**
Saturday	Σάββατο Sávvato	sa-**va**-to
Sunday	Κυριακή Kyriakí	keer-ee-a-ke**e**
January	Ιανουάριος Ianouários	ee-a-noo-**a**-ree-os
February	Φεβρουάριος Fevrouários	fev-roo-**a**-ree-os
March	Μάρτιος Mártios	mar-tee-os
April	Απρίλιος Aprílios	a-pree-lee-os
May	Μάιος Máios	ma-ee-os
June	Ιούνιος Ioúnios	ee-oo-nee-os
July	Ιούλιος Ioúlios	ee-oo-lee-os
August	Αύγουστος Avgoustos	av-goo-stos
September	Σεπτέμβριος Septémvrios	sep-tem-vree-os
October	Οκτώβριος Októvrios	ok-**to**-vree-os
November	Νοέμβριος Noémvrios	no-**em**-vree-os
December	Δεκέμβριος Dekémvrios	the-kem-vree-os